ESSENTIAL DEVELOPMENTAL BIOLOGY

Breno Lisboa

TAP

Toronto Academic Press

ESSENTIAL DEVELOPMENTAL BIOLOGY

Breno Lisboa

Toronto Academic Press
4164 Lakeshore Road
Burlington ON L7L 1A4
Canada
www.tap-books.com
Email: orders@arclereducation.com

ISBN: 978-1-77956-329-3

ABOUT THE AUTHOR

Breno Lisboa is a committed bioinformatician with a strong expertise in bioinformatics, biotechnology, synthetic biology and genetic engineering. Breno is now studying for a Doctorate in Bioinformatics at the Federal University of Minas Gerais (UFMG) in Belo Horizonte, Brazil. His research focuses on investigating the origins of life and the significance of microscopic terrestrial algae in the development of Earth's soils and the evolution of land plants. His study integrates evolutionary biology and ecology to provide new insights into these basic processes. Breno acquired expertise in the field of biotechnology via his Master's program at the Federal University of Bahia (UFBA). During his studies, he specifically concentrated on conducting structural analyses of house dust mites. His research efforts have made significant contributions to the advancement of hypoallergenic vaccinations and enhanced diagnostic techniques. He obtained a Bachelor's degree in Biotechnology from UFBA and conducted research at the Technical University of Munich. During his research, he used structural bioinformatics to investigate the immunological cross-reactivity between mite allergens and helminth proteins. Additionally, he achieved a gold medal in the iGEM 2022 competition, showcasing his inventive methodology and unwavering commitment to scientific superiority. Breno Lisboa's fervor for bioinformatics, together with his varied experiences and unwavering commitment to furthering scientific knowledge, establishes him as a proficient and inventive researcher in the realm of bioinformatics.

Table of Contents

List of Figures

List of Tables

PREFACE

Developmental biology stands as a cornerstone of modern scientific inquiry, offering an expansive framework that integrates multiple biological disciplines. It is an exciting and rapidly evolving field, providing scientists with the tools to investigate complex biological problems across various levels of organization. From the molecular and chemical levels to the cellular, tissue, organ system, ecological, and evolutionary levels, developmental biology encapsulates the breadth of biological phenomena.

At the molecular and chemical levels, developmental biology examines the minutiae of gene transcription and molecular interactions. For example, the study of globin genes, which are essential for hemoglobin production, involves understanding how these genes are transcribed and how transcription factors interact with DNA to initiate this process. This molecular approach provides insights into the fundamental mechanisms that govern gene expression and regulation, which are crucial for understanding broader biological functions.

Moving up to the cellular and tissue levels, developmental biologists explore how specific cells produce globin and the processes involved in the transport of globin mRNA from the nucleus to the cytoplasm. This level of study encompasses cell differentiation, signaling pathways, and the intricate choreography of cellular components during development. The cellular interactions and communications that guide tissue formation and function are vital for comprehending how organisms grow and develop from a single cell into complex multicellular entities.

At the organ and organ system levels, the focus shifts to the formation and branching of capillaries in different tissues and the instructions they receive to connect and form a functional vascular network. This aspect of developmental biology is intertwined with anatomy and physiology, revealing how organs and systems develop, maintain homeostasis, and adapt to changing conditions. Understanding these processes is pivotal for addressing medical conditions such as cancer, where normal developmental pathways are often hijacked.

Developmental biology also extends to the ecological and evolutionary levels, examining how differences in globin gene activation facilitate oxygen transfer from mother to fetus and how environmental factors influence the differentiation of red blood cells. These studies highlight the interplay between genetic regulation and environmental stimuli, shedding light on evolutionary adaptations and ecological dynamics. The insights gained from this research contribute to our understanding of biodiversity and the evolutionary pressures that shape life on Earth.

Thus, developmental biology is a dynamic field that synthesizes knowledge from molecular biology, physiology, cell biology, anatomy, cancer research, neurobiology, immunology, ecology, and evolutionary biology. It provides a comprehensive framework for understanding the complexity of biological development, making it indispensable for advancing our knowledge in any area of biology. As we continue to unravel the mysteries of development, this field promises to yield transformative insights that will drive scientific progress and innovation.

Content/Structure of the Book

This book is structured into ten chapters, each providing an in-depth exploration of key aspects of the field.

Chapter 1 is introductory chapter that lays the foundation by defining developmental biology and outlining its historical evolution. It discusses fundamental concepts such as cellular differentiation, morphogenesis, and the role of genes and the environment in shaping development.

Chapter 2 covers the genetic basis of development. It explores how genes control developmental processes, the principles of genetic regulation, and the methodologies used in developmental genetics. Key genetic pathways and model organisms commonly used in research are also covered.

In chapter 3, the focus shifts to the experimental techniques that have shaped our understanding of embryology. The chapter examines classical approaches, such as fate mapping and lineage tracing, alongside modern molecular techniques that reveal the intricate details of embryonic development.

Chapter 4 provides a comprehensive overview of the cell and molecular biology techniques employed in developmental biology. It includes discussions on imaging technologies, gene editing tools like CRISPR, and biochemical methods used to study cell signaling and gene expression.

Chapter 5 explores the processes by which individual cells organize into complex tissues. Topics include cell adhesion, extracellular matrix interactions, and the signaling pathways that guide tissue formation and maintenance.

Focused specifically on human development, Chapter 6 covers the stages from fertilization to early embryogenesis. It highlights the unique aspects of human development, including implantation, gastrulation, and the formation of primary germ layers.

An essential chapter for understanding organ formation, Chapter 7 reviews the methodologies used to study organogenesis and postnatal development. It covers *in vivo* models, organ culture systems, and advanced imaging techniques that allow for real-time observation of developing organs.

Chapter 8 provides a detailed look at the development of the nervous system. It discusses neural induction, the formation of neural circuits, and the role of growth factors and signaling pathways in shaping the nervous system.

Focused on organs derived from the mesoderm, such as the heart, muscles, and skeletal system, Chapter 9 examines the molecular and cellular events that guide the development of these critical structures.

The final Chapter 10 addresses the development of organs derived from the endoderm, including the liver, pancreas, and lungs. It details the signaling pathways and cellular interactions that drive the formation and differentiation of these organs.

This book provides a comprehensive and integrated approach to understanding developmental biology, combining classical knowledge with cutting-edge research and techniques.

Audience

This book is an invaluable resource designed for a diverse readership encompassing students, researchers, and professionals across various biological and biomedical disciplines. Undergraduate and graduate students will find the book's clear explanations and detailed illustrations particularly beneficial for gaining a solid foundation in developmental biology principles and techniques. For researchers, the comprehensive coverage of both classical and modern methodologies provides a robust reference for experimental design and interpretation, making it a crucial tool for those engaged in developmental genetics, embryology, and cell and molecular biology studies. Additionally, professionals in related fields such as medicine, veterinary science, and biotechnology will appreciate the book's insights into human and animal development, organogenesis, and tissue engineering, which are directly applicable to their work in clinical and applied research settings. Educators and instructors can also leverage the book as a comprehensive teaching aid, using its structured chapters to develop curricula and facilitate learning. The inclusion of cutting-edge research and current trends ensures that this book remains relevant in an ever-evolving field, making it a must-have for anyone looking to stay abreast of the latest developments in developmental biology. Whether used as a textbook, a reference guide, or a source of inspiration for new research, this book offers a wealth of knowledge that bridges fundamental concepts with advanced scientific inquiry, catering to a broad audience dedicated to understanding the complexities of life from its earliest stages.

Learning Features

- Start-of-chapter learning objectives.
- Key terms in each chapter.
- Summary Points at the end of each chapter.
- Review Questions at the end of each chapter.
- Case Studies and Activities in each chapter for solving real problems.
- References at the end of each chapter.

CHAPTER

1

Introduction to Developmental Biology

LEARNING OBJECTIVES

After studying this chapter, you will be able to:

- Understand a brief history of developmental biology;
- Discuss the scope of development;
- Focus on control and integration of development;
- Know how development works.

KEY TERMS FROM THIS CHAPTER

Blastogenesis	Carbohydrates	Cell
Cell differentiation	Cytoplasm	Deoxyribonucleic acid
Differentiation	Embryogenesis	Fertilization
Fertilization	Gametes	Gametogenesis
Genotype	Hemoglobin	Metamorphosis
Morphogenesis	Nucleus	Oogenesis

INTRODUCTION

Developmental biology is an ancient yet relatively new field of study; vastly understood based on the history of development. It was originated from the science of embryology; and began taking shape in 1950s; further becoming a recognized field of study in the 1970s. The learning of embryology and its application in the field of molecular embryology led to the gradual formation of developmental biology of a new field that is both the comprehensive study of this field and its advancement. Developmental biology has advanced greatly since the 1980s as a result of the growth of fields like genetics, cell biology, and molecular biology, as well as the application of numerous new research techniques. The appearance and formation of gametes, the process of fertilization, cellular differentiation, and morphogenesis which may include the reconfiguration and specialization of various cell groups during development- are the key topics, that should be covered in the field of study. There are certain other topics as well like the emergence of diverse cell types, the phenotypic characteristics of final organs, and the establishment of unique functions. Additionally, the expression, control, and regulation of genes at various developmental stages, the causal relationship between the genotypes and phenotype expression, the relationship between genotypic and phenotypic expression

1.1. A BRIEF HISTORY OF DEVELOPMENTAL BIOLOGY

From a molecular or genetic standpoint, developmental biology is a crucial branch of science that helps us comprehend how organisms, including humans, grow and develop. Many scientific streams are included in the field of developmental biology, such as genetics, molecular biology, embryology, the biology of regeneration, and metamorphosis. Study of advancements in developmental biology is crucial for in depth understanding of evolutionary aspects of human and animal lives, along with their cellular, metabolic and health profile.

1.1.1. Early Developmental Biology

The field of developmental biology is deeply rooted in ancient Greek and Roman philosophy, contribute to the earliest known observations in the field. The renowned Greek philosopher Aristotle, who flourished from 384 to 322 BC, made significant contributions to the early stages of developmental biology through his studies of the development of chick embryos. (Richardson & Keuck 2002).

1.1.2. The Advancement of the Microscope

Developmental biology advanced with the development of the microscope. Before its invention, the smallest objects that humans could see were about the width of the hair. The inner workings of the cell started to become clear with the aid of the microscope. Although, the microscope was created in 1590, it was not until much later that microscope research supported the advancements in the field of developmental biology significantly.

Notably, during the 17th and 18th centuries, early researchers Marcello Malpighi, an Italian biologist, and Antonie van Leeuwenhoek, a Dutch microbiologist, made significant contributions by using microscopes to examine embryos of different organisms.

1.1.3. Preformation Versus Epigenesis

With a better understanding over a period, the field of developmental biology was dominated by the controversy surrounding per formation versus epigenesis. Although, it was earlier believed that evolution is a gradual process of cellular growth and differentiation from simple to complex structure. The concept of epigenesis transformed the whole idea and highlighted the important role of the environment in the evolution process. With advancements in technology and deeper understanding, the fundamental ideas of epigenesis were increasingly verified.

1.1.4. 19th-Century Cell Theory

The development of cell theory by trailblazing scientists Matthias Schleiden and Theodor Schwann later in the 19th century cleared the path for our understanding of how organisms develop at the cellular level. Around this time, the notochord and the mammalian ovum were discovered by German scientist Karl Ernst von Baer, who is regarded as the father of embryology. Von Baer's groundbreaking work, "Mammalian Ovum and the Origin of Man," was published the same year that he discovered the ovum. Von Baer's extensive research led to the development of his laws of embryology, which improved our knowledge of mammalian development and reproduction (Bowler & Peter, 2003).

1.1.5. Evolution

Additionally, one of the most important scientists in the world who worked during the 19th century, Charles Darwin, made groundbreaking contributions. His theory of biological evolution by natural selection gave rise to a new field of study known as evolutionary biology. He not only founded this new field but also altered our understanding of human history by challenging Christian doctrine and providing a fresh perspective on what it means to be human. At this point, scientists realized the importance of embryology in developmental processes and merged it with Darwin's theory of evolution (Gilbert, Opitz, & Raff, 1996).

1.1.6. 20th-Century Genetics

The discovery of the DNA structure was arguably the most significant finding of the twentieth century. The double helix structure of DNA was discovered in 1953 by English physicist Francis Crick and American biologist James Watson, with crucial assistance from Rosalind Franklin's X-ray diffraction work. This research revolutionized developmental biology. Following this discovery, important genes involved in development were discovered, and the genetic rules governing living organisms began to be understood. Over the course of the 20th century, research on the role of genes in cell signaling, cell fate determination, and pattern formation, among other processes, deepened and significantly contributed to our understanding of developmental mechanisms.

1.1.7. 21st-Century Stem Cell Research

Scientists have investigated stem cells and their potential role in tissue regeneration at the close of the 20th and beginning of the 21st centuries. Stem cell research became a contentious issue in the 1990s, dividing both the public and experts. The controversy surrounding stem cell research serves as a contemporary illustration of the need for ethical considerations in science, especially in biological science. The world witnessed the successful cloning of its first mammal, Dolly the sheep, in 1996. Cloning has significant ramifications for developmental biology as it has allowed researchers to investigate the function of genetics, and consider how cloning might be used in medical endeavors (Arthur, July 2001).

1.1.8. Modern-Day Synthetic Biology

The advent of synthetic biology, an interdisciplinary field that aims to create biological functions or organisms with features that do not occur naturally, has led to some of the most recent developments in the field of developmental biology. This field of seeks to design biological systems with specific functions in mind. These advancements could eventually impact various fields, such as artificial intelligence, bioinformatics, materials science, education and research, healthcare and pharmaceuticals, agriculture and food production, environmental sustainability, energy production, and more.

1.1.9. Future Directions

The rapidly growing field of developmental biology promises many more advancements and discoveries. Progress in related fields such as genetics, imaging, and bioinformatics will continue to drive the advancements of developmental biology. In the future, we can expect to gain deeper insights into similarities and differences between human and animal brains, with the potential goal of improving healthcare. As our technology advances and our understanding of cellular processes deepens, it is crucial to consider the ethical implications of altering natural processes to ensure that the knowledge gained from developmental biology is always applied safely and appropriately.

1.2. THE SCOPE OF DEVELOPMENT

All living things, even the most basic, are made up of two parts that German biologist August Weismann identified in the late 1800s as the "germ plasm" and the "soma." The germ plasm contains the essential components, or genes, that are passed down from one generation to the next, while the soma contains the body that may be created as the organism is developed. In contemporary terminology, DNA (deoxyribonucleic acid) is the distinguishing characteristic of Weismann's germ plasm. Encoded within the intricate structure of its molecule are the instructions required for the synthesis of the organism's other compounds and their proper assembly into structures. The soma is made up of all these additional substances (fats, proteins, carbohydrates, and so forth) arranged in a way that allows the organism to function metabolically. Therefore, all of the processes involved in carrying out the instructions found in DNA are included in biological development. Only the proper executive machinery, the first part of which is supplied by the cell that carries DNA into the next generation (in the case of plants and animals, by the fertilized egg cell; in the case of viruses, by the infected cell), can carry out those instructions. When life histories exceed a minimal level of complexity, new mechanisms of protein synthesis are brought into functional condition and the genetic instructions are gradually implemented, resulting in modifications to the executive machinery itself. Understanding the interaction between genetic instructions and the mechanisms that carry them out is the fundamental challenge of developmental biology.

In the language of genetics, the term "phenotype" refers to the functional organisms that are created by the genetic instructions, passed down from one generation to the next in the genes. The generation of phenotypes is, therefore, the process of biological development. The development of a generation's phenotype is dependent on the way certain aspects of the previous generation's phenotype (e.g., egg cell), trigger the new organism's genotype's instructions to be interpreted.

1.2.1. Types of Development

There are numerous distinct modes of development in the entire domain of organisms, the most significant of which can be categorized as pairs of opposing types.

1.2.1.1. Quantitative and Qualitative Development

Development can be limited to a quantitative shift in a system that remains essentially unchanged, usually an increase. The process of qualitative development entails changing the structure of the system. It is challenging to locate pure instances of the basic devek.

When an animal or plant reaches a structure with all of its organs, it approximates it; after that, it only seems to grow in size, or quantitatively. This would be a time of basic development. But a closer look almost always reveals that there is also some qualitative change occurring in the system. For example, a human newborn already has all of its organs. However, during the subsequent developmental stage leading up to adulthood, there are processes of maturation that involve both qualitative and quantitative changes in addition to growth. In certain simple plants and animals, quantitative development can be seen in perhaps its most basic forms. For instance, when starved, flatworms may shrink in size but reappear in size when given enough nourishment; these are examples of quantitative changes. However, even in these situations, it is discovered that the constituent organs may experience partial loss rather than just a reduction in size.

1.2.1.2. Progressive and Regressive Development

Since most plants' and animals' regular developmental processes result in system expansions, increases in size and complexity, and the introduction of new elements, they can be viewed as progressive. As previously mentioned, some organisms may experience regressive changes in size and complexity when exposed to unfavorable environments. For some organisms, such regressive changes are a typical aspect of their life cycle. These species are characterized by the development of a relatively complex structure that allows for motility early in the life cycle, followed by the adoption of a form of life that does not require motility. One such instance is the group of marine crustaceans known as barnacles, in which the egg first develops into a motile larva that quickly settles down and forms a strong attachment to a solid underwater surface. The barnacle then transforms into its well-known stationary form after losing many of the organs that made up the motile phase.

There are several more instances, especially in groups where the adults take on a parasitic lifestyle, particularly in the digestive tract or other tissues of a host animal, from which they only need to absorb their food without moving or have the right organs to catch prey. In these situations, the early developmental stage is marked by a move toward more complex forms, which is followed by a regression phase during which a large number of these organs may disappear. In this phase of regression, specific elements within the organism (i.e., those involved in operating as a sessile or parasitic form) may experience progressive development concurrent with the regress of the other organs.

1.2.1.3. Single-Phase and Multiphase Development

The most recognizable organisms, such as humans, develop in a single phase; the organs that emerge early in life remain throughout the entirety of the organism. Numerous animal species go through one or more larval stages where they develop to adapt to a life that differs from that of an adult. Among them, the common frog is arguably the most well-known. The egg first becomes a tadpole, which can swim thanks to its large, muscular tail. Eventually, the tadpole experiences a metamorphosis, or change in form. This includes the limbs growing and the tail resorbing. While the organs continue, they go through relatively extensive progressive changes, while the rest of the tadpole's body experiences less significant changes during this time. The transition between the larval and adult forms may be considerably more pronounced in other animals. For example, a sea urchin's egg first develops into a small larva called the pluteus, which is entirely different from that of an adult. Nearly all of the pluteus's structures vanish during metamorphosis, and the five-rayed adult emerges from a tiny larval rudiment. Before the adult form emerges, there may be successive larval stages in other groups of marine invertebrates.

While fundamentally distinct from the multiphased development recently discussed in animals, plants generally seem to display a type of development that is related to it in general. We refer to this as the "generational alternation." Like most higher animals, most higher plants are diploid (2n), meaning that each of their cells contains two sets of similar chromosomes. However, during sexual reproduction, diploid cells go through a reduction division to create the haploid precursors of the sex cells, or i.e., they are composed of a single set of chromosomes. These cells in animals give rise to the sex cells (egg and sperm), which combine during fertilization. Before the haploid cells in plants develop into functional sex cells, they go through a few developmental stages. The term "haploid generation" refers to the offspring of this development."

The majority of higher plants have very little haploid development, meaning that haploid individuals only have a small number of nuclei—a small number connected to the female side of the egg and a small number connected to the male side of the pollen tube. However, the haploid development may be much more extensive in some lower plants, like mosses and ferns, and result in the formation of fairly sizable separate plants. In these situations, a species gives rise to two types of individuals, each of which is the result of distinct developmental processes under the same genotype. This can be likened to the multiphase development of animal larvae. While the adult and larval forms of an animal species share the same chromosomal makeup, the situation in plants is characterized by the two forms of the organism—haploid and diploid—having different chromosomal constitutions.

1.2.1.4. Structural and Functional Development

These two categories are two aspects of all biological development processes and can only be conceptually separated for the sake of description convenience. They cannot be viewed as a pair of opposites, as were the previous pairs in this list. Function is the biological system's ability to perform tasks. These functions include walking, swimming, eating, digesting, and other bodily functions. Typical cell functions include breathing, contracting, sending nerve impulses, secreting hormones, and so forth. Additionally, all molecular functions rely on the synthesis of enzymes, which are encoded by specific genes. All of an organism's components that are able to perform specific tasks locally within its body and arranged in a specific spatial pattern are collectively referred to as its structure. For instance, contractile cells are grouped together to form muscles, and other cells are grouped together to form skeleton elements; there are distinct spatial relationships between the muscles and the skeletal elements. Function and structure are two aspects of development that are not incompatible with one another. Conversely, it is evident that the appropriate structural relationships and functions of cell systems are necessary for the higher-level functions to function. Secretion or nervous conduction fundamentally depend on the appropriate structural relationship of the subcellular components, even at the most fundamental cellular or molecular levels. However, it is frequently convenient to concentrate discussion on one or both of these two aspects of development; for example, without worrying about structural issues right now, a study into the developmental processes that lead to the production of insulin or hemoglobin by a particular type of cell may be conducted. Alternatively, the emphasis might be on the outcomes of a specific process that turns a mass of cells into a normal hand with five fingers. The structural elements of this kind of investigation are crucial.

1.2.1.5. Normal and Abnormal Development

When several fertilized eggs of a particular species are given the right circumstances to grow at all, they will remarkably frequently mature into remarkably similar adult organisms. They can withstand a fair amount of conditions, and the final products are remarkably complete in terms of similarity. Indeed, there are valid reasons to acknowledge what needs to be regarded as typical development. Given that plants grown from a given batch of seed under various environmental conditions frequently exhibit significantly greater variation than is typically found among animals, the situation may be more pronounced in plants than in animals. However, even within plants, the variations brought about by various growing environments typically amount to only numerical variations in the quantity and size of organs like leaves and flowers; thus, an individual can be classified as either well or poorly developed instead of normally or abnormally developed. Rarely do two distinct conditions cause a plant to develop in remarkably different ways, neither of which can be categorized as aberrant or normal. For example, the emergent leaves of some aquatic plants have a different shape than the underwater-developing leaves. In these situations, a plant can go through two normal stages of development.

Of course, exposing a developing system to stimuli not typically found in a typical environment—like specific chemicals—can result in the production of aberrant organisms. Deviations from normal developmental processes can also be caused by the presence of uncommon genes. Such abnormalities can be considered to be the result of incomplete completion of normal developmental processes in the great majority of cases. A functional abnormality in an adult is when the system is unable to produce a specific enzyme or type of functional cell; a structural abnormality is when certain component elements appear unusually or are arranged in patterns that are not fully realized. It is very uncommon to find examples where the abnormality is the formation of a new elemental structural pattern or the addition of a new enzyme not produced in normal development.

Carcinogenesis, or the formation of tumors, is one significant form of development that, from some perspectives, can be seen as an exception to the general rule that aberrant development is almost always retrogressive. A shift in a cell population's developmental behavior is a component of carcinogenesis. At first, it frequently results in the cells losing some of their former structural and functional qualities. However, it is frequently followed by the assumption of new characteristics, which, despite the fact that they may be harmful to the host animal, must be viewed as a progressive type: for example, thc cells frequently grow and multiply more quickly than noncancerous cells. Additionally, the character and arrangement of the cells within the tumor may change in a series of ways. From the perspective of development, each of these traits can be considered progressive. It would appear that the organs produced during a species' normal development actually exhaust all of its genotype's potentialities for the production of orderly functional structures, given the extreme rarity of cases of abnormal development that result in progressive changes. The only anomalous developments that seem to be possible are either the displacement of normal organs, deficiencies in the execution of normal processes, or the start of progressive but highly disordered processes, such as the formation of tumors.

1.2.2. General Systems of Development

1.2.2.1. Development of Single-celled Organisms

Development in viruses is just the assembly of these component units. Activities in viruses involve the production of units for constructing new virus or phage particles, assisted by the machinery of a host cell.

An organism in the next higher grade of biological organization is made up of just one cell. Numerous unicellular algae generate distinct cell types known as gametes, which are representative of the sex cells. These cells can combine during fertilization to form zygotes, which are fertilized eggs that go through a brief developmental stage. However, reproduction occurs in many other single-celled organisms through the straightforward division of an original cell into two daughter cells. Development in these forms typically occurs during the subdivision process. It entails the division of the parent cell into two smaller cells after it has been remodeled. Of course, the division of cells in higher organisms must also involve something similar. Nonetheless, the cell in many single-celled organisms is made up of several distinct components that are arranged in very specific ways, making the remodeling process visible and easy to witness. This is the case, for example, with ciliated protozoans, in which the cortex is equipped with numerous, precisely patterned appendages, such as cilia, and frequently with additional structures like gullets or mouths. During cell division, these structures are replicated in two identical but smaller copies. This does not necessarily mean that there aren't any more developmental processes that could occur. For example, the regeneration of removed parts happens quite independently of cell division.

1.2.2.2. Open and Closed Systems of Development

The general systems of development in multicellular plants and multicellular animals differ significantly. Certain cell groups in plants maintain their ability to develop into many different types of cells during the course of the plant's life. These areas, called meristems, appear as a cylindrical sheath encircling the stem and at the tips of growing branches and roots. They are made up of quickly dividing cells that can group together to form buds, which can then sprout new stems, leaves, flowers, or roots. In contrast, the majority of animals lack any unique areas that preserve their embryonic characteristics. In most cases, the entire egg and the group of cells that were taken straight from it participate in the processes of development and become components of the growing embryo. Certain cells in some forms that go through several larval stages have their development interrupted early on and are set aside to continue developing later to form a different type of larva or to form the adult once the larval stages are finished. Some insects' imaginal buds serve as an illustration. Since these buds can only carry out the developmental processes necessary to produce the specific late-larval or adult structure for which they have been set aside, it is not possible to consider the cells within them to have fully embryonic characteristics, similar to those of plant meristems. Therefore, in general, plants maintain their embryonic characteristics, having the ability to essentially start over and complete the entire developmental process. In this way, their development is "open." However, most animals do not have these kinds of persistently embryonic cells, so their development can be described as "closed." The major developmental processes that give rise to the general form of the organism cannot be attributed to neoblasts, as meristems are responsible for the development of plants. However, there may be some exceptions to this in very simple forms, such as flatworms, where specific cells known as neoblasts seem able to participate in any type of development; these cells are usually scattered throughout the body.)

1.2.2.3. Blastogenesis Versus Embryogenesis

A second system of development exists in some animals, distinct from the "closed" embryonic system discussed in the previous section. In its most advanced form, this system creates a new organism by remodeling a part of the parental body in the absence of eggs or sperm. A section of the body may grow incredibly quickly in an adult hydra, a microscopic aquatic animal. Its cells differentiate into different cell types and mold into the constituent organs to form a new individual that is exactly like the parent. Early on, the collection of cells exhibiting this behavior is known as a bud, or blastema. These cells may seem almost identical to other body cells before they are activated, and they exhibit no embryonic potential similar to that of plant meristems. Certain insects, reptiles, and amphibians are among the higher organisms in which incomplete but relatively extensive new developments of a similar kind may occur. However, they need an injury to trigger them, which could mean removing a portion of the normal body. A new development to regenerate, or replace, the missing part is typically the outcome. An initial phase of these regenerative processes is the development of a blastema, which is a collection of rapidly dividing cells that exhibits minimal signs of cellular specialization. The data suggests that they might be formed of cells close to the site of the injury rather than, as was previously believed, from surviving embryonic cells dispersed throughout the adult body. These cells become capable of developing into most of the tissues needed to replace the parts removed by the injury, and they also lose their typical adult characteristics.

The typical process of development from a fertilized egg is called embryogenesis; development from a blastema, or blastogenesis, differs greatly from this process. Tissues that appear sequentially during embryonic development may form simultaneously in blastogenesis, with no apparent sequential relationship. However, very little is currently known about the mechanisms underlying the differentiation of the different tissues within the blastema from one another. It's possible that these mechanisms resemble those in embryonic development more than meets the eye.

1.2.3. Constituent Processes of Development

1.2.3.1. Growth

As previously mentioned, growing systems

typically get larger as they develop, at least in the beginning. This phenomenon is referred to generally as "growth." It consists of two primary components: (1) a rise in cell size and (2) a rise in cell number through cell division. In certain instances, these two processes may take place quite independently of one another; for example, cells in specific quickly expanding tissues (e.g., the connective tissue or blood-forming systems in vertebrates) may grow significantly in quantity while the size of the cells stays roughly the same. Conversely, in certain organs (e.g., insects' salivary glands) the cells may grow significantly without changing in number, with each cell becoming enlarged or hypertrophied. In these massively enlarged cells, gene duplication frequently occurs, increasing the nucleus's DNA content. However, no cell division occurs; instead, the nucleus remains a single body with a multiplied set of chromosomes, or "polyploid,".

Nonetheless, an organ's ability to grow is frequently dependent on increases in both the quantity and size of its cells. It is still unclear how important these two processes are in relation to one another. The size of the Drosophila fruit fly's wings is one instance that has been thoroughly researched. Since each wing cell bears a single hair that can be seen and counted in straightforward microscopic preparations, the number of cells in each wing can be easily ascertained. It has been discovered that there is an accommodation of factors: if there are a lot of cells, they might be a little bit smaller than usual, keeping the wing's overall size mostly constant.

The fact that the idea of growth is a quantitative concept linked to an ill-defined entity may present the concept with the greatest theoretical challenges. Growth is defined as an increase in size, but what kind of size is an increase in? Is it still considered growth when a cell or organ gains volume only through the absorption of water or the laying down of a mineral substance like calcium carbonate?

1.2.3.2. Morphogenesis

Morphogenesis, as previously mentioned, is the term used to describe all the processes by which components of a developing system acquire a distinct shape or take up specific relative positions in space. It could be thought of as the development architecture. Unlike processes of differentiation, which only require chemical reactions, morphogenetic processes involve the movement of components of the developing system across space, thereby involving physical forces. Although in practice the physical and chemical processes of development normally proceed in close connection, for purposes of discussion it is often convenient to make an artificial separation between them.

Living organisms possess an immense diversity of different types of structures. They are found in all sizes, from the trunk of an elephant to the organelles that are only visible under an electron microscope inside a cell. The vast array of processes that give rise to these structures still lacks a satisfactory classification.

1. **Morphogenesis by Differential Growth:** An organism's different organs and regions may grow in size at different rates after they are initiated. The overall shape of the body will change as a result of these processes of differential growth. Such processes are extremely common in animals, especially in later stages of development. They play a significant role in the morphogenesis of plants, where the rates at which certain component

elements—like stems, lateral shoots, and the vein and intervein material in leaves—grow determine the overall shape of the plant, the shape of individual leaves, and so on. Hormones play a major role in these growth processes in both plants and animals. Internal components of individual cells most likely always have an impact as well.

Differential growth can result in dramatic changes to an organism's overall shape, but these changes are likely to be regarded as purely surface-level because they merely alter a fundamental pattern established by other processes. For example, in a plant, the primary pattern is set by the way the lateral buds surround the central growing stem; the question of how these buds proceed to grow in relation to the stem—however remarkable the outcome—is secondary.

2. **Morphogenetic Fields:** Many essential mechanisms for the creation of patterns (e.g., the arrangement of lateral buds in developing plants) take place inside regions or three-dimensional masses of tissue that don't reveal the locations of the pattern's constituent parts until they really do. These tissue masses that exhibit a pattern have been referred to as "fields." German writers first used this term in the early 1900s, drawing comparisons between biological morphogenetic fields and physical phenomena like magnetic or electromagnetic fields. The behavior of the developing system is described by the biological field; it is not explained. The system functions as if every cell or subunit had "positional information" indicating where it is in the field and a set of instructions defining the developmental behavior corresponding to that position.

Several attempts have been made to explain the nature of the instructions that go along with the positional information. The gradient hypothesis is the most well-established and well-known of these. Many fields have a region that is "dominant" in some way, giving the impression that the field is structured around it. It is hypothesized that a substance or activity is highly concentrated in this area and that the concentration decreases gradually throughout the remainder of the field. The primary flaw in the hypothesis is that no one has been able to satisfactorily identify the variables distributed in the gradients as of yet. Investigations into the possibility that they are gradients of metabolic activity have always encountered problems that can only be resolved by defining metabolic activity in a way that turns the hypothesis into a circle, with metabolic activity being defined as that which is distributed in the gradient.

Regarding position data, a new proposal has been put forth lately. Negative feedback control systems are typically involved in the majority of cell processes. These systems often oscillate, or fluctuate, on a regular basis. It's possible that every aspect of cell metabolism is essentially oscillatory in nature, with the cycle of cell division and growth serving as merely one illustration of a much larger phenomenon. It is probable that the substances causing these oscillations consist of diffusible molecules that have the ability to affect the behavior of adjacent cells. It is not difficult to imagine the possibility that certain localized areas with higher frequency or greater amplitude oscillations could serve as centers from which waves radiate outward in all directions. It has been proposed that phase differences between two or more of these trains of transmitted oscillations can be used to specify positional information.

Stochastic (random) variations may be amplified in some types of field phenomena. The area and amplitude of chance variation on either side of an initial condition of equilibrium can increase in systems with multiple substances and suitable rates of reaction and diffusion. This means that the processes could result in a pattern of distinct areas that are dispersed according to boundary conditions.

3. **Morphogenesis by the Self-assembly of Units:** Interaction between units that have properties that allow them to fit together in a particular way can result in complex structures. This is especially suitable for the basic level of morphogenesis in molecules or cells. Larger molecules, like those of tropocollagen or protein subunits in general, can assemble themselves into complexes whose structure is dependent on localized and directed intermolecular forces. Units such as the atoms of carbon, hydrogen, oxygen, nitrogen, and so on can assemble themselves into orderly molecular structures. Although it appears that relatively large entities, like the units that combine to form bacterial flagella or the head structures of bacteriophages, are capable of orderly self-assembly, little is known about the chemical forces that give rise to the interunit bonds.

Aggregates of cells can host processes that are broadly classified with self-assembly. The cells themselves are the self-assembling units. Cells of one or more distinct types can come together in assemblages and interact and agglomerate. In these situations, it is frequently observed that the initially isolated cells have a tendency to adhere to one another—at first, essentially randomly and regardless of their characteristics—but that later, they rearrange into multiple regions made up of cells of the same kind. The assortment sometimes combines cells from the same organ, sometimes from the same species, when the cells in the original collection differ in two different ways, such as species and organ of origin. Cells from two different species of amphibia sort out into groups from the same species more or less independently of organ type, but mixes of chick and mouse cells reassort themselves into groups derived from the same organ.

It is likely that the application of this morphogenetic process to the formation of structures during normal development is limited, occurring only in a few tissues (e.g., the connective system) do cells ever go through a free stage where they are not in close proximity to one another, and cells that are not related typically do not mix together to the point where reassortment processes are necessary. The results that can be obtained from the different behaviors of cells that are in constant close contact with one another are what can be used to explain the normal morphogenetic processes of both plants and animals. Numerous writers have demonstrated how dramatic morphogenetic changes can occur within a mass of cells that stay in contact but experience variations in the strength of adhesion between adjacent cells, surface area relative to cell volume, and other parameters.

1.2.3.3. Differentiation

To put it simply, differentiation is the act of becoming unique. There are two different kinds of differentiation if morphogenesis in relation to biological development is taken out as a component for independent study. A portion of a developing system will change over time in the first type; for example, a portion of the mesoderm begins as embryonic cells with minimal internal features, progressively developing striated myofilaments, and eventually maturing into a fully formed muscle fiber. In the second type, time is not as important as space. For example, other cells in the same mass of embryonic mesoderm may begin to surround themselves with an external matrix before finally turning into cartilage. Histogenesis is the process by which the distinctive characteristics of adult tissues are produced during differentiation in development. Regionalization is the process by which a initially homogeneous mass of tissue is divided into distinct regions. This process is known as differentiation in space.

Many new protein species are synthesized during histogenesis in accordance with a suitable schedule. The easiest to characterize are proteins like myosin and actin that form in muscle cells at a relatively late stage of histogenesis. Genes regulate the synthesis of proteins, so the issue of histogenesis basically boils down to the genetic processes that govern protein synthesis.

The emergence of differences between initially homogeneous or nearly homogeneous masses is the focus of regionalization. It serves as an introduction to histogenesis, which moves on in different directions in the various areas that are marked out in this way. Certain processes covered under morphogenesis must be connected to the ways in which the various regions develop their unique, contrasting characteristics. Regionalization does not necessarily require alterations to the overall spatial shape of the tissues that are undergoing it, in contrast to morphogenesis. Regionalization is more akin to the kind of process that has been explained by field theories.

1.3. CONTROL AND INTEGRATION OF DEVELOPMENT

In order to sustain themselves, living things function as cohesive wholes. By describing the mechanisms involved in each action, the action itself can be explained. However, this does not explain how different tasks are coordinated and carried out in a variety of ways on its own. This relies on a particular kind of mechanism known as a control mechanism.

1.3.1. Phenomenological Aspects

The tendency of all developmental systems to produce a normal outcome despite injuries or abnormalities that may have impacted the system in previous stages is one of their most remarkable traits. Only injuries sustained during a specific, limited period of development can often, if not always, be fully compensated for. The system is said to be able to regulate or return to normal during these times.

Homeostasis, or regulatory mechanisms, is a common term used to discuss developmental regulation. When thrown off course, many systems—including biological ones—have a propensity to revert to their original equilibrium. By definition, a developing system is one that is constantly changing over time, evolving from an initial stage, like a fertilized egg, through different larval stages to adulthood, and then onto senescence, following a predetermined time trajectory. In these systems, regulation takes place to a future segment of the time trajectory rather than, as in homeostasis, back to an initial stable equilibrium. The process can be best described by the term "homeorhesis," which refers to the restoration of a flow.

An additional significant phenomenological feature of development is that the final state attained is not monolithic, but rather can be dissected into several distinct organs and tissues. As a result, the system's overall time trajectory can also be broken down into a number of component trajectories, each of which leads to a different end product that can be identified later on. The fact that the various time trajectories diverge from one another relatively abruptly during a brief period of development—typically well before any visible signs of divergence can be seen microscopically or by any other available means of analysis—was one of the key discoveries of the early experiments on developing systems. The most striking and significant illustration of this came from research on how an amphibian egg develops during gastrulation, or the creation of a hollow ball of cells. At this point, the embryo's upper hemisphere will stay on the surface and grow in size to encompass the entire embryo, while the lower hemisphere will be forced inward (invaginated) to give rise to the mesoderm and endoderm. The nervous system will develop from about one-third of the upper hemisphere, and the skin will develop from the remaining portion. A process occurs whereby a portion of the upper hemisphere enters a trajectory toward neural tissue and another part enters a trajectory leading to epidermal development during the time of these morphogenetic movements of invagination and expansion. Determining the developmental pathways occurs rather quickly, during a stage in which the cells from the two distinct regions superficially resemble each other. In actuality, only experiments can prove that the determination occurs. Any portion of the hemisphere can develop into skin or neural tissue prior to it happening. Each component can only grow into one or the other of these options after it has occurred.

It is evident that a sufficient theory of development must take into consideration both the mechanisms through which a system develops along its proper time trajectory and the characteristics of the processes by which those trajectories diverge and become fixed or determined in the developing cells.

Many cell generations can be used to transmit the determined state. In *Drosophila* flies, this transmission is demonstrated. In the early stages of development, *Drosophila* imaginal buds—small packets of cells—separate from the embryo's main body. They endure throughout the larval stage and, when prompted to do so by the hormones secreted during pupation, transition into the distinct characteristics of the adult. These pupation hormones vanish from the adult insect's body, and although transplanted imaginal buds undergo numerous cell generations within the adult's body cavity, they do not differentiate into the corresponding adult organ's specific tissues. However, the cells can be reinserted into a larva that is ready to pupate after many generations of proliferation; as a result, they submit to the pupation hormones and differentiation takes place. The cells have managed to preserve their ability to determine which adult organ they will differentiate into upon the release of pupation hormones, even after many generations of multiplication.

To date, efforts to determine the determining agent have not proved fruitful. Nonetheless, a significant general conclusion has been drawn from studies conducted on amphibian eggs: only during a specific developmental stage, when the cells in the upper half of the egg are halfway between developing into neural tissue and into skin, can a decision be made. They are currently regarded as "competent" for either of these developmental stages. Various external agents have the ability to switch them into any of the possible pathways while they are in this state, and only while they are. A situation like this can be compared to one where the cells were neutral or featureless and needed an outside substance to give them the ability to develop into skin or nerve tissue. This would imply that information or instructions had to be added to the reacting cells from the outside. A circumstance like this is not typical of biological development. External agents simply serve as a releaser, activating a process in which all the necessary information is already present in the cells involved, whether in complex organisms like bacteria or highly developed ones like amphibians.

1.3.2. Analytical Aspects

In the first part of this century, it was discovered that these developmental phenomena actually existed. Biologists at the time lacked a clear understanding of the essential ideas required to explain development. Developmental biologists, also known as embryologists, tried to explain their findings by using vague concepts like "potencies" or "organ-forming substances," or by making reference to characteristics of cells that are real enough but clearly complex and fundamentally secondary in nature, like the ability of cell surfaces to absorb certain substances differently, or the adhesiveness of cells. Developmental biologists only gradually realized the significance of the genetic demonstration that almost all the instructions needed to build a new organism are found in the genes that unite during fertilization, and that the small amount of additional information, mostly found in the ovum, is itself a result of genetic instructions provided in the mother's body during ovulation production. Therefore, comprehending how these components interact to create more intricate mechanisms that give rise to the cellular or tissue behaviors of various developing system types is one of the core challenges of the theory of development.

For example, a large number of genes must be active to regulate the synthesis of specific proteins during the development of the neural system in vertebrates. Approximately 20 to 30 genes have been shown to be actively involved in the development of a *Drosophila* wing, and many more are undoubtedly involved as well. All these genes, however, work together to form a network of interactions and feedback loops that ultimately produce a product with many different components at precisely defined concentrations. Additionally, the developmental process that leads to this final product needs to be stabilized or buffered so that, in the event that it deviates from its normal trajectory at an early stage, it reverts to a later stage. The discovery that genes are the fundamental building blocks of development suggests that tens, if not hundreds, of genes are involved in a stabilized time trajectory. One of the long-standing philosophical debates regarding the nature of development—between preformation and epigenesis—has finally been settled by the realization that biological development is essentially an expression of the tightly controlled activities of genes. The former postulated that the system already contained a representative of each organ that would eventually make an appearance at the beginning of development, for example, in the fertilized egg. On the other hand, the proven theory of epigenesis postulated that entities that appeared later were created during development.

According to the current understanding of epigenesis, certain entities with clearly defined properties are present during the early stages of development, specifically the genes. However, these do not directly represent the later formed organs, which are the result of the properties of gene groups gradually interacting and unfolding.

A fundamental issue in contemporary developmental biology is the nature of "determination," which necessitates an understanding of how genes are "primed" to activate in response to the right stimulus. Since some genes may be activated by the activity of the earlier active ones, the state of priming presumably needs to apply to a sizable number of genes, though maybe not all of those that will be involved in the stabilized, or buffered, time trajectory. Furthermore, the priming needs to be able to endure cell division and be passed down through numerous generations of cell growth. There haven't been many specific recommendations made about the mechanism yet. One is known as the "masked messenger" theory and holds that although the primed genes are already generating the messenger RNAs (ribonucleic acid molecules) that control protein synthesis in the cell, these messengers are somehow deactivated or unable to activate the machinery responsible for synthesizing proteins. However, rather than being direct, the arguments in favor of this hypothesis are circumstantial. There is direct evidence against it in some cases, such as the *Drosophila* imaginal buds case. Another theory, which is possibly more appealing but far less specific, holds that some of the substantial amounts of repeated DNA that are known to exist in the cells of higher organisms play a role in the determination or priming process. However, there is currently no compelling theory in biology that explains how genes function.

At first glance, it seems that actual differentiation is better understood than initial determination. In order for differentiation to occur, specific genes that code for different proteins must be synthesized under control. Of course, a lot is known about the mechanisms governing how genes function to direct the synthesis of proteins in simple organisms like bacteria and viruses. It is easy to think that comparable mechanisms govern how genes in higher organisms synthesize and function. Sadly, despite a thorough search, no single instance of a precisely similar controlling system has ever been found in higher organisms. It's possible to argue that the way in which the activity of these genes is regulated at later stages cannot really be fully explained until the mechanism of "priming" genes at the time of determination is better understood.

1.4. HOW DEVELOPMENT WORKS

Understanding how an organism develops—from a single cell to an ordered collection of cells that are then programmed to become specialized for particular tasks at predetermined times—is the goal of developmental biology.

1.4.1. Gametogenesis

Meiosis is the process by which gametogenesis, or the production of sperm and eggs, occurs. Meiosis is a process in which the paired chromosomes in the nucleus and the chromatids produced at an earlier phase of the cell's life cycle are separated by two cell divisions. Meiosis results in haploid cells that have half of each pair of chromosomes that diploid cells typically have. The process of producing sperm is known as spermatogenesis, and the process of producing eggs is known as oogenesis (Saitou, Mitinori, Hayashi, & Katsuhiko, 2021).

1.4.1.1. Spermatogenesis

Spermatogenesis takes place in the wall of the seminiferous tubules, where spermatozoa are found in the tube lumen and stem cells are located at the tube's periphery. The diploid, undifferentiated cells are located directly beneath the tubule's capsule. These stem cells, known as spermatogonia (plural: spermatagonium), undergo mitosis, whereby one of the progeny develops into a sperm cell and the other produces the sperm that will proliferate the following generation (Figure 1.1).

An entity known as a primary spermatocyte initiates meiosis. Following the initial meiotic division, a haploid cell known as a secondary spermatocyte is created. Because it is haploid, this cell needs to go through one more meiotic division. The cell created at the conclusion of meiosis is known as a spermatid; when it enters the tubule lumen and begins to develop a flagellum, it is known as a sperm cell. Every primary spermatocyte that undergoes meiosis yields four sperm.

Although they are dormant, stem cells are present from birth until the start of adolescence and are deposited during gestation. Gonadotropic hormones produced by the anterior pituitary during adolescence activate these cells and result in the production of viable sperm. This persists until old age.

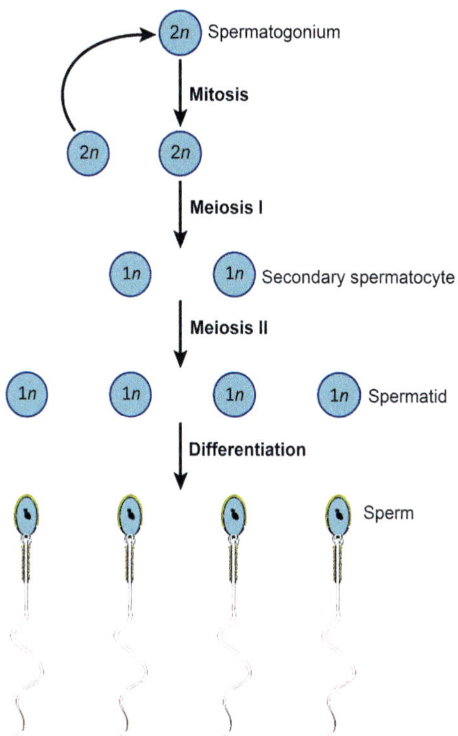

Figure 1.1. *During spermatogenesis, four sperm result from each primary spermatocyte.*

Source: https://courses.lumenlearning.com/wm-biology2/chapter/gametogenesis/.

1.4.1.2. Oogenesis

The outermost layers of the ovaries are where oogenesis takes place. Similar to the process of producing sperm, oogenesis begins with a germ cell known as an oogonium (plural: oogonia). However, this cell divides through mitosis to multiply, ultimately giving rise to one or two million cells in the developing embryo. As seen in Figure 1.2, the cell that initiates meiosis is referred to as a primary oocyte. This cell will initiate the first meiotic division and enter the first prophase, where its development will be stopped. All upcoming eggs are in the prophase stage when they are born. Many follicles in an ovary develop during adolescence as a result of anterior pituitary hormones. As a result, the first meiotic division of the primary oocyte is completed. When the cell divides unevenly, only one set of chromosomes and a small amount of cytoplasm go to the other cell, while the majority of the cellular material and organelles go to one cell, known as a secondary oocyte. The second cell is known as a polar body, and it typically perishes. At the metaphase II stage, there is another meiotic arrest. This secondary oocyte will be released during ovulation and move via the oviduct in the direction of the uterus. In the event that the secondary oocyte becomes fertilized, the cell proceeds to meiosis II, generating a second polar body and a fertilized egg that has 46 chromosomes total, half of which come from the sperm.

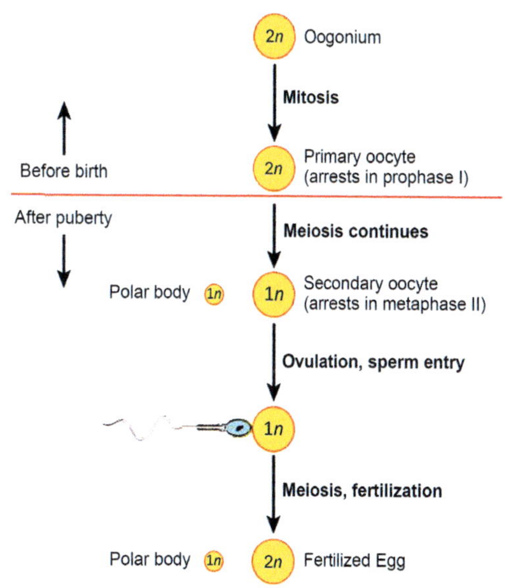

Figure 1.2. *The process of oogenesis occurs in the ovary's outermost layer.*

Source: https://courses.lumenlearning.com/wm-biology2/chapter/gametogenesis/.

Prior to puberty, egg production is stopped during meiosis, but individual cells continue to produce eggs during each menstrual cycle. Every meiotic process results in the production of one egg, with the excess chromosomes and chromatids entering polar bodies that degenerate and are subsequently reabsorbed by the body.

1.4.2. Early Development

The growth and formation of the human embryo is known as human embryonic development, or human embryogenesis. It is distinguished by the early developmental stages of the embryo's cell division and cellular differentiation processes. The biological process of human body development involves the transformation from a single-celled zygote to an adult human. When a sperm cell enters and successfully fuses with an egg cell (ovum), fertilization takes place. The germinal stage of development then starts when the genetic material from the sperm and egg combines to form the single cell zygote. In humans, the first eight weeks of development are referred to as embryonic development; at the start of the ninth week, the embryo is called a fetus. There are 23 stages in eight weeks.

The study of this development in the first eight weeks following fertilization is known as human embryology. The typical gestational period, or 40 weeks of pregnancy, is approximately nine months.

The period of time from fertilization through the early embryo's development and the completion of implantation in the uterus is known as the germinal stage. It takes roughly ten days for the germinal stage. The zygote starts to split during this phase, a process known as cleavage. After that, a blastocyst develops and implants in the uterus. The processes of neurulation and organogenesis come after the next stage of embryogenesis, known as gastrulation, during which the three germ layers of the embryo form through a process known as histogenesis.

The fetus has a more complete set of developing organs and more distinguishable external features than the embryo. Throughout the entire process of embryogenesis, changes in cell growth, differentiation, and gene expression occur in a coordinated spatial and temporal manner. Other species go through a nearly similar process, particularly chordates.

1.4.2.1. Germinal Stage

1. **Fertilization:** After the spermatozoon enters the ovum successfully, fertilization occurs when the two sets of genetic material carried by the gametes fuse together to form the zygote, which is a single diploid cell. Usually, one of the fallopian tubes' ampullas is where this happens. The 23 chromosomes from the ovum nucleus and the 23 chromosomes from the sperm nucleus make up the combined genetic material carried by the male and female gametes in the zygote. Before the mitotic division that results in the formation of an embryo with two cells, the 46 chromosomes go through modifications.

   Three mechanisms facilitate successful fertilization and serve as checks to guarantee species-specificity. First, chemotaxis, which guides sperm motion toward the ovum, is involved (Marlow & Florence, 2020). Second, there is an adhesive compatibility that takes place between the egg and the sperm. The third stage of the acrosomal reaction occurs once the sperm is attached to the ovum. An acrosome, which caps the front of the spermatozoan head and contains digestive enzymes to dissolve the zona pellucida and permit entry, is then formed. When sperm enter, calcium is released, preventing other sperm cells from entering (Singh & Vishram, 2013). The zona reaction is a parallel reaction that happens in the ovum. In order to prevent polyspermy, this results in the release of cortical granules that release enzymes that break down sperm receptor proteins (Standring & Susan, 2015). Additionally, the granules change the zona pellucida and fuse with the plasma membrane to stop additional sperm entry.

2. **Cleavage:** When the zygote divides into two cells during mitosis, the cleavage process begins. As this mitosis proceeds, the initial two cells split into four, eight, and so forth cells. Every division takes twelve to twenty-four hours. The zygote experiences cleavage without experiencing an overall increase in size, and it is large in comparison to other cells. This indicates that the ratio of nuclear to cytoplasmic material rises with each subsequent subdivision (Forgács, Newman, & Stuart, 2005). The dividing cells, known as blastomeres (*blastos* Greek for sprout), aggregate into an undifferentiated sphere that is initially contained within the ovum's zona pellucida. Eight blastomeres begin to compact after they have formed. They start to form gap junctions, which allow them to grow in an integrated manner and synchronize their reactions to environmental stimuli and physiological signals. The solid ball of cells inside the zona pellucida is called a morula when there are about sixteen of them.

3. **Blastulation:** The process of forming the blastocyst, known as blastulation, begins with cleavage. Cells divide into an inner cell mass and an outer layer of cells known as the trophoblast. The distinct trophoblasts, which are the individual outer blastomeres, blend together with additional compaction. They remain contained within the pellucid zone. The fluid that the cells will eventually secrete is contained within the structure thanks to this compaction. The center mass of cells polarizes at one end and differentiates into embryoblasts. Gap junctions are created when they close together and help with cellular communication. The structure that results from this polarization is now known as the blastocyst. It leaves a cavity called the blastocoel. (This is known as the blastula in animals other than mammals). Fluid is secreted into the blastocoel by the trophoblasts. The blastocyst subsequently grows larger and breaks through the zona pellucida, which eventually dissolves. This phenomenon, known as zona hatching, occurs just prior to the implantation process on the sixth day of embryo development. Proteases released by the blastocyst cells aid in the hatching of the human embryo by breaking down the zona pellucida's proteins and creating a hole. The blastocyst then experiences a rise in internal pressure as a result of its regular expansion and contractions;

this causes the hole to enlarge and, at last, the blastocyst can emerge from this stiff envelope. The fetal portion of the placenta will develop from the outer trophoblast layer, whereas the inner cell mass will give rise to the pre-embryo, the amnion, the yolk sac, and the allantois. The conceptus, which at this point has entered the uterus, is the embryo plus its membranes. In the end, the zona pellucida completely vanishes, exposing the trophoblast cells and enabling the blastocyst to adhere to the endometrium, where it will implant. At the start of the second week, the hypoblast and epiblast, the bilaminar germ disc's two major layers, begin to form. The trophoblast and the embryoblast will both differentiate into two sub-layers. The inner cells will differentiate into the hypoblast layer, which will encircle the epiblast layer. Together, these layers will form the embryonic disc, from which the embryo will grow (Sadler, Langman, & Jan, 2012).

The trophoblast will also differentiate into two sub-layers: the syncytiotrophoblast, located inside the endometrium, and the cytotrophoblast, which is located in front of it. Subsequently, a supplementary layer known as the exocoelomic membrane or Heuser's membrane emerges, enveloping both the primitive yolk sac and the cytotrophoblast. The syncytiotrophoblast will develop and go through a stage known as the lacunar stage, during which some vacuoles will form and fill with blood over the course of the next few days. The hypoblastic flat cells that make up the exocoelomic membrane, which coats the interior of the cytotrophoblast to form the primitive yolk sac, are the first steps in the development of the yolk sac. The syncytiotrophoblastic cells erode the endothelial lining of the mother's capillaries, forming the mother sinusoids. From these, blood enters the trophoblastic lacunae and flows through them, initiating the uteroplacental circulation. The chorionic cavity will then be formed by newly formed yolk sac cells that are situated between the trophoblast and the exocelomic membrane. These cells will eventually give rise to extra-embryonic mesoderm.

Towards the end of the second week of development, some trophoblast cells enter the syncytiotrophoblast and form rounded columns. We refer to these columns as primary villi. Simultaneously, additional migrating cells, smaller than the primitive yolk sac, form a new cavity called the secondary or definitive yolk sac within the exocelomic cavity.

4. **Implantation:** The endometrial lining changes into a secretory lining following ovulation in order to make room for the developing embryo. It thickens, becomes increasingly vascular, and its secretory glands elongate. Now referred to as the decidua, the lining of the uterus (or womb) generates a large number of large decidual cells in its increased interglandular tissue. The trophoblast is the outer layer formed by the arrangement of blastomeres in the blastocyst. Next, the trophoblast divides into two layers: the cytotrophoblast,

which is the inner layer, and the syncytiotrophoblast, which is the outer layer. The syncytiotrophoblast is a syncytial layer devoid of cell boundaries, while the cytotrophoblast is the source of dividing cells and contains cuboidal epithelial cells.

The embryonic portion of the placenta is formed when the syncytiotrophoblast uses projections of chorionic villi to implant the blastocyst in the decidual epithelium. After the blastocyst is implanted, the placenta forms, attaching the embryo to the uterine wall. The decidua in question is known as the decidua basalis; it constitutes the maternal portion of the placenta and is located between the blastocyst and the myometrium. Enzymes with hydrolytic properties that break down the epithelium help with the implantation. Human chorionic gonadotropin, a hormone that promotes the release of progesterone from the corpus luteum, is also produced by the syncytiotrophoblast. Progesterone helps the uterus to develop a thick layer of capillaries and blood vessels so that the growing embryo can get oxygen. To feed the developing embryo, the uterus releases sugar from glycogen that has been stored in its cells. The embryo's blood vessels are contained in the dividing villi. Nutrient exchange occurs between terminal or free villi and other villi. A thin connecting stalk that eventually becomes the umbilical cord, which connects the placenta to the embryo, connects the embryo to the trophoblastic shell. The decidua's arteries are remodeled to enhance the flow of maternal blood into the placenta's intervillous spaces, facilitating gas exchange and the embryo's nutrition transfer. The embryo's waste products will spread throughout the placenta (Campbell et al., 2006).

The uterine wall is penetrated by the syncytiotrophoblast, which causes the embryoblast, the inner cell mass, to develop concurrently. The embryonic stem cells that give rise to all tissues and organs are found in the inner cell mass. These stem cells are pluripotent, meaning they can differentiate into any of the three germ layer cells.

5. **Embryonic Disc:** The upper layer of the embryoblast, known as the epiblast, and the lower layer, known as the hypoblast, combine to form an embryonic disc. The disc is stretched between the yolk sac and the area that will become the amniotic cavity. Columnar cells make up the epiblast, which is located next to the trophoblast, while cuboidal cells make up the hypoblast, which is located closest to the blastocyst cavity. The amniotic cavity is formed when the epiblast migrates downward and away from the trophoblast. Amnioblasts, which are developed from the epiblast, form the cavity's lining. The exocoelomic cavity, or yolk sac, is formed by the hypoblast being forced downward. As they move along the inner cytotrophoblast lining of the blastocoel, certain hypoblast cells secrete extracellular matrix. Heuser's membrane, also known as the exocoelomic membrane, is made up of these hypoblast cells and extracellular matrix. It covers the

blastocoel to form the yolk sac, also known as the exocoelomic cavity. The extraembryonic reticulum is disrupted when hypoblast cells migrate along the outer edges of the reticulum to form the extraembryonic mesoderm. The reticulum soon develops pockets that eventually come together to form the chorionic cavity (extraembryonic coelom).

6. **Gastrulation:** Around the seventeenth day (week 3) after fertilization, gastrulation begins with the appearance of the primitive streak, a linear collection of cells formed by the migrating epiblast. Through the primitive streak, which creates bilateral symmetry, the process of gastrulation reorganizes the two-layer embryo into a three-layer embryo and also gives the embryo its distinct head-to-tail and front-to-back orientation. In front of the primitive streak, which serves as the neurulation organizer, a primitive node, also known as a primitive knot, forms. The center of the primitive node develops a depression called a primitive pit, which links to the notochord directly below. The neural plate, which forms the foundation of the nervous system, is induced to form by the node, which originated from the epiblasts of the amniotic cavity floor.

   The primitive streak from ectodermal tissue, which thickens and flattens into the neural plate, will be opposite the neural plate when it forms. In that area, the epiblast descends into the streak at the primitive pit's location, where the mesoderm-forming process known as ingression takes place. During this ingression, mesenchymal stem cells—multipotent stromal cells with the ability to differentiate into different cell types—become epithelial cells as they migrate into the primitive streak from the epiblast. After being forced aside, the hypoblast forms the amnion. The epiblast continues to grow and gives rise to the mesoderm, a second layer. The bilaminar disc is now a trilaminar disc, the gastrula, as the epiblast has now differentiated into the three germ layers of the embryo.

   The ectoderm, mesoderm, and endoderm are the three germ layers; they form as three flat discs that overlap. All of the body's structures and organs are derived from these three layers through the processes of somitogenesis, histogenesis, and organogenesis. Cells that develop between the epiblast and endoderm form the mesoderm, whereas epiblastic cells that invade and migrate to the hypoblast form the embryonic endoderm. Generally speaking, the epiblast is the source of all germ layers. The outermost layer of skin, the central and peripheral nervous systems, the eyes, the inner ear, and numerous connective tissues are all derived from the upper layer of ectoderm. The heart, the start of the circulatory system, the bones, muscles, and kidneys are all derived from the middle layer of mesoderm. The lungs, intestine, thyroid, pancreas, and bladder will all begin to develop from the inner layer of endoderm.

   After ingression, a blastopore forms on one side of the embryo where the cells have invaded. This blastopore deepens to become the archenteron, which is the first stage of the gut's development. The gut tunnels through the embryo to the other side, where the opening becomes the mouth, and the blastopore, like in all deuterostomes, becomes the anus. Now that the digestive tube is working, gastrulation is finished, and neurulation can start.

7. **Neurulation:** The gastrula is now known as the neurula because the ectoderm produces neural and epithelial tissue after gastrulation. The neural plate, which originated from the ectoderm as a thickened plate, keeps growing in width and begins to fold upward into neural folds at its ends. The folding process that turns the neural plate into the neural tube, known as nerulation, happens in the fourth week of pregnancy. They fold in the neural plate along a shallow neural groove that has developed as a dividing median line. When the folds meet and close together at the neural crest, this gets deeper as they continue to grow in height. In order to form the paraxial mesoderm, which gives rise to the somitomeres that, during the process of somitogenesis, will differentiate into somites that will form the sclerotomes, syndetomes, myotomes, and dermatomes to form cartilage and bone, tendons, dermis (skin), and muscle, cells must migrate through the most cranial part of the primitive line. Cells that migrate from the middle region of the primitive line comprise the intermediate mesoderm, which gives rise to the urogenital tract. Extraembryonic mesoderm is formed by cells that migrate through the most caudal portion of the primitive line, while other cells migrate through the caudal part and form the lateral mesoderm.

   The embryonic disc is round and flat at first, but it gradually stretches to form a narrow caudal end and a wider cephalic portion. The primitive line begins at the tip and extends cephalic, returning caudally 18 days after fertilization until it vanishes. The germ layer exhibits distinct differentiation at the start of the fourth week in the cephalic portion and at the end of the fourth week in the caudal portion. Neuropores in the cranium and cauda gradually shrink until they shut entirely (by day 26), creating the neural tube (Carlson & Bruce, 1999).

1.4.3. Growth and Death

Growth is the process through which an organism's cell size and number increase over its life.

1.4.3.1. The Process of Growth

Growth rarely happens at random. Instead, it proceeds in accordance with a plan that ultimately establishes the individual's dimensions and form. Certain areas of the organism, like the layers of cells that divide and enlarge close to the tip of a plant shoot, may be the only places where growth occurs. Alternatively, as in the case of the human embryo, the cells involved in growth may be widely dispersed throughout the body of the organism. In the latter instance, there are distinct differences in the rates of cell division and size growth. Adult forms demonstrate that the growth patterns of both plants and animals are predetermined and regular. However, in some organisms—most notably, the slime molds—no regular pattern of growth takes place, leading to the formation of a formless mass of cytoplasm. The rate at which an organism's constituent parts grow could have a significant impact on its capacity for environmental adaptation, and as a result, could influence evolution. For example, a fish with a modified fin would have an easier time adapting to terrestrial locomotory life than a fish without the modification. This would be because the fleshy parts of the fin would grow faster. Without the fin's disproportionate growth, which eventually came about due to haphazard mutations in the genetic code, it may not have been possible for limbs to evolve through natural selection.

1.4.3.2. Types of Growth

1. **In Cells:** A developing organism's increased size and shape are determined by the size and quantity of cells that comprise the individual. Mitosis, a precise cellular reproductive mechanism, is responsible for the increase in cell number. The genetic material-containing chromosomes are replicated in the nucleus during mitosis, and the doubled chromosomes are subsequently evenly distributed to the two daughter cells, with one copy of each chromosomal type going to each daughter cell. Before the ends of the dividing cell split apart, each end receives a full complement of chromosomes. This is known as cytokinesis, or the pinching off of the cell membrane in animal cells; in plant cells, a new cellulose wall grows in between the newly formed cells.

   The mother cell frequently doubles in size during the stage of cell life that precedes the actual division of chromosomes. Consequently, a cycle of cell division and growth is created. Cell division, in which the cytoplasmic mass and chromosomes are transferred to the daughter cells, comes after cell growth, which is characterized by an increase in cytoplasmic mass, chromosome number, and cell surface. That being said, cell-division cycles do not always result in an increase in cytoplasmic mass. For instance, in the early stages of an embryo's development, the original, typically very large egg cell divides into thousands of smaller cells through a series of repeated cell divisions without any growth intervals in between. The typical pattern of growth and mitosis does not begin until the embryo is able to obtain food from its surroundings.

2. **In Plants:** One crucial difference between plant and animal growth is the extensive size increase that most plant cells experience without accompanying cell division. Daughter cells that result from cell division behind the root or shoot tip of a plant may experience significant volume expansions. This is achieved by the cells taking up water, which is kept in a central chamber known as a vacuole. The pressure created by the intake of water, along with other elements, presses against the cellulose walls of the plant cells, lengthening, girdling, and stiffening (turgor) the plant as a whole. Plants experience a significant increase in size following cell division, which is mostly due to an increase in cell water content rather than a

significant increase in dry weight.

Numerous cells that go through the cycles of growth and division are dispersed throughout the mass of the very young, developing plant embryo. However, the ability for cell division is limited to cells in specific areas known as meristems once the positions of the root tip, shoot tip, and embryonic leaves are established. Just beneath the surface of the developing root is one meristematic center, where all increases in the primary root's cell count take place. A portion of the daughter cells stay at the tip of elongation and keep dividing. The lengthening of other daughter cells that remain in the root allows the new root to pierce deeper into the soil. The growth shoot of higher plants follows the same general pattern, with cell elongation taking place behind the meristematic center and the formation of the stem and leaf cells coming from a restricted meristematic region at the tip. The vascular strands of phloem and xylem, which transport water from the soil to the leaves and sugar from the leaves to the rest of the plant, are the tissues that the young seedling primarily develops its cells associated with. These cells have the capacity to divide once more, producing new cell material needed for the vascular strands and the development of a woody covering. Hence, the growth of higher plants—*i.e.,* those aspects involving both the pattern of stems, leaves, and roots and the increase in bulk—results primarily from cell division at the meristem followed by a secondary increase in size because of water uptake. These processes take place all during the growing season of the plant.

3. **In Animals:** Animal growth is more temporally constrained than plant growth, but cell division occurs more widely throughout the organism's body. The developing embryo possesses a widely distributed capacity for cell division, despite regional variations in the rate of division. Size growth occurs quickly during the embryonic stage, continues at a slower pace in the juvenile stage, and then stops altogether. Even after there is no longer an increase in the overall size of the body, cell division and size growth nevertheless continue. The post-juvenile increase in cell number is essentially a replacement phenomenon because cell death balances these events. Mammals can only grow to a certain height when their long bones stop accumulating bone and dividing into new cells.

Humans have an unusually long juvenile growth period; most higher animals reach adult size shortly after the end of embryonic development. Certain organ systems experience minimal cell division and growth following birth; for example, all of a female's germ cells—which are the precursors to her egg cells—are formed at birth. In a similar vein, by the end of the embryonic stage, every nerve cell in the brain has developed. The nervous system grows larger as a result of nerve fiber expansion and the deposition of a fatty insulating substance along their length. While nerve fiber outgrowth in animals reflects a true increase in the amount of cytoplasm and cell surface and not just an uptake of water, the largest increase in size of nerve cells occurs, as in plant cells, after the cessation of cell division.

Certain organs continue to have the capacity for growth and cell division for the duration of the animal's life. For instance, the liver constantly produces new cells to replace aging and dying ones. While the liver is a single organ where cells divide and grow, other organs have a unique cell population known as stem cells that are still able to divide. The marrow of the long bones is the only place where the cells that make the circulating red blood cells of mammals can be found. They create a population of dividing cells that never stops, taking the place of the red blood cells that eventually die and vanish from the bloodstream. Different body parts can have quite different rates of cell division and growth. One of the main elements that determines an organism's shape is this differential increase in size.

1.4.3.3. Normal and Abnormal Growth

1. **Tumors:** Growth irregularities and tumors can arise from improper growth regulation. Hepatomas, or liver tumors, may develop if there is an abnormal rise in the number of liver cells. The absence of the typical growth patterns and rates is, in fact, one characteristic of malignant tumors, or cancers. Malignant tumor cells possess not only atypical growth rates but also modified adhesive characteristics that facilitate their easy separation from the tumor. This allows the cells to metastasize, or spread to different parts of the body, and grow in unexpected places. Usually, tumor growth occurs outside of the original organ and results in the death of the organism. Growth rates of tumors can differ greatly. They can grow at a pace that is comparable to normal cell division in adult tissues, or they can grow extremely slowly. In addition to an accelerated rate of cell division, tumors also exhibit aberrant growth patterns. Large nodules may form from the newly formed tumor cells, which are disorganized and not integrated into the organ's structure. There might not be any health issues with these aberrant growths (e.g., moles) or could have catastrophic consequences, as in the case of pressure on the brain brought on by a tumorous mass of the meningeal covering of the brain.
2. **Regeneration:** Tumors are not the only abnormal growths. When a tree is partially burned, the vascular strands that are exposed are covered again by cells beneath the bark. Growth might not be typical, and a noticeable scar or new bark growth is noticeable. Likewise, in the event that a mammal sustains severe skin damage, the organism experiences no physiological difficulties despite the repair being abnormal and imperfect. Many organisms are able to regenerate, or grow back, damaged or missing body parts to varied degrees of perfection. Salamanders have amazing regeneration abilities; if they lose an eye or limb, they can grow new ones. Humans can regenerate portions of their livers, and lizards can grow new tails. The causes of the variations in regenerative capacities amongst

animals are still a puzzling puzzle with significant practical implications. In the event that regeneration takes place, some specialized cells typically lose their specialized traits and go through a phase of faster cell division; the newly formed cells then respecialize into the tissues of the original body part. Sometimes, dormant tissues in plants that have had their tops pruned can regenerate into new meristematic centers, which then sprout new shoots.

3. **Compensatory Growth:** Animal organs frequently exist in pairs, and when one is lost, the other enlarges to accommodate the increased demand. For example, when a human has one of their two kidneys removed, the other grows larger. This phenomenon, known as a compensatory reaction, can manifest as either hypertrophy (an increase in cell size), hyperplasia (an increase in the rate of cell division), or both. The kidney's compensatory response is mostly caused by an increase in cells, but the number of individual filtration units, or glomeruli, does not increase. Cell division therefore results in an increase in glomeruli's size but not in their total number. Animal endocrine organs, which secrete hormones, are responsible for some of the most remarkable instances of cell size increases. For instance, stimulation of the thyroid gland can result in a dramatic increase in the size of the gland's individual cells.

1.4.3.4. Factors That Regulate Growth

1.4.3.4.1. Environmental Factors

1. **Temperature:** An organism's living environment has a significant impact on how quickly and how much it grows. Environmental influences could be physical (e.g., atmospheric pressure, temperature, radiant energy, and chemical. The temperature sensitivity of organisms and the cells that make them up is remarkable; as the temperature drops, the biochemical reactions required for life occur more slowly. A 10°C (18°F) drop in temperature slows metabolism at least twice and frequently more. Cell division and the expansion of secondary meristematic tissue beneath the bark are two factors that contribute to trees' increased width. Cell division and enlargement may completely stop in the winter months, but they resume in the spring. Three factors affect this sporadic growth: water, light, and temperature. If there is a cold spring, if there are obstacles preventing sunlight from reaching the plant, or if there is a drought, the amount of growth may be significantly reduced. Since the spacing between growth rings of varying sizes has been correlated with known periods of cold and drought, the width of the growth rings visible on the surface of the cut tree trunk actually provides a partial history of climatic conditions, allowing for reliable archaeological dating of various structures, such as the timbers used in Indian pueblos in the southwestern United States.

   Moreover, animals with warm or cold blood are impacted by temperature. Many warm-blooded (e.g., bears) and cold-blooded (e.g., frogs) vertebrates cease growing during the cold winter and simply enter an inactive or dormant state, which is characterized by a very low rate of metabolism. When it gets cold, animals that don't go dormant have higher demands for food consumption to keep their bodies warm. If food is scarce, this use of food energy could restrict the amount of energy that can be used to grow larger.

2. **Pressure:** Air pressure is probably not very important in regulating growth because it is mostly constant everywhere except in the mountains. Since it is well known that increases in hydrostatic pressure obstruct cell division, pressure increases at the ocean's depths could be substantial. Deep-sea fish tissues, which have received little research up to this point, must have evolved to withstand these kinds of pressure effects. The peculiar forms of some conifers that grow along coastlines exposed to strong prevailing winds are clear indications that movements of the terrestrial atmosphere, or winds, may have an impact on growth patterns in trees and shrubs.

3. **Light:** Light is the physical factor with the most dramatic and best understood role out of all of them. Many of light's effects on plant growth are clear-cut and immediate. Photosynthesis, the chain of chemical reactions in green plants that converts carbon dioxide and water into carbohydrates and is ultimately responsible for all life, is propelled by light energy. Green plants with inadequate light either die off or experience growth retardation. However, light also has significant indirect effects. Small amounts of a pigment known as phytochrome, which can take two forms, are found in green plants. Red light (660 millimicrons, or mμ; 1 mμ = 3.937×10^{-8} inch) is absorbed by one form. Plants that contain this pigment absorb red light, which transforms the pigment into a different form that absorbs far-red light (730 mμ). This far-red form can then be converted back to the original form that originally absorbed red light. These transformations have striking effects; red light, for instance, promotes leaf growth, chloroplast development, red flower coloration, and spore germination but inhibits stem elongation and lateral root formation. The formation of flowers can also be influenced by red and far-red light cycles.

Even though they are less evident, light's effects on animals can still be significant. One such effect is how light affects the development of some animals' reproductive systems. Some birds appear to start developing their sex organs (gonads) in the spring when the days get longer and there is more light. Strangely, cells deep in the brain are sensitive to the tiny amounts of light that pass straight through the bird's thin skull. This means that the eyes are not the receptors for the light signal that triggers the endocrine system to start the growth of gonads.

Most animals show cyclic activity, or rhythms, in various important physical (*e.g.*, movement) and chemical (*e.g.*, respiration) events that are essential to the individual. Short bursts of light are frequently used to control these rhythms.

1.4.3.4.2. Chemical Factors

The gases in the atmosphere and the water, minerals, and nutritional value of food are examples of chemical elements that are significant in the environment.

For photosynthesis to occur, plants need carbon dioxide, water, and sunlight; a drought can impede or even kill a plant. The consequences of air pollution—e.g., hydrocarbons, carbon monoxide, and nitrogen oxides—are recognized to negatively impact the development and reproduction of both plants and animals.

Minerals and trace amounts of elements like zinc, magnesium, and boron are necessary for both plants and animals. Plants receive their nitrogen and phosphorus from the soil in the form of nitrates and phosphates. Insufficient levels of any given nutrient in the soil lead to subpar crop yields and stunted plant growth. Water, oxygen, and other elements from the environment are necessary for animals. Animals must get their sugars from their diet, either directly from plants they eat or indirectly through other animals they eat that have also eaten plants, as they are unable to synthesize sugars from carbon dioxide. Either growth is retarded or death occurs if this food is of poor quality or quantity.

Animals require trace amounts of a class of molecules known as vitamins, which have a range of chemical structures. Because animals are unable to synthesize all the vitamins they need, they must obtain the vitamins through their diet, either from plants or from other animals that are able to do so. Vitamin deficiencies can cause stunting, malformation, disease, or even death during growth because specific vitamins are required for certain critical metabolic reactions.

1.4.3.4.3. Internal factors

The environment provides the organism with the building blocks for growth, but the organism also controls its own growth. Mutations can have an impact on events like the rate and location of cell division and the extent of cell enlargement because the size and form of plants and animals are determined by genetics. However, the exact mechanism by which these factors, which are the final arbiters of growth, are regulated in individual cells remains unknown.

Hormones represent a significant class of intrinsic growth regulators. The main hormone in plants, auxin, is produced in the leaves and travels to other parts of the plant through exact mechanisms that are still unclear. These mechanisms regulate various processes in the plant, including the elongation of plant cells. The plant cell's hard cell wall is altered by auxin in some way, making it more flexible. As a result, the cell is forced to expand due to internal pressure. Other hormones found in plants, such as gibberellins and cytokinins, which affect the rate of cell division in meristems, might also be involved in this process. Gibberellin alone can encourage certain dwarf plants to reach normal size.

Hormones are also essential for animal growth. Because of its profound and pervasive effects on growth, one hormone produced by the pituitary gland at the base of the brain is known as growth hormone. In pre-adolescents, a growth hormone shortage causes dwarfism, while an excess of the hormone—often brought on by a tumor—causes gigantism. If an excess of growth hormone is produced after the long bones can no longer grow—*i.e.,* post-adolescence—a disease called acromegaly, which is characterized by increases in the size of the hands and feet and broadening of facial features, results. Growth retardation in children is also brought on by a thyroid hormone deficiency.

The gonads' growth is regulated by a complex interaction between the sex hormones secreted by the pituitary gland.

The gonads then go on to produce the hormones estrogen, progesterone, and testosterone in females and testosterone in males. These hormones regulate the growth of the vocal cords in males, the enlargement of the mammary glands in females, and body hair in humans. Though they are essential for growth, the precise mechanism by which sex hormones and growth hormones work has not been determined.

Apart from their capacity to synthesize growth-regulating factors, plants and animals also clearly exhibit sophisticated mechanisms for combining and controlling the production of hormones; i.e., for normal growth, the right hormones are produced in the right amounts at the right times in the right places.

While the growth of many plants, including trees, is continuous throughout their lives, the growth of individual organisms is not; e.g., a species' leaves reach a maximum size and are unable to enlarge. Animals stop growing completely after the juvenile stage, with the exception of replacement. Genetic mechanisms most likely set limits for both total body size and organ size. Although the exact causes of an organism's growth restriction are unknown, evidence suggests that the liver releases bloodstream-based protein molecules that have the ability to restrict an organ's growth. According to one theory, an organ could therefore create chemicals that function to restrict its own growth, creating a feedback loop. Certain regions of the mammalian nervous system depend on a protein known as nerve-growth factor for growth. The growth of sympathetic nerve fibers is extensive and abnormal if there is an excess of the nerve-growth factor. The sympathetic nerves wither and disappear if the nerve-growth factor is removed from the body, which can be done by injecting an antibody against the factor. It may eventually be possible to find additional subtle growth regulatory substances that are particular to different organ systems.

1.4.3.5. The Dynamics of Growth

1.4.3.5.1. Measurement of Growth

There has long been interest in the mathematical study of the rate of growth. The principle of cell division—that one cell gives rise to two daughter cells—forms the basis of it. As a result, a geometric series would theoretically increase the number of cells by one cell producing two, four, eight, sixteen, and so forth. In actuality, however, the rate of growth is not constant and eventually decreases due to environmental factors or innate genetic limitations. Therefore, when growth is plotted against time on a graph, the curve representing the growth of cell populations and organisms is typically S-shaped, or sigmoid. The rising portion of the curve is explained by the increase in cell number brought about by cell division; at the curve's plateau, the rate of cell division falls. In general, the growth of organisms can be modeled by the S-shaped growth curve. The early, intense growth (referred to as log growth) in the rising phase of the growth curve falls on a straight line when plotted against time on a logarithmic scale.

The differential equation $v = dW/dt\ (1/W)$, where W is the weight at any given time, t, and v is the growth rate, can be used to define the rate of growth. A value for relative increase—the weight increase in relation to the starting mass of the growing substance—is obtained from the solution of this equation. Insect larvae are the animal that most closely resembles a constant rate of growth. In the majority of animals, growth rate decreases with increasing size and age. The growth rates of various components of entire organisms differ, even though the S-shaped growth curve fairly accurately depicts the growth of populations of single cells, such as bacteria or cells of higher organisms in tissue culture—the growth of cells of tissues from organisms in a sterile nutrient environment. Allometry is the relationship between an organism's growth in one part and its growth in another. The basic relationship of allometric growth can be expressed as follows: $y = bx^k$, where x is the size of one organ, y is its size, b is a constant, and k is the growth ratio. Unfortunately, despite the fact that these mathematical tools have made it possible to describe the differential growth of various organism parts in great detail, they have not shed light on the physical and chemical processes that regulate growth rate.

1.4.3.5.2. The Study of Growth

Despite the elusive nature of the chemical, physical, and genetic bases of growth, growing tissues in a sterile nutrient environment has provided valuable insights into the process. The cells will start to grow again in culture, frequently at a logarithmic rate of increase, even if the source of the tissue is an organ that has completely stopped growing, such as the phloem of a plant or the nervous system of an animal. Thus, it is possible to draw the conclusion that the organism as a whole imposes limitations on each cell's capacity for reproduction, and that these limitations are what prevent cells from reaching their full potential for growth. The sigmoid-shaped growth curve, however, results from the eventual slowing of cell growth, even in tissue culture. In tissue culture, cells that undergo rapid growth typically lose the capacity to perform the specialized function unique to their original organ; cartilage cells, for instance, can no longer synthesize cartilaginous matrix if they divide quickly. Evidence suggests that certain types of specialized cells can be maintained in tissue culture for very long times and still retain the ability to carry out specialized biosynthesis, suggesting that the apparent loss of specialized function in tissue culture cells may not be primarily the result of a mutual exclusivity of growth and differentiation. This phenomenon of apparent despecialization has been a topic of great theoretical interest: are rapid growth and specialization mutually exclusive activities?

One reason for the slowdown in tissue-culture cell growth is the medium's critical components running out. However, even with frequent medium replacements, the growth rate decreases as soon as a layer of cells densely packs the bottom of the culture dish—a phenomenon known as contact inhibition of growth. It is thought that a signal that slows down cell division is produced by cells that are so close to one another that they are always in contact. The growth rate of seemingly identical cells in tissue culture also varies greatly. For example, when skin cells are cultured, some of them might divide every eight hours, while other cells that are similar might only divide every 36 hours. Tissue culture, which allows cells to grow in a controlled environment, provides a wealth of opportunities for researching the basic mechanisms governing cell growth and, by extension, the growth of organisms and populations.

1.4.3.6. Cell Death

The event of a biological cell ceasing to function is known as cell death. This could be the outcome of diseases, localized injuries, or the demise of the organism that the cells are a part of. It could also be the natural process of old cells dying and being replaced by new ones, as in programmed cell death. Necrosis is a non-physiological process that happens as a result of infection or injury, whereas apoptosis, also known as Type I cell death, and autophagy, also known as Type II cell death, are both types of planned cell death (Nirmala & Lopus, 2020).

1.4.3.6.1. Types

1. **Programmed Cell Death:** An intracellular program is responsible for cell death in cases of programmed cell death (PCD) (Engelberg-Kulka et al., 2006). PCD is implemented through a controlled process that typically offers benefits to an organism at various stages of its life cycle. For instance, in a developing human embryo, the differentiation of fingers and toes happens as a result of apoptosis in the cells separating the fingers. PCD plays essential roles in the tissue development of metazoa (multicellular animals) and plants.

- **Apoptosis:** The mechanism through which multicellular organisms may experience programmed cell death (PCD) is called apoptosis. Cell death and morphology changes that are characteristic are caused by biochemical processes. Blebbing, cell shrinkage, nuclear fragmentation, chromatin condensation, and chromosomal DNA fragmentation are some of these alterations. It is now believed that, in a homeostatic setting, cells are positively induced to commit suicide in a developmental context; the lack of specific survival factors may act as a trigger for suicide. The morphology and even the biochemistry of these suicide pathways seem to vary somewhat; while some take the route of "apoptosis," others take a more broad route that leads to deletion; in any case, both are typically driven by genetic and synthetic factors. While there is evidence that some "apoptosis" symptoms, like endonuclease activation, can be artificially induced without initiating a genetic cascade, true apoptosis and planned cell death most likely need to be genetically mediated. It's also becoming more obvious that signals from the right growth or survival factors are necessary to achieve a balance between mitosis and apoptosis, which are toggled or linked in some way (Bowen, 1993).
- **Autophagy:** Cytoplasmic autophagy is typified by the formation of sizable vacuoles that consume organelles in a particular order before the nucleus is destroyed. Often called autophagy, macroautophagy is a catabolic process that breaks down excess or damaged organelles, aberrant protein aggregates, and bulk cytoplasmic contents via autophagosomal-lysosomal degradation. In addition to being linked to physiological and pathological processes like development, differentiation, neurodegenerative illnesses, stress, infection, and cancer, autophagy is typically triggered by nutritional deprivation.
- **Other Variations of PCD:** There are now additional known pathways for programmed cell death (Kroemer & Martin, 2005). These alternate pathways to death, known as "non-apoptotic programmed cell-death" (or "caspase-independent programmed cell-death"), are just as effective as apoptosis and can

serve as primary PCD or as a backup plan.

Among these are excitotoxicity, ferroptosis, an iron-dependent type of cell death, anoikis, which is nearly identical to apoptosis except for its induction, cornification, a type of cell death unique to the eyes, and Wallerian degeneration. Plant cells experience specific PCD processes that are comparable to autophagic cell death. However, both plants and metazoa share a number of common PCD characteristics.

The combination of the Fas receptor (Fas, CD95) and the Fas ligand (FasL, CD95 ligand) results in activation-induced cell death (AICD), which is a predetermined form of cell death. It is brought on by repeatedly stimulating particular T-cell receptors (TCRs) and contributes to the preservation of peripheral immune tolerance. Therefore, autoimmune diseases could result from a change in the process. Stated differently, AICD functions as the antagonistic regulator of activated T cells. One type of unintentional, passive cell death that is frequently regarded as a fatal injury is ischemic cell death, also known as oncosis. The process is typified by swelling of the nucleus and cytoplasm, cytoplasm vacuolization, and mitochondria (Vitale, Galluzzi, Castedo, & Kroemer, 2011).

An oncosuppressive mechanism known as "mitotic catastrophe" can result in cell death when cells enter mitosis too soon or improperly. When cancer cells are exposed to ionizing radiation and numerous other anti-cancer treatments, this is the most frequent way for the cells to die. A type of cell death known as immunogenic apoptosis or immunogenic cell death is brought on by certain cytostatic drugs, including radiotherapy, photodynamic therapy (PDT), oxaliplatin, and bortezomib. Pyroptosis is a highly inflammatory type of programmed cell death that primarily happens when intracellular pathogens are infected. In myeloid cells, it is likely to be a component of the antimicrobial response.

Combining elements from various cell death pathways, PANoptosis is a distinct inflammatory cell death pathway. Necroptosis, apoptosis, and pyroptosis taken separately cannot explain the entirety of the biological effects in PANoptosis. Multifunctional macromolecular complexes known as PANoptosomes control PANoptosis. Cell death brought on by the phagocytosis of a live cell (i.e., eaten) by another cell, usually a phagocyte, which causes the engulfed cell to die and undergo digestion. Cells that are pathogenic, cancerous, old, damaged, or have more cells than needed can undergo phagoptosis.

2. **Necrotic Cell Death***:* Necrosis, which can take many different forms, is a type of cell death that happens when a cell sustains significant damage from outside factors like trauma or infection. A cell swells during necrosis, and then the cell membrane spontaneously bursts, releasing the contents of the cell. Subsequently, these cell contents frequently cause adjacent cells to become inflamed (D'Arcy, June 2019). Necroptosis is a type of

programmed necrosis that has been identified as a substitute for programmed cell death. It is proposed that when endogenous or exogenous factors, such as viruses or mutations, block the apoptosis signaling, necroptosis can act as a backup cell-death mechanism to apoptosis. Death receptors, such as the tumor necrosis factor receptor 1, are linked to necroptotic pathways. Morphology was once used to classify cell death, but in recent years, molecular and genetic conditions have taken its place.

CASE STUDY

ABNORMAL CELL DIVISION

Kim, who is seven years old, is fighting leukemia, a blood cell-related cancer, just like the young kids in Figure 1. As blood cells are made in the bone marrow, leukemia typically begins there. Depending on the type of leukemia, it can affect other types of blood cells in addition to producing abnormal blood cells, most commonly white blood cells. Over time, the patient's normal blood cells are replaced by the aberrant blood cells. Symptoms like exhaustion, recurrent infections, and easy bruising or bleeding may result from this. Although leukemia can be fatal, there are thankfully some treatment options that can extend life or even potentially reverse the disease's course.

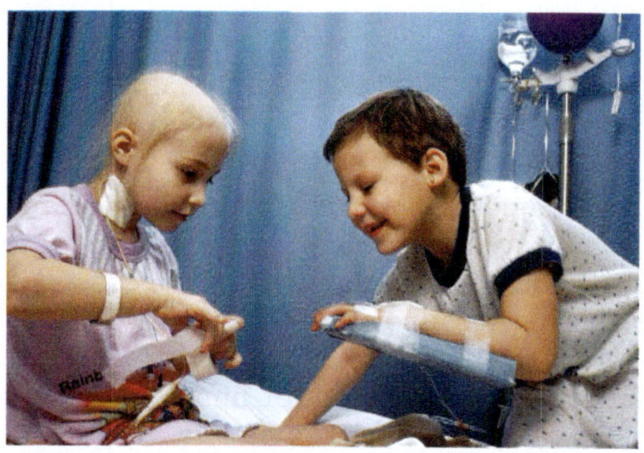

Figure 1. *Pediatric patients receiving chemotherapy.*

Source: https://bio.libretexts.org/Courses/Butte_College/BC%3A_BIOL_2_-_Introduction_to_Human_Biology_(Grewal)/Text/07%3A_Cell_Reproduction/7.1%3A_Case_Study%3A_Genetic_Similarities_and_Differences.

Kim's parents have been informed by his doctors that although he has received chemotherapy to eradicate the cancerous cells, it is insufficient. To replace his aberrant bone marrow with healthy bone marrow, Kim requires a bone marrow transplant. Although they must first undergo testing to determine whether they are a compatible match, his family members are eager to donate bone marrow to him.

Bone marrow transplants necessitate far more precise matching between donor and recipient than blood transfusions, where finding a compatible blood donor is usually not too difficult. Their cells' surface proteins, known as human leukocyte antigens (HLAs), must be similar in multiple ways. In Figure 2, one kind of HLA protein is shown. Depending on their unique genes, different people have different kinds of HLA proteins, or markers. In most cases, the prospective bone marrow donor and recipient have eight to ten HLA markers tested and compared. For a match to be made, there must be at least six or seven identical HLA markers between them.

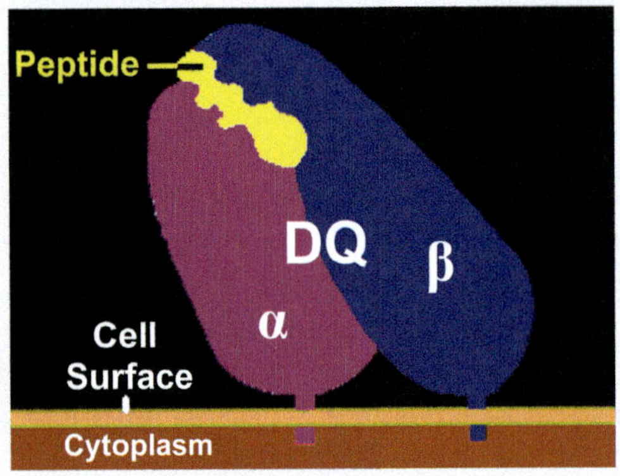

Figure 2. *An illustration of a human leukocyte antigen protein, HLA-DQ, attached to the surface of a cell, showing its α (pink) and β (blue) chains.*

Source: https://bio.libretexts.org/Courses/Butte_College/BC%3A_BIOL_2_-_Introduction_to_Human_Biology_(Grewal)/Text/07%3A_Cell_Reproduction/7.1%3A_Case_Study%3A_Genetic_Similarities_and_Differences.

The patient's body may reject the bone marrow transplant if the match is poor, or the transplanted bone marrow may create immune cells that assault the patient's body. For bone marrow donation to be both safe and successful, a good match between donor and recipient is essential. Because they share many of the same genes from their parents, full siblings are often the best match for bone marrow donation. After testing, Kim's sister discovers that she is not a match for him. Since there is only roughly a 25% chance that a sibling will have the same HLA, this is not particularly surprising. After being tested, his parents and other family members are also unmatched. Kim has to become one of the seventy percent of patients who have to search for a bone marrow donor outside of their family.

Conclusion

Genetically, humans are far more alike than different from one another. On Earth, there is 99.9% genetic similarity between any two individuals. However, in certain situations, such as bone marrow donation for the treatment of leukemia, the tiny 0.1% difference can have a significant impact. In a family where there is sexual reproduction, these variations are passed down. A recipient's and a donor's genes that code for human leukocyte antigen (HLA) proteins must be genetically compatible. Immune system antibodies will recognize the antigens as "non-self" and attack the transplanted cells if a patient receives a bone marrow transplant from a donor who has different types of HLAs than the patient. Furthermore, because bone marrow generates immune cells, antibodies present in the transplanted tissue may use the same mechanism to target the patient's own cells.

It is not always easy to find a good HLA match, even between full siblings, as you have also learned. This results from random assortment and crossing over of a single person's gametes, which causes genetic variations. When two distinct gametes fertilize, the variations multiply. Kim needs to meet someone who is not related to him. Approximately 14,000 Americans are required to search a bone marrow registry for a compatible donor each year. These registries keep track of potential donors' HLA types, such as the woman pictured above. She is taking a DNA sample from her cheeks. If a compatible match is found at all, it may take months or even years.

Kim needs to use chemotherapy to halt the production of aberrant WBCs in the interim. Chemotherapy is the use of medications, also known as "anticancer drugs," to treat cancer by destroying cancer cells. Nowadays, unlike targeted therapy, the term "chemotherapy" usually refers to cytotoxic drugs that generally affect rapidly dividing cells. Chemotherapy drugs interfere with cell division in various possible ways, e.g., with the duplication of DNA or the separation of newly formed chromosomes. Chemotherapy may cause damage to healthy tissue, particularly to tissues with a high rate of replacement (e.g., stomach lining). Patients may experience loss of taste buds, lining of the digestive system, and hair follicles as a result of these side effects. Following chemotherapy, these cells often heal themselves. Drugs that interact better with one another than they do separately are sometimes administered in combination. Since most chemotherapy regimens are administered in combinations, this is known as "combination chemotherapy."

CLASS ACTIVITY

CELLS

The Cell as a System: This activity enables students to explore the components of a cell and how they work together as a system.

Objectives: Students will identify major cell components; know structures and functions of components; understand how the parts of a cell interact together.

SUMMARY

- Developmental biology is an important field of science that allows us to understand how organisms, including humans, grow and develop from a molecular or genetic perspective.
- The most recent developments in the field of developmental biology have involved the emergence of synthetic biology – an interdisciplinary field that aims to create biological functions or organisms with features that do not occur naturally. The goal of this field of science is to engineer biological systems that have specific capabilities.
- Development may amount to no more than a quantitative change (usually an increase) in a system that remains essentially unaltered. Qualitative development involves an alteration in the nature of the system.
- The normal processes of development in the majority of plants and animals may be considered progressive since they lead to increases in size and complexity and to the addition of new elements to the system. As already indicated, some organisms, when placed in adverse conditions, may undergo regressive changes, both in size and complexity. Such regressive changes are a part of the normal life history of certain organisms.
- If a number of fertilized eggs of a given species are provided with conditions that enable them to develop at all, they will, with extraordinary regularity, develop into exceedingly similar adult organisms. The range of conditions they can tolerate is rather wide, and the similarity of the end products surprisingly complete.
- In viruses, activities consist in the production, aided by the machinery of a host cell, of units for building new virus or phage particles: development is simply the assemblage of these constituent units.
- One of the most striking characteristics of all developmental systems is a tendency to produce a normal end result in spite of injuries or abnormalities that may have affected the system in earlier stages. In many cases, perhaps in most, only injuries inflicted during a certain restricted period of development can be fully compensated for. During such periods the system is said to be capable of regulation or the restoration of normality.
- The existence of these developmental phenomena was realized in the first third of this century. During this period, biologists had no clear notion of the fundamental concepts needed to explain development.
- Gametogenesis, the production of sperm and eggs, takes place through the process of meiosis.
- Spermatogenesis occurs in the wall of the seminiferous tubules, with stem cells at the periphery of the tube and the spermatozoa at the lumen of the tube. Immediately under the capsule of the tubule are diploid, undifferentiated cells. These stem cells, called spermatogonia, go through mitosis with one offspring going on to differentiate into a sperm cell and the other giving rise to the next generation of sperm.

- Oogenesis occurs in the outermost layers of the ovaries. As with sperm production, oogenesis starts with a germ cell, called an oogonium (plural: oogonia), but this cell undergoes mitosis to increase in number, eventually resulting in up to about one to two million cells in the embryo.
- Human embryonic development or human embryogenesis is the development and formation of the human embryo. It is characterized by the processes of cell division and cellular differentiation of the embryo that occurs during the early stages of development.
- Growth, the increases in cell size and number that take place during the life history of an organism.
- Cell death is the event of a biological cell ceasing to carry out its functions. This may be the result of the natural process of old cells dying and being replaced by new ones, as in programmed cell death, or may result from factors such as diseases, localized injury, or the death of the organism of which the cells are part.

REVIEW QUESTIONS

1. Who is the founder of developmental biology?
2. What is the historical foundation of developmental biology?
3. What is the significance of developmental biology?
4. What is the scope of development biology?
5. What are the features of developmental biology?
6. What is gametogenesis and its types?
7. What is an early stage of development organism?
8. What is the process of growth?

REFERENCES

1. Arthur, W. (2001). Developmental drive: An important determinant of the direction of phenotypic evolution. *Evolution & Development, 3*(4), 271–278. https://doi.org/10.1046/j.1525-142x.2001.003004271.x.
2. Bowen, I. D. (1993). Apoptosis or programmed cell death? *Cell Biology International, 17*(4), 365–380. https://doi.org/10.1006/cbir.1993.1075.
3. Bowler, P. J. (2003). *Evolution: The History of an Idea*. University of California Press.
4. Campbell, N. A., Williamson, B., & Heyden, R. J. (2006). *Biology: Exploring Life*. Pearson Prentice Hall.
5. Carlson, B. M. (1999). *Human Embryology & Developmental Biology* (1st ed.). Mosby, Inc.

6. D'Arcy, M. S. (2019). Cell death: A review of the major forms of apoptosis, necrosis, and autophagy. *Cell Biology International, 43*(6), 582–592. https://doi.org/10.1002/cbin.11137.

7. Engelberg-Kulka, H., Amitai, S., Kolodkin-Gal, I., & Hazan, R. (2006). Bacterial programmed cell death and multicellular behavior in bacteria. *PLOS Genetics, 2*(10), e135. https://doi.org/10.1371/journal.pgen.0020135.

8. Forgács, G., & Newman, S. A. (2005). Cleavage and blastula formation. In *Biological Physics of the Developing Embryo* (p. 27). Cambridge University Press.

9. Gilbert, S. F., Opitz, J. M., & Raff, R. A. (1996). Resynthesizing evolutionary and developmental biology. *Developmental Biology, 173*(2), 357–372. https://doi.org/10.1006/dbio.1996.0032.

10. Kroemer, G., & Martin, S. J. (2005). Caspase-independent cell death. *Nature Medicine, 11*(7), 725–730. https://doi.org/10.1038/nm1263.

11. Marlow, F. L. (2020). *Maternal Effect Genes in Development*. Academic Press.

12. Nirmala, J. G., & Lopus, M. (2020). Cell death mechanisms in eukaryotes. *Cell Biology and Toxicology, 36*(2), 145–164. https://doi.org/10.1007/s10565-019-09496-2.

13. Richardson, M. K., & Keuck, G. (2002). Haeckel's ABC of evolution and development. *Biological Reviews of the Cambridge Philosophical Society,* 77(4), 495–528. https://doi.org/10.1017/s1464793102005948.

14. Sadler, T. W., & Langman, J. (2012). Chapter 3: Primera semana del desarrollo: De la ovulación a la implantación. In S. Seigafuse (Ed.), *Langman, Embriología Médica* (pp. 29–42). Lippincott Williams & Wilkins, Wolters Kluwer.

15. Saitou, M., & Hayashi, K. (2021). Mammalian *in vitro* gametogenesis. *Science, 374*(6563), eaaz6830. https://doi.org/10.1126/science.aaz6830.

16. Singh, V. (2013). *Textbook of Clinical Embryology* (E-book). Elsevier Health Sciences.

17. Standring, S. (2015). *Gray's Anatomy E-Book: The Anatomical Basis of Clinical Practice*. Elsevier Health Sciences.

18. Vitale, I., Galluzzi, L., Castedo, M., & Kroemer, G. (2011). Mitotic catastrophe: A mechanism for avoiding genomic instability. *Nature Reviews Molecular Cell Biology, 12*(6), 385–392. https://doi.org/10.1038/nrm3115.

19. Zagris, N. (2022). Aristotle (384–322 BC): The beginnings of embryology. *The International Journal of Developmental Biology, 66*(1–2–3), 5–8. https://doi.org/10.1387/ijdb.220040nz.

CHAPTER

2

Approaches to Development: Developmental Genetics

LEARNING OBJECTIVES

After studying this chapter, you will be able to:

- Discuss the concept of developmental genetics;
- Explain the sex chromosomes;
- Focus on cloning of genes;
- Learn about transgenes;
- Discuss the gene duplication.

KEY TERMS FROM THIS CHAPTER

Achondroplasia
Genes
Homologues
Organisms
Cell biology
Homeobox genes
Multicellular
Developmental Genetics

INTRODUCTION

The study of developmental genetics focuses on how genes control an organism's growth, differentiation, and tissue patterning. It uses concepts from developmental biology, molecular biology, and genetics to understand the complex processes that transform a single fertilized egg into a multicellular, complex organism. Understanding how genes encode information that controls the formation of structures and functions within an organism is at the heart of developmental genetics. The instructions needed to synthesize essential proteins and RNA molecules for development are found in genes. These molecules orchestrate cell division, differentiation, and morphogenesis through precise molecular interactions that set off cascades of molecular events.

Gene regulation is one of the core ideas of developmental genetics. Gene expression can be regulated both temporally and spatially; genes are not always active. DNA regulatory sequences function as switches, activating or deactivating genes in response to environmental or developmental cues. For example, during embryonic development, a particular gene may be exclusively expressed in cells that will eventually form muscle tissue, whereas a different gene may only be expressed in nerve cells. The final form of the organism is shaped in large part by the precise timing and location of gene expression. Diseases or abnormalities in development can result from gene mutations or modifications to regulatory sequences. Investigating these mutations sheds light on how genes normally function and what part they play in development. For instance, mutations in the HOX genes, which are essential for figuring out how an animal's body is patterned, can cause drastic changes to the way the body is structured.

Developmental genetics has been transformed by technological breakthroughs. Scientists can precisely edit genomes using methods like CRISPR-Cas9, which allows them to study the roles of particular genes during development. Mapping gene expression patterns across tissues and developmental stages has been made easier by high-throughput sequencing technologies, which have also produced detailed insights into gene regulatory networks. Additionally, model organisms are essential to the study of developmental genetics. In comparison to human beings, certain organisms, such as fruit flies (*Drosophila melanogaster'*), zebrafish (Danio rerio), and mice (Mus musculus), have well-characterized genomes and simpler developmental processes. They are effective systems for researching the *in vivo* regulation and function of genes.

The goal of developmental genetics is to identify the molecular processes that enable a single fertilized egg to develop into a complex organism. In addition to expanding our knowledge of fundamental biological concepts, researchers that can better understand how genes govern developmental processes also open up new avenues for application in biotechnology, medicine, and agriculture. The field of developmental genetics promises more discoveries that will impact our comprehension of life's basic processes as technologies continue to progress.

2.1. CONCEPT OF DEVELOPMENTAL GENETICS

The study of developmental genetics employs genetics to understand how a multicellular organism's fertilized egg develops into an adult. The majority of our knowledge is derived from research on mice, zebrafish, roundworms, and drosophila, or fruit flies. These "model" organisms can be experimentally made mutant, or mutagenized, and the effects of this genetic modification can be studied in organisms as old as adults and as young as early embryos. Naturally, such research is unimaginable for humans. However, the fact that the majority of the genes regulating development have been conserved throughout evolution has been extremely helpful to scientists. Certain genes that have been isolated from worms and flies have homologues in mice and humans. Studies of gene function support this conservation of gene structure: a mutation in a fly can be corrected by using the human homologue of a gene that is active during fly development.

This remarkable conservation says two things:

- Some genes controlling development are hundreds of millions of years old;
- What is learned about development in model organisms is often transferable to humans.

Geneticists started creating drosophila mutants that disrupted normal development in the 1970s. Maternal-effect genes, segmentation genes, and pattern-formation genes were the three gene categories they discovered to have an impact on these flies' early development. They then connected these genes' mode of action to the phenotypes that resulted from their mutations. The pioneering research of these pioneers was followed by the ever-growing body of developmental geneticists.

2.1.1. Developmental Mutants

Genetic variations known as developmental mutants result in organisms that, when compared to people of the wild type, display aberrant or altered developmental patterns. These mutants are essential to the study of developmental genetics because they shed light on the functions of particular genes in regulating body patterning, organ formation, and tissue differentiation, among other aspects of development. Researchers can identify genes that are crucial for particular developmental processes and learn more about the molecular mechanisms underlying normal development by examining developmental mutants. A well-known illustration of developmental mutants originates from research conducted on the fruit fly *Drosophila melanogaster.'* The fruit fly's short generation time, high number of progeny, and well-characterized genome have made it a model organism in genetics for more than a century. Drosophila mutations can cause a broad variety of phenotypes that impact internal organs, appendages, and body segments. These phenotypes can provide important information about how genes function during development.

For example, the identification of the HOX genes, which are essential for establishing the anterior-posterior axis during development, was made possible by the discovery of the Antennapedia (Antp) mutant in Drosophila. The antennae of Antp mutants become legs, emphasizing the function of the Antp gene in determining segment identity along the body axis. Likewise, Drosophila mutants affecting the eyeless (ey) gene have contributed to our understanding of the molecular mechanisms underlying eye development. Fly mutants that cause flies to either have no eyes or drastically reduced eye structures show how important ey is to the development of eyes.

Apart from Drosophila, significant insights into developmental genetics have also been gained from mutants in other model organisms such as mice (Mus musculus) and zebrafish (Danio rerio). For instance, because of their genetic resemblance to humans, mouse mutants have proven invaluable in the study of mammalian development. In mice, mutations in genes like Sonic Hedgehog (Shh) have demonstrated their critical roles in the development of the limb and neural tube; these discoveries have implications for our understanding of developmental disorders in humans. Because of their transparent embryos and quick external development, zebrafish mutants have proven especially helpful for researching vertebrate development. The genetic mechanisms governing the formation and segmentation of the tail during embryogenesis have been clarified by mutations in zebrafish genes such as spadetail (spt).

Targeted mutant production in a variety of organisms has accelerated thanks to technological advancements in genome editing, such as CRISPR-Cas9. With CRISPR-Cas9, scientists can precisely alter particular genes and closely examine the phenotypic alterations that arise. The ability to produce mutants with previously unheard-of speed and accuracy has transformed developmental genetics. Researching developmental mutants has potential applications in biotechnology and medicine in addition to improving our understanding of fundamental biological concepts. Finding the genes responsible for human developmental disorders can be facilitated by research findings in developmental genetics. For instance, human congenital malformations affecting limb development can be caused by mutations in HOX genes, which are similar to those studied in Drosophila.

Additionally, developmental mutants are useful resources for studying the role of genes in stem cell biology and tissue regeneration. Researchers may be able to treat birth abnormalities and repair damaged tissues by using their understanding of how genes regulate developmental processes. In order to understand the genetic basis of organismal development, developmental genetics research relies heavily on the developmental mutants. Researchers are still working to understand the intricate webs of genes and pathways that coordinate development from a single cell to a fully formed organism. This work is being done through the analysis of mutants in model organisms and the use of cutting edge genetic technologies.

2.1.2. Gene Classes

2.1.2.1. Maternal-Effect Genes

Development is under genetic control even before the zygote forms. We can infer this from the fact that very little gene transcription occurs in the first few cell divisions after fertilization and that three types of mutations in female flies have distinct and repeatable effects on the development of the early embryo. As we've previously stated, the egg has reserves of nutrients that are essential to the zygote's survival. Furthermore, the egg stores spindle proteins, nucleotides, ribosomes, and enzymes needed for initial rounds of cell division. These products derived from mothers include proteins that are essential for the early stages of development. These proteins are transferred from the egg's cytoplasm into the zygote and developing embryo. The zygotic genes that control axis formation—which distinguishes between the head and tail, or anterior and posterior, and the top and bottom, or dorsal and ventral—are regulated by these maternal-effect gene products.

2.1.2.2. Segmentation and Pattern Formation Genes: Homeobox Genes

Every organism has a specific time when its own genes become active and replace those of the mother. This transition takes place in humans between the four and eight-cell stage. The basic body plan is established by two types of events. Initially, the embryo is divided into a linear, antero-posterior series of progressively smaller segments by the interaction of the segmentation genes (Figure 2.1). Then, each segment is given a specific developmental fate by the pattern-formation genes. (These unduly simplistic statements depict profound events: various gene hierarchies interacting both within and between themselves to enable the numerous processes that must take place during embryogenesis, including cellular migration, differentiation, and proliferation.

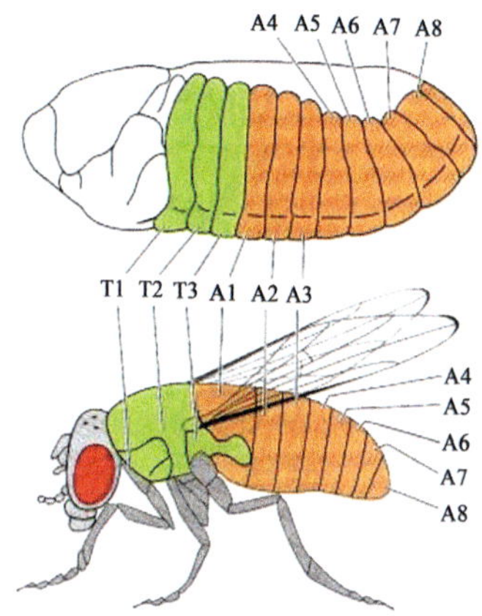

Figure 2.1. *Early development of Drosophila melanogaster' embryo: (A) immediately after fertilization; (B) about 9 hours later. Note prominent division of embryo into a linear series of segments. Segmentation results from the action of many genes, which have homologues in humans.*

https://ars.els-cdn.com/content/image/3-s2.0-B9780123852120000159-f15-04-9780123852120.jpg.

The homeobox genes are a hierarchy of genes essential to pattern formation and segmentation. The aforementioned segments' distinct structures—such as the head structures from the most anterior segment and the thoracic features from the mid-segments—are determined by these genes. When the mutant flies were first discovered, they had remarkable phenotypes, like wings growing from the head where the antennae should have been or legs growing from the head where the eyes should have been. Studies to isolate and characterize these genes were prompted by these peculiar phenotypes. Thanks to this work, a class of genes known as the homeobox genes and their mutants, known as homeotic mutants, were identified. The major conclusions of this work are worth enumerating:

- Multicellular organisms from flies to humans contain a collection of homeobox genes. Flies have 8 of them; humans have 38. The fly genes are abbreviated HOM; the mouse genes, Hox; the human ones, HOX.
- Homeobox genes have arisen by a process of gene duplication and translocation. All HOM genes are located on one fly chromosome. Human HOX genes have been distributed to four chromosomes (numbers 2, 7, 12, and 17) (Figure 2.2).
- There is striking similarity in DNA and amino acid sequences in homeobox genes within a species and between species. The genes contain a conserved 180-bp region of DNA coding for a 60 amino acid region of the homeobox protein. This conserved region is called the homeodomain.
- The linear order of homeobox genes in one species, such as the fly (Figure 2.2), remains relatively the same in other species as different as humans. Remarkably, these genes are expressed temporally and spatially along the embryo's anterior/posterior axis in the order of their chromosome position. That is, the genes found closest to the 3′ end of the cluster (Figure 2.2), *a1* and *b1* in the mouse) are expressed first. Their expression starts in the most anterior region and extends posteriorly as development proceeds. Then the second genes (*a2* and *b2*) are turned on and are expressed a bit more posteriorly but extending tail-ward. Finally, the most 5′-located gene is expressed, but only in the tail region. Each body segment, then, has its own array (or orchestra) of homeobox gene expression. The HOM, Hox, and HOX gene clusters represent copies of an ancestral gene cluster that likely existed before insects and mammals diverged.
- The proteins encoded by HOM or HOX genes are transcription factors that bind to enhancer or promoter regions of other genes and either turn them on or turn them off. There is still much to be learned about the specific genes modulated by homeobox-derived transcription factors.
- Homeobox genes are involved prominently in embryonic development. Mutational studies have clearly established their role in development of the brain, the limbs, and the vertebrae.
- A rare and unusual human birth defect is caused by a mutation in a single HOX gene (d13). Three families have been described in which hand and foot abnormalities were noted. Fusion and duplication of digits occurred in each affected patient. This condition is inherited as an autosomal dominant trait with incomplete penetrance.

Homeobox genes are but one class of genes regulating segmentation and pattern formation. We discuss them rather than others because more is known about them and because our knowledge about them is representative of the action of other gene classes.

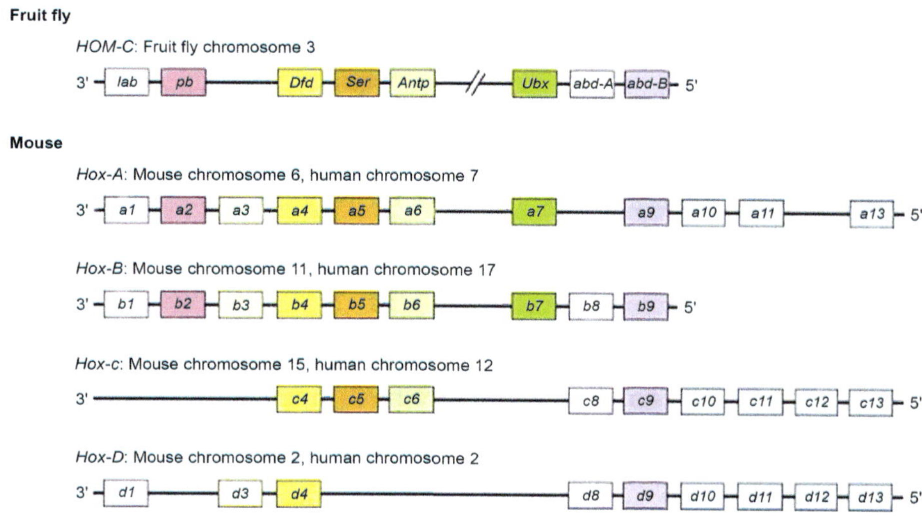

Figure 2.2. *The homeobox gene family in drosophila and vertebrates. In the fly, eight genes make up the HOM-C family. In mice and humans, 38 genes, distributed to 4 chromosomes, constitute the family called Hox in mice and HOX in humans. Color-coding of the eight HOM-C genes in flies depicts each gene's domain of activity in forming an adult fly. The 3⊠-to-5⊠ order of the genes on the chromosome matches the anterior to posterior activity of the genes. In mice and humans, Hox and HOX genes have been dispersed to four chromosomes, but their 3⊠-to-5⊠ order has been conserved on each chromosome.*

https://ars.els-cdn.com/content/image/3-s2.0-B9780123852120000159-f15–05–9780123852120.jpg.

2.1.3. Signaling Molecules

In multicellular organisms, communication between cells is essential for performing a variety of tasks. They communicate with one another by means of chemical signals. Numerous signaling molecules have been identified, including peptides and proteins, derivatives of amino acids and fatty acids, steroids and retinoids, and dissolved gases. These signaling molecules are frequently taken from one cell and recognized by a target cell that is nearby or far away. A reaction is started once the signaling molecule attaches to the target cell, usually through a cell surface receptor. Such cell-to-cell communication is essential for the development of the embryo. All components of signaling pathways, including signal synthesis, cell transport, target cell receptors, and metabolic pathways responding to signals, are governed by genes. We'll talk about two signaling modalities that are especially pertinent to people.

2.1.3.1. Fibroblast Growth Factor

Ten distinct fibroblast growth factor (FGF) loci, which encode proteins that promote the growth of various cell types, are found in the human genome. These factors are present in many different types of cells, despite the fact that they were first identified in fibroblasts. They play a number of roles in the development of the embryo, including the formation of the midbrain, spinal cord, facial structures, limb buds, and nerve cell processes (axons and dendrites). They control the closure of the skull bones and the growth of long bones later in development. FGFs bind to target cells' FGF receptor (FGFR) proteins to send a signal. The FGFR protein then initiates growth-promoting pathways.

There have been numerous reports of human FGFR and FGF gene mutations. All of these mutations are inherited as Mendelian traits and result in a range of birth defects unique to the affected locus. The most common abnormalities caused by these mutations are in the limbs and face. The most common type of human dwarfism, achondroplasia, is brought on by a mutation in the FGFR3 gene and will be covered in more detail later. Different conditions, some more severe than achondroplasia, are brought on by other mutations in FGFR3. All in all, mutations in one of the four FGFR genes cause over 100 distinct birth defects. Serious deformities of the face, limbs, and skull are frequently the result of these mutations.

2.1.3.2. Sonic Hedgehog Genes

Hedgehog is the name given to one of the many flies mutations that affect development because of its peculiar appearance and small size. It bears the name of a character from a video game. We now understand that pattern formation depends on this gene locus, which goes by the name hh. The morphogen, or protein, that the hh locus encodes is a protein. The creation of new structures during development is referred to as morphogenesis. Hedgehog protein is one of the morphogens that is essential to the development of the embryo because it diffuses through the tissues and creates concentration gradients of the protein that lead neighboring cells to develop differently. The most well-studied class of loci encoding morphogens is the hh gene family.

Humans are among the many other organisms that contain homologues of hh. In humans, several distinct hedgehog gene loci have been found. The most researched is known as Sonic Hedgehog (SHH). Human notochord and neural tube cells secrete the SHH protein, which creates a gradient that induces and organizes brain and spinal cord cells. Similarly, the asymmetrical formation of the digits in the developing limb is caused by secretions of SHH.

Human SHH mutations are uncommon. They could result in severe birth defects or ones that are hardly noticeable (Figure 2.3). Severe affected individuals (Figure 2.3A) show failure of brain and mid-face development, a condition known as holoprosencephaly. These babies are characterized by the absence of the forebrain, closely spaced eyes, and clefts of the lip and palate. However, the same mutation within the same family could produce a phenotype that is very different. The mother of the child in Figure 2.3A is the woman in Figure 2.3B. Her only symptom is the loss of one central incisor tooth. It appears likely that other genes have modified the expression of the SHH mutation, but this hypothesis raises the intriguing and as-yet unanswered question: Which modifying genes, and which actions do they modify? How can we account for this widely variable expressivity?

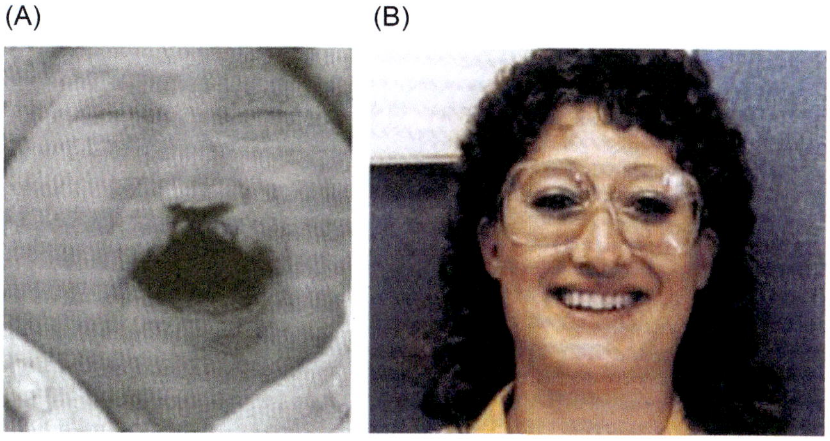

Figure 2.3. *Variable expression of identical mutations in the SHH gene. (A) An infant has a small head, closely spaced eyes, missing nose bones, and a cleft lip. (B) The mother of the infant shown in (A). She has the same SHH mutation as her child, yet her only clinical manifestation is absence of a single tooth—a central upper incisor.*

https://ars.els-cdn.com/content/image/3-s2.0-B9780123852120000159-f15–06–9780123852120.jpg.

2.1.4. Programmed Cell Death (Apoptosis)

The opposite process, known as programmed cell death, or apoptosis, is essential to both embryonic development and cell growth. Unlike the process that eliminates damaged or dead cells brought on by trauma, infection, or other environmental insults, apoptosis is a controlled event. Apoptosis is a methodical process that results in the specific molecules' orderly disintegration and the death of the designated cell. Apoptotic cells contract, causing fragmentation and clumping of their nuclear chromatin in the vicinity of the nuclear membrane. Ultimately, the dead cells are consumed by nearby cells or circulating scavenger cells.

Apoptosis is a necessary aspect of human existence. Our hands have a paddle-like shape when they are still developing. Then, to form fingers, apoptotic events eliminate rows of cells. Our four-chambered organ, the heart, is formed through remodeling during the process of apoptosis. In the brain, far more cells are created during the embryonic and fetal stages than are present at birth, and the extra cells are systematically eliminated. All of this begs the question, "How is apoptosis controlled?" and, predictably, the answer is that suicide (or cell death) genes regulate cell suicide. These genes were originally discovered in the roundworm Caenorhabditis elegans, and other organisms have homologs. Over a dozen genes in the worm either stimulate or inhibit apoptosis. There are known to be over 25 such genes in vertebrates. This "cassette" of death genes responds to signals coming from both inside and outside of cells. A subset of these genes produce endonucleases, which break down chromosomal DNA, and proteases, which break down proteins. Cell surface receptors or transcription factors comprise the remaining ones.

2.2. SEX CHROMOSOMES

Sex chromosomes are chromosomes that contain the genes that identify a person's sex. The sex chromosomes of humans are a normal pair of allosomes found in mammals. They are not the same as autosomes in size, shape, or behavior. Members of an allosome pair may differ from one another, in contrast to autosomes, which are found in homologous pairs in diploid cells whose members have the same form. In 1905, Nettie Stevens and Edmund Beecher Wilson made separate discoveries regarding sex chromosomes. But Stevens is acknowledged as having found them before Wilson.

As a result of sex determination, people acquire traits that enable them to be classified as either male, female, or, in certain situations, hermaphrodites. For example, in the solid nematode C and other species. elegans, sexual characteristic differences can be minuscule; in fact, in C. elegans, the existence of a testis as opposed to an ovotestis is the only sexual trait that sets them apart. The phenotypic variations between the sexes can be fairly notable in other species. Think about a tom turkey's amazing plumage and display in contrast to a female turkey's relatively plain features. Mammals can be easily identified as female or male based on a variety of internal and external phenotypes, behaviors, and metabolic differences.

The discovery of sex chromosomes in the early 1900s marked the first significant advancement in our understanding of sex determination. Through careful examination of the chromosomes of male and female insects, researchers found that while most chromosomes were present in equal amounts in both sexes, there were one or two extra chromosomes that were not equally represented in either. Over the years, analyzes of more species have shown that most animals' sex determination is mostly caused by chromosomal differences.

2.2.1. Chromosomal Mechanisms of Sex Determination in Insects

Given that insects are the most diverse class of organisms on the planet, a significant diversity in their sex determination mechanisms is not surprising (Saccone et al., 2002). Nonetheless, the majority of insects share dimorphic sex chromosomes that allow for cytological differentiation with most other animals.

Being heterogametic means that an animal has two different sex chromosomes, which allows it to produce two different kinds of gametes. On the other hand, only one kind of gamete can be produced by members of the homogametic sex. Males typically have one X and one Y chromosome, while females have two X chromosomes in humans and many other animals. In butterflies and moths (order Lepidoptera), the system is inverted; males are homogametic and females are heterogametic. In Lepidoptera, the sex chromosomes are called W and Z. W chromosome development is typically linked to the emergence of female traits. ZO develops into females and ZZ into males when the W chromosome is missing. For certain species, developing as a female does not even require possessing a W chromosome. In the absence of the W chromosome, a moth known as *Talaeporia tubulosa* regulates sex determination based on the surrounding temperature. The Z chromosome travels to the outer pole in colder climates, where it results in an increase in the production of male offspring. In warm climates, the Z chromosome is located on the inner spindle and more female eggs are produced (Traut et al., 2007). This system reflects the adaptive advantage of producing more female offspring when the environment is favorable (warm) and there is a greater chance that resources will be available for their future reproduction.

A single chromosome (XX/XO) sex determination system is also used by some grasshoppers; in this system, males are classified as XO since they have only one sex chromosome. Because they generate two distinct types of gametes, men are the heterogametic sex. In some mosquito genera, the two sexes are chromosomally indistinguishable, further reducing the system of chromosomal sex determination. It is believed that in this homogametic group, a dominant male-determining factor determines sex.

2.2.2. Sex Determination in Drosophila

Our understanding of heredity has been greatly aided by the fruit fly *Drosophila melanogaster''s* sex chromosomes. Consequently, it might surprise you to learn that fruit flies identify sex using a comparatively uncommon method. Actually, the X:A ratio—that is, the ratio of the number of X chromosomes to the number of sets of autosomes—determines sex in Drosophila (Cline & Meyer, 1996). Which sex-specific transcription pattern is initiated depends on the balance between male- and female-determining factors encoded on the autosomes and on the X chromosome. Therefore, XY and XO flies are male, and XX, XXY, and XXYY flies are female. Because of their dosage compensation mechanism, flies cannot survive with more than two copies of an X chromosome. (Dosage compensation describes the mechanisms by which animals balance the quantity of gene products produced in males and females from X-linked genes. All of the X chromosomes in Drosophila are still active, in contrast to those in mammals, and flies regulate the amounts of X-linked gene products by doubling the expression of the X chromosome in males. Aneuploidy, a condition that severely upsets cell equilibrium, is caused by an extra copy of the X chromosome, which houses nearly one-third of the genes found in flies.).

There are various other ways in which the process of determining sex in Drosophila is different from that of mammals. First, there is no indifferent period; sex determination happens right away at fertilization. Furthermore, each cell in the embryo detects the X:A ratio, which causes either the female- or male-specific pattern of transcription. Hormones have no role in sex-specific traits. The results of microarray experiments show that there are significant sex-specific variations in gene expression. Indeed, it was discovered that sex-specific expression biases were present in about 30% of Drosophila genes (Parisi et al., 2004).

2.2.3. Sex Determination in Mammals

Sex in placental mammals is determined by the Y chromosome. Male and female cells typically have an X and a Y chromosome, while female cells typically have two X chromosomes. Sometimes people are born with sex chromosome aneuploidies, in which case their sex is always determined by whether or not they have a Y chromosome. As a result, people with the karyotypes 45,X and 47,XXX are female, and people with the karyotypes 47,XXY and 47,XYY are male. X inactivation and the relatively gene-poor human Y chromosome allow humans to tolerate supernumerary numbers of sex chromosomes. While the Y chromosome has been implicated in the determination of mammalian sex since the early 1900s, scientists were unable to pinpoint the specific Y chromosome region responsible for this process until 1959 (McLaren, 1991). Afterwards, scientist David C. Page examined the chromosomes of sex-reversed XX men, a unique subset of people who resemble men but carry two X chromosomes rather than one X and one Y chromosome. Page found that sex-reversed males carried genes from a 140-kilobase region on the short arm of the Y chromosome through DNA hybridization with probes corresponding to different regions of the Y chromosome (Figure 2.4). This area had most likely undergone a translocation where it was moved to the X chromosome (Page et al., 1985). Following investigations, this area was narrowed down (McLaren, 1991), and it was discovered that the master regulator of sex determination was a single gene known as the sex-determining region of the Y, or SRY. So, all that is needed to cause male development is the existence of this particular Y chromosome region (Koopman et al., 1991).

In human embryos, the SRY gene encodes a unique transcription factor that activates a testis-forming pathway at about week seven of development. Before this time, the embryonic gonad is "indifferent," meaning that it is capable of developing into either a testis or an ovary (Figure 2.5).

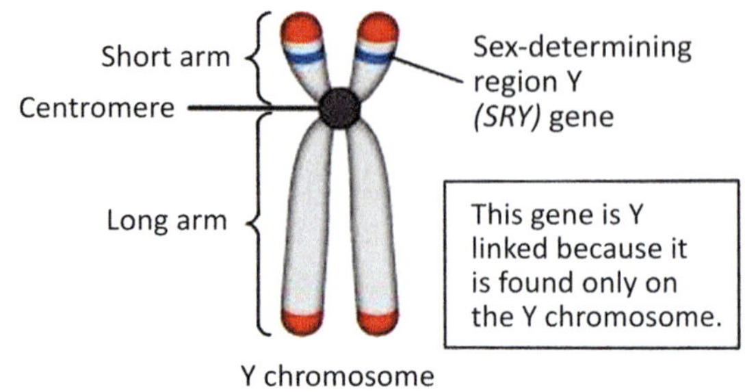

Figure 2.4. *Location of the sex-determining region Y (SRY) gene on the Y-chromosome. The SRY gene (blue band) on the male Y chromosome regulates sex determination in mammals.*

https://www.nature.com/scitable/content/ne0000/ne0000/ne0000/ne0000/117894590/7_2_1.jpg.

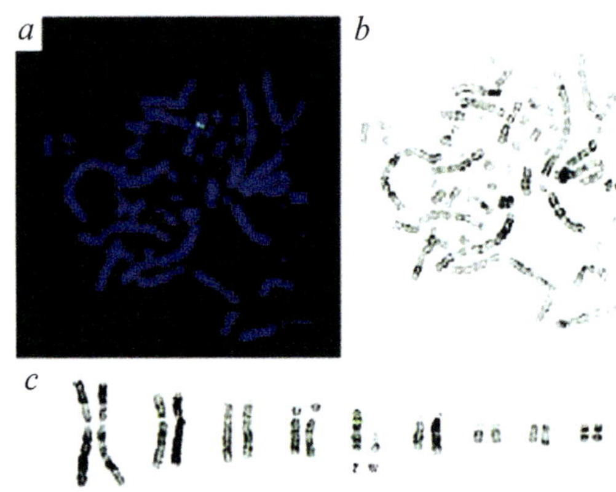

Figure 2.5. *FISH mapping of DMRT1 on chicken Z chromosome. a. Hybridization of DMRT1 cosmid to a female chicken metaphase spread. b. DAPI banding of the same metaphase spread, converted into G-like bands. c. G-banded karyotype of chicken macrosomes hybridized with DMRT1 (green spot on Z chromosome).*

https://www.nature.com/scitable/content/31393/ng0399_259b-f2_MID_1.gif.

Likewise, the early embryo has two systems of ducts, Wolffian and Müllerian ducts, which are capable of developing into the male and female reproductive tracts, respectively. Once the SRY gene product stimulates the indifferent gonad to develop into a testis, the testis begins producing two hormones, testosterone and anti-Müllerian hormone, or AMH. Testosterone and one of its derivatives, dihydrotestosterone, induce formation of other organs in the male reproductive system, while AMH causes the degeneration of the Müllerian duct. In females, who do not contain the SRY protein, the ovary-forming pathway is activated by a different set of proteins. The fully developed ovary then produces estrogen, which triggers development of the uterus, oviducts, and cervix from the Müllerian duct.

2.2.4. Sex Determination in Birds: Z and W Chromosomes

The Z and W chromosomes in birds determine sex, with females making up the heterogametic sex. It was quickly discovered that the mechanisms by which birds and mammals determine their sex differed significantly. Scientists looked for homologues of other genes necessary for the formation of testes in mammals after working with chickens and failing to locate a counterpart of the SRY gene needed for the determination of testis in mammals. The DMRT1 gene on the chicken Z chromosome was found as a result of these studies (Nanda et al., 1999).

The DMRT1 gene product is an essential component of the SRY-initiated testis-forming pathway in mammals. Even in the presence of SRY, two copies of the DMRT1 gene are required for testis development. Mammals have two copies of DMRT1 because it is located on an autosome, chromosome 9 in particular. In contrast, since DMRT1 is located on the Z chromosome, only male chickens have two copies of it. The results of an *in situ* hybridization experiment in which a female chicken's chromosomes were hybridized with a DRMT1 probe are displayed in Figure 2.5. Under a microscope, the chromosomes can be seen in Figure 2.5b, and Figure 2.5a demonstrates that the DRMT1 probe only hybridizes to one area on the female Z chromosome. Since these chromosomes are in metaphase, hybridization has been seen on their two sister chromatids. A partial karyotype with specific chromosomes is displayed in Figure 2.5c. The Z chromosome is noticeably bigger than the W chromosome, as is easily observed.

It is still unclear what exactly sets off the testis-forming pathway in chickens. The requirement for female development of two distinct genes on the W chromosome, FET1 and ASW, sets birds apart from mammals. FET1 is expressed in the gonads prior to the time of sexual differentiation, although its exact function is unknown (Smith & Sinclair, 2004). Similar to mammals, chickens have an indifferent gonad until approximately the fourth day of development, at which point an ovary or a testis begin to form. The development of other sex-specific traits is then regulated by hormones. But unlike mammals, chickens need estrogen earlier in the process of determining their sex—in fact, it's essential for the development of the ovary. It is actually possible to turn genetically male chickens into females by injecting estrogen into the eggs when they are still developing sensitively. There are other animals besides birds that use the ZZ-ZW method to determine sex. A comparable method of determining sex has been found in certain fish and amphibians, as well as in reptiles.

2.2.5. Sex Determination and Environment

As was previously mentioned in the discussion of *Talaeporia tubulosa*, sex determination can occasionally be significantly influenced by environmental factors. Insects are not an exception; in certain groups of reptiles, temperature also plays a significant role in determining mating behavior. For instance, the temperature at which eggs are incubated determines sex in most turtle species and crocodilian reptiles. This theme has multiple variations. When eggs are incubated below 22°C or above 28°C, they develop into females in the alligator snapping turtle (*Macroclemys temminckii*), whereas eggs incubated at intermediate temperatures primarily develop into males. Similar biphasic dependence on temperature is seen in American alligators, although the curve is shifted toward higher temperatures. All females of the European pond turtle Trachemys scripta are produced at incubation temperatures above 30°C, while all males are produced at temperatures below 25°C. Equal numbers of males and females are produced at 28.5°C. In the European turtle, Emys orbicularis, the temperature-dependent portion of the sex-determination pathway has been extensively investigated. Within E. orbicularis, the synthesis of the enzyme aromatase—which changes androgens like testosterone into estrogens—appears to be the crucial temperature-dependent component. The sex ratio is biased toward more females at higher temperatures because of increased aromatase activity, which produces more estrogens. It is evident that evolution has produced a variety of strategies for producing distinct sexes when comparing the various mechanisms for determining sex among species. Because sexual reproduction adds new genetic variability to a population with every generation, it has enormous adaptive value for a species. In most species, chromosomes are decisive, but environmental factors still add extra complexity to the developmental process.

2.2.6. Maternal and Zygotic

Different roles are played by maternal and zygotic genes in the early development of many organisms, especially during embryogenesis. These gene groups play a key role in coordinating the intricate processes that develop an egg into a multicellular organism.

2.2.6.1. Maternal Genes

During oogenesis, or the formation of eggs, the mother deposits maternal genes into the egg. They are translated and transcribed into proteins and mRNA, which are then stored in the cytoplasm of the egg. Before the zygotic genome is fully activated, maternal mRNA and proteins are crucial for supplying the first instructions and resources required for early embryonic development. When the embryo depends only on the mother's mRNA and proteins in the early stages of development, the mother's contribution is especially significant. The polarity of eggs, the development of body axes, and the formation of early embryonic structures are all regulated by maternal factors. For instance, in Drosophila, maternal morphogens such as bicoid and nanos establish anterior-posterior polarity in the developing embryo.

After fertilization, maternal mRNA molecules usually remain stable and last for different amounts of time. This guarantees a steady supply of vital proteins until zygotic transcription takes over. Maternal gene products gradually deteriorate or become less abundant as the zygotic genome becomes active and starts producing its own transcripts.

2.2.6.2. Zygotic Genes

Unlike maternal genes, zygotic genes are transcribed from the embryo's own genome following fertilization. The embryo starts to rely on its own genetic machinery to drive development at zygotic genome activation (ZGA). For the embryo to develop cell fate specification and differentiation and to become independent of maternal contributions, this transition is necessary. The complex processes of cell division, differentiation, and morphogenesis that occur later in embryonic development are controlled by zygotic genes.

These comprise genes involved in patterning, organ formation, and differentiation specific to particular tissues. For instance, zygotic genes regulate the development of neural tissues, skeletal structures, and organs in vertebrate embryos. Zygotic gene expression is tightly regulated in terms of timing. The zygotic genome is either completely dormant or very slightly active at first. Transcription factors and signaling pathways cause the activation of zygotic genes in particular spatial and temporal patterns when maternal gene products decrease.

2.2.6.3. Interplay and Regulation

The coordinated development process depends on the interaction of maternal and zygotic genes. Zygotic genes take over to drive more specialized and precise developmental processes, while maternal genes set the initial framework and supply the necessary materials for early embryonic development. Errors in the equilibrium or operation of maternal and zygotic genes may result in aberrant development or the death of the embryo. Investigations utilizing model organisms such as mice, zebrafish, and Drosophila have yielded important insights into the interactions and regulation of developmental programs by these genes. Different stages of gene expression during embryonic development are represented by the maternal and zygotic genes. Zygotic genes control the later phases of organogenesis and differentiation, whereas maternal genes supply the initial direction and resources for early development. Deciphering the functions and control of these genes is crucial to comprehending the intricacies of embryonic development and its consequences for well-being and illness.

2.2.7. Genetic Pathways

Genetic pathways are complex webs of genes and the products of those genes that work together to control different biological functions in cells and organisms. These pathways are essential for growth, metabolism, responding to stimuli, and preserving cellular homeostasis. Clarifying how genetic information is translated into phenotypic traits and how disruptions in these pathways can result in disease requires an understanding of genetic pathways.

2.2.7.1. Components of Genetic Pathways

Genetic pathways consist of several key components that work together to transmit signals and execute biological functions:

1. **Genes:** Genes within a pathway encode proteins or non-coding RNAs that directly or indirectly participate in the pathway's function. These genes can be organized sequentially, forming linear pathways, or they can interact in more complex networks.
2. **Transcription Factors:** Transcription factors are proteins that regulate gene expression by binding to specific DNA sequences (promoters or enhancers) and either activating or repressing transcription of target genes. They play pivotal roles

in determining when and where genes are expressed, thereby controlling developmental processes and cellular responses.

3. **Signaling Molecules:** Signaling molecules, such as hormones, growth factors, and cytokines, transmit extracellular signals to the cell interior. These molecules bind to receptors on the cell surface, initiating intracellular signaling cascades that often involve phosphorylation events and activation of downstream effectors.
4. **Effector Proteins:** Effector proteins are the downstream targets of signaling pathways that carry out specific functions within the cell. They include enzymes involved in metabolism, structural proteins, and regulators of cell cycle progression or apoptosis.
5. **Feedback Mechanisms:** Genetic pathways often include feedback loops to regulate their own activity. Negative feedback loops dampen the initial signal, preventing overstimulation or prolonged activation. Positive feedback loops amplify signals, leading to robust responses in certain contexts.

2.2.7.2. Types of Genetic Pathways

Genetic pathways can be categorized based on their functions and the processes they regulate:

1. **Developmental Pathways:** These pathways control the spatial and temporal patterning of tissues and organs during embryonic development. Examples include pathways involved in body axis formation, organogenesis, and limb patterning.
2. **Metabolic Pathways:** Metabolic pathways regulate the biochemical reactions that convert nutrients into energy and building blocks for cellular processes. Examples include glycolysis, the citric acid cycle, and fatty acid synthesis pathways.
3. **Signal Transduction Pathways:** These pathways transmit signals from the cell surface to the nucleus or other cellular compartments, regulating gene expression and cellular responses to environmental stimuli. Examples include the MAPK (mitogen-activated protein kinase) pathway and the PI3K-Akt pathway.
4. **Cell Cycle and Apoptosis Pathways:** These pathways govern the progression of cells through the cell cycle stages (G1, S, G2, M) and regulate programmed cell death (apoptosis) to maintain tissue homeostasis and prevent aberrant proliferation.

2.2.7.3. Techniques for Studying Genetic Pathways

Advances in molecular biology and genetics have enabled researchers to study genetic pathways with increasing precision:

1. **Gene Knockout and Knockdown:** Using techniques such as CRISPR-Cas9 or RNA interference (RNAi), researchers can selectively disrupt or silence specific genes within pathways to investigate their functions.
2. **Transcriptomics:** High-throughput RNA sequencing (RNA-seq) allows researchers

to profile gene expression patterns across different conditions or developmental stages, providing insights into how genetic pathways are regulated.

3. **Proteomics:** Mass spectrometry-based proteomics enables the identification and quantification of proteins involved in genetic pathways, elucidating protein-protein interactions and post-translational modifications.
4. **Bioinformatics and Systems Biology:** Computational approaches are used to model and simulate genetic pathways, integrating data from multiple sources to predict pathway dynamics and behavior.

2.2.7.4. Clinical Relevance

Gaining an understanding of genetic pathways is essential to improving our understanding of disease mechanisms and creating tailored treatments. Diseases like cancer, metabolic disorders, and aberrant development can result from pathway dysregulation. Treating these conditions by restoring normal pathway function through drug or gene therapy targeting specific nodes within pathways shows promise.

The complex molecular networks that control biological processes in cells and organisms are represented by the genetic pathways. Technological developments and interdisciplinary methods continue to expand our knowledge of these pathways, providing new perspectives on basic biological concepts and their possible uses in biotechnology and medicine.

2.2.8. Genetic Mosaics

When an organism has cells with distinct genotypes, it is said to be genetically mosaic. This is usually the result of genetic mutations that happen after fertilization. This mosaic pattern can appear in different parts of the organism, ranging from individual cells within tissues to distinct organs or regions. Genetic mosaicism sheds light on the dynamic nature of genetic variation and its effects, playing a vital role in development, disease, and evolution.

2.2.8.1. Mechanisms of Genetic Mosaicism

Genetic mosaicism can arise through several mechanisms:

1. **Somatic Mutations:** Somatic mutations occur in non-germline cells after fertilization and can lead to mosaicism within tissues and organs. These mutations can result from errors in DNA replication, exposure to mutagens, or defects in DNA repair mechanisms. The extent of mosaicism depends on the timing and nature of the mutation during development.
2. **Chromosomal Mosaicism:** Chromosomal mosaicism involves variations in the number or structure of chromosomes within cells. For example, some individuals may have cells with aneuploidies (abnormal chromosome numbers) due to errors in cell division during early embryogenesis.
3. **X-Chromosome Inactivation:** In females, one of the X chromosomes in each

cell is randomly inactivated during early development. This process, known as lyonization, leads to mosaicism in gene expression between cells, contributing to the diverse phenotypes observed in females with X-linked disorders.

2.2.8.2. Developmental Implications

Genetic mosaicism can have profound effects on development and phenotype:

1. **Tissue-Specific Effects:** Mosaic mutations can affect specific tissues or organs depending on when and where the mutation occurs. This can result in localized abnormalities or variations in traits within the same individual.
2. **Functional Compensation:** In some cases, mosaic mutations may be compensated for by unaffected cells or tissues, allowing for relatively normal development despite the presence of genetic variation.
3. **Clinical Manifestations:** Mosaicism can lead to variable clinical presentations in individuals with genetic disorders. For example, individuals with mosaic forms of conditions like neurofibromatosis or tuberous sclerosis may exhibit milder symptoms compared to those with non-mosaic forms.

2.2.8.3. Detection and Study

Advances in genetic technologies have enhanced the detection and study of genetic mosaicism:

1. **Next-Generation Sequencing (NGS):** NGS allows for high-resolution analysis of genetic variation within tissues and organs, enabling the detection of low-frequency mosaic mutations that may have clinical implications.
2. **Single-Cell Sequencing:** This technique allows researchers to analyze the genomes of individual cells, revealing mosaic mutations and their distribution within tissues with unprecedented detail.
3. **Imaging Techniques:** Imaging modalities such as fluorescence *in situ* hybridization (FISH) and immunohistochemistry can visualize mosaicism at the tissue and cellular levels, providing insights into spatial patterns of genetic variation.

2.2.8.4. Clinical and Evolutionary Significance

Understanding genetic mosaicism is crucial for clinical genetics and evolutionary biology:

1. **Diagnosis and Counselling:** Knowledge of mosaicism can impact genetic counselling and clinical management of individuals with genetic disorders, influencing decisions regarding prognosis, treatment options, and recurrence risks.
2. **Evolutionary Adaptation:** Mosaicism contributes to genetic diversity within populations and may play a role in evolutionary adaptation by providing a reservoir of genetic variation upon which natural selection can act.

One intricate and dynamic facet of genetic variation found in organisms is represented by genetic mosaicism. It emphasizes how crucial it is to take into account genetic heterogeneity at the cellular and tissue levels in both disease mechanisms and developmental processes.

Genetic mosaicism's roles in health, disease, and evolution are becoming more and more clear as a result of advances in genomic and genetic technologies.

2.2.9. Screening for Mutants

In genetics and biotechnology, searching for mutants is a basic procedure used to find people or organisms that have genetic variants that cause noticeable phenotypic changes. These mutations may arise naturally, as induced changes, or as specific adjustments brought about by genetic engineering methods. Finding mutants is a crucial step towards understanding gene function, researching genetic pathways, and creating new features for biomedical or agricultural uses.

2.2.9.1. Methods and Approaches

1. **Phenotypic Screening:** Phenotypic screening involves visually or functionally assessing organisms or cells for specific traits or behaviors that differ from the wild-type or control counterparts. For example, in a population of fruit flies (*Drosophila melanogaster'*), mutants may be screened for altered eye color, wing morphology, or behavior patterns.
2. **Genetic Markers:** Genetic markers, such as visible markers (e.g., eye color or coat color) or selectable markers (e.g., antibiotic resistance genes), are used to identify mutants that have undergone genetic modification. These markers aid in distinguishing mutant individuals from wild-type organisms.
3. **Forward Genetics:** In forward genetics approaches, researchers induce mutations randomly across the genome using mutagens such as chemicals (e.g., EMS – Ethyl Methane Sulfonate) or irradiation (e.g., X-rays). Mutagenized populations are then screened for individuals exhibiting desired phenotypic traits, followed by genetic mapping to identify the underlying mutations.
4. **Reverse Genetics:** Reverse genetics involves starting with a known gene of interest and generating mutants with specific disruptions or alterations in that gene. Techniques such as gene knockout (complete deletion), knockdown (partial reduction), or point mutations (single base changes) are used to create targeted mutations. These mutants are then screened for phenotypic changes associated with the gene of interest.
5. **High-Throughput Screening:** Advances in automation and robotics have enabled high-throughput screening of large mutant libraries. Automated systems can process thousands to millions of samples, accelerating the identification of mutants with desired traits in fields such as drug discovery, agricultural biotechnology, and functional genomics.

2.2.9.2. Applications

1. **Biomedical Research:** Screening for mutants is essential for understanding the genetic basis of human diseases and identifying potential therapeutic targets. Mutant animal models, such as mice with targeted gene disruptions, are used to study disease mechanisms and evaluate novel treatments.

2. **Agricultural Biotechnology:** Mutant screening is employed in crop improvement programs to develop varieties with enhanced traits such as disease resistance, improved yield, or nutritional content. Mutagenesis techniques are used to generate genetic diversity, and screening identifies mutants with desirable agronomic traits.
3. **Functional Genomics:** Mutant screening contributes to deciphering gene function and regulatory networks in various organisms. By studying mutants with altered phenotypes, researchers gain insights into the roles of specific genes in development, metabolism, and response to environmental cues.
4. **Drug Discovery:** Screening for mutants in model organisms or cell lines can reveal targets for drug development. Mutants with altered responses to drugs or toxins provide valuable information on drug mechanisms and potential side effects.

2.2.9.3. Challenges and Considerations

1. **Off-Target Effects:** Induced mutations or genetic modifications may have unintended effects on unrelated genes or pathways, requiring careful validation and characterization of mutant phenotypes.
2. **Ethical Considerations:** The use of animal models and genetic manipulation raises ethical concerns regarding animal welfare, genetic integrity, and potential environmental impacts.
3. **Validation and Reproducibility:** Phenotypic changes observed in mutants must be validated through rigorous experimentation to confirm causality and reproducibility.

Mutant screening is a powerful tool in genetics and biotechnology that makes it possible to identify genetic variants that affect an organism's traits and behaviors. The use of mutant screening is becoming more widespread in a variety of domains, from applied biotechnology and medicine to basic research, thanks to developments in screening methodologies and genetic manipulation techniques.

2.3. CLONING OF GENES

Researchers frequently use gene cloning in molecular biology labs to make duplicates of a specific gene for use in genotyping, sequencing, mutagenesis, and heterologous protein expression, among other downstream uses. A DNA fragment of interest is transferred from one organism to a self-replicating genetic element, like a bacterial plasmid, in the conventional method of gene cloning. These days, this method is frequently applied to isolate long or poorly understood genes and protein expression. Polymerase chain reaction (PCR) is a more modern method for amplifying a target gene. As previously mentioned, the benefit of PCR over conventional gene cloning is the shorter time required to produce a pure sample of the desired gene. However, only genes with predefined sequences can be amplified using PCR for gene isolation. This is the reason that before PCR can be carried out for additional analysis, many unknown genes must first undergo gene cloning and sequencing.

2.3.1. Gain-and Loss-of-Function Experiments

In molecular biology and genetics, gain-of-function and loss-of-function experiments are essential methods for examining the functions of particular genes or gene products in biological processes. In these experiments, the expression of a gene of interest is altered to either increase or decrease its function. This enables researchers to understand the gene's normal function, regulatory mechanisms, and possible roles in disease.

2.3.1.1. Gain-of-Function Experiments

Gain-of-function experiments aim to enhance the activity or expression of a gene beyond its normal levels. This can be achieved through various methods:

1. **Overexpression:** Inserting extra copies of a gene or using strong promoters to drive increased transcription and translation of the gene. This leads to higher levels of the gene product (protein or RNA) within cells or organisms.
2. **Gene Activation:** Using techniques like CRISPR activation (CRISPRa) to target and activate specific genes without altering their DNA sequence. CRISPRa uses a catalytically inactive Cas9 protein (dCas9) fused to transcriptional activators to induce gene expression.
3. **Insertional Mutagenesis:** Inserting DNA sequences (such as transposons) near or within a gene can disrupt regulatory elements or enhance gene expression, resulting in gain-of-function phenotypes.

Gain-of-function studies are useful for understanding how a gene functions in situations where it might typically express itself at low levels or under particular circumstances. They can shed light on the gene's downstream effects on cellular functions, the molecular pathways it regulates, and its possible involvement in the development of disease.

2.3.1.2. Loss-of-Function Experiments

Loss-of-function experiments involve reducing or eliminating the function of a gene to study its effects on cellular or organismal phenotype. Common approaches include:

1. **Gene Knockout:** Deleting or disrupting the gene sequence using techniques like CRISPR-Cas9. This results in a complete loss of gene function and allows researchers to study the consequences of gene absence.
2. **RNA Interference (RNAi):** Introducing double-stranded RNA (dsRNA) or short hairpin RNA (shRNA) molecules that target and degrade mRNA transcripts of the gene of interest, leading to reduced gene expression.
3. **Antisense Oligonucleotides:** Using synthetic oligonucleotides complementary to the mRNA sequence of the gene to inhibit translation or induce mRNA degradation.

Experiments involving loss of function are essential for comprehending the role that genes play in developmental processes, disease mechanisms, and biological processes. They aid in identifying genes that may be used as therapeutic targets and clarifying the phenotypic effects of gene inactivation.

2.3.1.3. Applications and Insights

Both gain-of-function and loss-of-function experiments have broad applications across biomedical research:

1. **Drug Target Validation:** Identifying genes that, when targeted, affect disease processes. Gain-of-function screens can identify genes that, when overexpressed, drive disease progression, while loss-of-function screens can pinpoint genes whose inhibition leads to therapeutic benefits.
2. **Developmental Biology:** Studying the roles of genes in embryonic development, tissue differentiation, and organogenesis using gain-of-function and loss-of-function approaches to manipulate gene expression patterns.
3. **Cancer Research:** Identifying oncogenes (genes that promote cancer) through gain-of-function experiments and tumor suppressor genes (genes that inhibit cancer) through loss-of-function experiments, providing insights into cancer biology and potential therapeutic strategies.

Gain-of-function and loss-of-function studies are essential resources for deciphering the intricate relationships between gene regulation and function. They make it possible for scientists to control the levels of gene expression, identify gene interactions, and investigate the molecular foundations of illnesses and biological processes. The precision and scalability of these experiments are being improved by advances in genetic engineering technologies, creating new opportunities for research and innovative therapeutics.

2.4. TRANSGENES

A transgene is a gene that has been naturally transferred from one organism to another, or has been created through one of several genetic engineering methods. A process called transgenesis, which involves the introduction of a transgene, can alter an organism's phenotype. A transgene is a section of DNA that has been isolated from one organism and inserted into another, carrying a gene sequence. This non-native DNA segment may change the normal function of the transgenic organism's genetic code or preserve the transgenic organism's capacity to produce RNA or protein. Usually, the DNA is integrated into the germ line of the organism. In higher vertebrates, for instance, this can be achieved by injecting the foreign DNA into the fertilized ovum's nucleus. Using this method, strains of laboratory mice are regularly given human disease genes or other genes of interest to investigate the function or pathology associated with those specific genes.

A transgene's construction necessitates the assembly of several key components. A stop sequence, an exon, a protein coding sequence (typically derived from the cDNA for the protein of interest), a promoter—a regulatory sequence that controls where and when the transgene is active—and an exon are required components of a transgene. Usually, these are assembled into a bacterial plasmid, and transgenes with previously identified roles are used to select the coding sequences.

There are numerous uses for transgenic or genetically modified organisms in research, whether they are fungi, bacteria, or viruses. It has been possible to breed transgenic fish, plants, insects, and mammals, including humans. In certain nations, transgenic plants—like corn and soybeans—have supplanted wild varieties in agriculture (e.g., the US). Since 2001, transgene escape for GMO crops has been shown to be persistent and invasive. Transgenic organisms raise moral concerns and could compromise biosafety.

Transgenesis is a basic genetics and molecular biology technique that entails inserting foreign genetic material into an organism's genome. Using this method, scientists can model human diseases, investigate the function of genes, and produce genetically modified organisms (GMOs) that possess desired traits. Transgenesis has provided tools to manipulate and comprehend genetic information in living organisms, revolutionizing biotechnology, agriculture, and biomedical research.

1. Methods of Transgenesis

i. **Microinjection:** This method is commonly used in vertebrate species such as mice and fish. It involves injecting foreign DNA, typically in the form of a plasmid or a transgene, directly into the pronucleus of a fertilized egg. The injected DNA integrates randomly into the host genome, and the resulting embryos are implanted into surrogate mothers for development.

ii. **Viral Vectors:** Viral vectors, such as retroviruses or adenoviruses, can be engineered to deliver transgenes into target cells or tissues. These vectors exploit the natural ability of viruses to insert their genetic material into host genomes, allowing for efficient transgene integration and expression.

iii. **Electroporation:** In this method, cells are exposed to brief electrical pulses that create temporary pores in the cell membrane. DNA molecules are then introduced into the cells, and some of the DNA integrates into the genome of the target cells. Electroporation is widely used in cell culture studies and in generating transgenic plants.

iv. **CRISPR-Cas9-Mediated Transgenesis:** The CRISPR-Cas9 system can be used not only for genome editing but also for targeted transgene insertion. By co-delivering Cas9 protein and guide RNAs along with donor DNA, researchers can achieve precise integration of transgenes into specific genomic loci.

2. Applications of Transgenesis

i. **Biomedical Research:** Transgenesis is instrumental in creating animal models of human diseases, allowing researchers to study disease mechanisms and test potential therapies. For example, transgenic mice can be engineered to express genes associated with cancer, neurodegenerative disorders, or cardiovascular diseases.

ii. **Drug Discovery:** Transgenic animals serve as valuable models for testing the efficacy and safety of new drugs. They enable researchers to evaluate drug metabolism, pharmacokinetics, and toxicology in a living organism that closely mimics human physiology.

iii. **Agricultural Biotechnology:** Transgenic plants are engineered to express traits such as herbicide resistance, pest resistance, improved nutritional content, or drought tolerance. These traits can enhance crop yield, reduce reliance on chemical pesticides, and address agricultural challenges related to climate change and food security.

iv. **Biopharmaceutical Production:** Transgenic animals, such as goats or cows, can be engineered to produce therapeutic proteins in their milk or eggs. This approach offers a cost-effective method for large-scale production of biopharmaceuticals.

3. Ethical and Regulatory Considerations: The extensive application of transgenesis gives rise to moral questions about the treatment of animals, the environment, and human health. In numerous nations, regulatory bodies have instituted protocols and structures to guarantee the conscientious application of genetically modified organisms and to evaluate possible hazards prior to their discharge into the surroundings or utilization in food manufacturing.

To sum up, transgenesis is a strong and adaptable method that has transformed biological study and applications across a range of domains. Transgenesis continues to propel advances in biotechnology, agriculture, and medicine by permitting the introduction and study of particular genetic traits in organisms, providing chances for improvements in human health and sustainable development.

2.4.1. Other Gain-of- Function Techniques

Apart from conventional approaches such as overexpression and activation, a number of contemporary gain-of-function (GOF) techniques have surfaced in the literature with distinct benefits for investigating gene function, disease mechanisms, and therapeutic development. These methods use cutting-edge molecular tools and technologies to precisely control gene expression and activity.

Here are some notable examples:

1. **CRISPR Activation (CRISPRa):** CRISPR activation utilizes the CRISPR-Cas9 system to enhance gene expression without altering the DNA sequence. Instead of inducing double-strand breaks (like in gene editing), CRISPRa uses a catalytically inactive form of Cas9 (dCas9) fused to transcriptional activators (e.g., VP64, p65AD, or others). When targeted to a specific gene promoter or enhancer region, dCas9 activates transcription by recruiting transcriptional machinery and initiating RNA synthesis. This technique enables researchers to upregulate genes of interest in a targeted and reversible manner, offering insights into gene function, pathway regulation, and cellular responses.
2. **Optogenetics:** In optogenetics, light-sensitive proteins (opsins) that are genetically encoded into cells or organisms are used to control biological processes. Researchers can control the activation or inhibition of particular signaling pathways, ion channels, or gene expression in response to light stimulation by fusing opsins to target proteins. The ability to precisely control both temporal and spatial aspects of cellular activities makes this technique useful for researching behavior, disease mechanisms, and neuronal circuits. Optogenetics can be used in gain-of-function applications to clarify the roles of intracellular signaling pathways and neuronal firing in health and disease.
3. **Chemical Inducible Systems:** Chemically inducible systems use tiny molecules to reversibly control protein activity or gene expression. These systems usually consist of protein domains or engineered promoters that react to particular ligands, like hormones (e.g., systems based on estrogen receptors) or artificial inducers (e.g., systems that are rapamycin-inducible). Researchers can exert precise temporal control over the activation or suppression of biological processes by creating constructs in which protein function or gene expression is contingent upon ligand binding. Chemically induced systems are useful research instruments for signaling pathways, developmental transitions, and dynamic cellular processes.
4. **Transcriptional Enhancer Mimicry:** Using this strategy, targeted genes or pathways are activated by creating artificial DNA sequences that resemble natural transcriptional enhancers. Researchers can control gene expression levels in a cell-specific or inducible way by creating synthetic enhancers that react to endogenous transcription factors or external stimuli. Through the combination of reporter genes or selectable markers with synthetic enhancers, scientists can identify enhancer variations that maximize gene expression for biotechnological or therapeutic purposes.

The enhanced precision, temporal control, and versatility that these sophisticated gain-of-function techniques provide when modifying gene expression and cellular processes make them a valuable addition to conventional methods. They keep pushing the boundaries of biological research, drug development, and biotechnology innovation, opening doors for new therapeutic approaches and a deeper understanding of gene function.

2.4.2. Targeted Mutagenesis

Targeted mutagenesis is a potent molecular biology technique that allows for precise modifications to particular genomic loci, which makes it easier to study gene function and develop genetically engineered and therapeutic treatments. Mutagenesis is an important laboratory technique whereby DNA mutations are deliberately engineered to produce libraries of mutant genes, proteins, strains of bacteria, or other genetically modified organisms. To investigate a genetic locus, process, or product in detail, mutations can be made to any of the gene's constituent parts, regulatory components, or gene products. Mutations can result in mutant proteins with intriguing characteristics or improved, unique functions that could find application in the commercial sector. It is also possible to create mutant strains that are useful in real-world situations or that enable research into the molecular underpinnings of specific cell functions.

There are a lot of mutagenesis techniques used today. When mutations were first created artificially in the lab using techniques like UV irradiation, they were completely random in nature. Specific regions or sequences of the genome cannot be targeted by random mutagenesis; however, more targeted changes are possible with the development of site-directed mutagenesis. Since 2013, the prokaryotic viral defense system that underpins the CRISPR/Cas9 technology has made it possible to edit or mutate a genome *in vivo*. In circumstances where random mutagenesis is not helpful, site-directed mutagenesis has shown to be effective. Combinatorial and insertional mutagenesis are two more mutagenesis methods. Non-random mutagenesis can be used to engineer proteins, clone DNA, and study the effects of mutagens. Additionally, it has medical uses that include curing diseases like beta thalassemia, treating HIV and cancer, and assisting immunocompromised patients.

2.5. GENE DUPLICATION

During molecular evolution, one of the main mechanisms by which new genetic material is produced is gene duplication, also known as chromosomal duplication or gene amplification. Any duplication of a gene-containing region of DNA can be classified as it. Gene duplications can result from a variety of mistakes made by the machinery that replicates and repairs DNA, as well as from accidental capture by self-centered genetic elements.

Ectopic recombination, retro transposition events, aneuploidy, polyploidy, and replication slippage are common causes of gene duplications. An organism can acquire more than one copy of a gene or genomic region through a process known as gene duplication. This phenomenon affects genetic diversity, adaptation, and the emergence of novel traits. It is fundamental to evolutionary processes and can have a variety of biological consequences.

1. Mechanisms of Gene Duplication

i. **Unequal Crossing Over:** During meiosis, homologous chromosomes may misalign, leading to unequal crossing over. This results in one chromosome gaining an extra copy of a gene or genomic segment, while the other loses one.

ii. **Retro position:** Retrotransposons, which are mobile genetic elements capable of reverse transcription and integration into new genomic locations, can duplicate genes by creating extra copies in different chromosomal locations.

iii. **Whole Genome Duplication (Polyploidy):** Entire sets of chromosomes can be duplicated during evolution, resulting in polyploid organisms (e.g., tetraploid, hexaploid). This process can occur in plants and some animals, leading to increased genetic redundancy and potential for evolutionary innovation.

2. Biological Implications

i. **Functional Redundancy and Divergence:** Initially, duplicated genes often maintain similar functions due to redundancy. Over time, mutations can accumulate in one or both copies, leading to functional divergence. This divergence may result in specialization of each gene copy to perform distinct roles or acquire new functions.

ii. **Evolutionary Innovation:** Gene duplication provides raw genetic material for evolutionary experimentation. Duplicated genes can evolve new functions or regulatory patterns that contribute to phenotypic diversity and adaptation to changing environments. Examples include the evolution of new enzyme functions or the development of complex traits like sensory perception or immune response.

iii. **Disease and Genetic Disorders:** Gene duplication events can also lead to genetic disorders when they disrupt normal gene dosage or function. For instance, duplications in certain chromosomal regions are associated with developmental disorders, neurodegenerative diseases, and cancer.

3. Methods of Detection and Study

   i. **Comparative Genomics:** By comparing genomes of different species, researchers can identify orthologous genes (genes derived from a common ancestor) and paralogous genes (genes derived from duplication events within a species or lineage).

   ii. **Experimental Approaches:** Laboratory techniques such as gene knockout, RNA interference (RNAi), and functional assays are used to study the effects of gene duplication on phenotype and biological function.

2.5.1. Mechanisms of Duplication

2.5.1.1. Ectopic Recombination

Unequal crossing-over, which happens during meiosis between misaligned homologous chromosomes, is the cause of duplications. The degree to which two chromosomes share repetitive elements determines the likelihood that it will occur. This recombination results in a reciprocal deletion and a duplication at the exchange site. Usually, sequence similarity at the duplicate breakpoints—which result in direct repeats—mediates ectopic recombination. Transposable elements, which are repetitive genetic elements that are frequently present at duplication breakpoints in plants and mammals, provide one source of repetitive DNA that can aid in recombination.

2.5.1.2. Replication Slippage

Short genetic sequences may duplicate due to a replication slippage error in DNA replication. DNA polymerase starts to copy DNA during replication. Replication stalls when the polymerase separates from the DNA at some point during the process. The replicating strand is aligned incorrectly and accidentally duplicates the same section multiple times when the polymerase reattaches to the DNA strand. Repetitive sequences also frequently aid in replication slippage, which only needs a small number of similar bases.

2.5.1.3. Retro-Transposition

On occasion, retrotransposons—primarily L1—can affect cellular mRNA. Retrogenes are created when transcripts are reverse transcribed to DNA and randomly inserted into the genome. The resulting sequences frequently contain poly(A) sequences that are also integrated into the genome and typically lack introns. When compared to their parental gene sequences, many retrogenes exhibit altered gene regulation, which can occasionally lead to the emergence of new functions. Chromosome evolution can be influenced by retrogenes that migrate between distinct chromosomes.

2.5.1.4. Aneuploidy

When nondisjunction at a single chromosome results in an abnormal number of chromosomes, the condition is known as aneuploidy. In mammals, aneuploidy frequently causes spontaneous abortions, or miscarriages, which can be harmful. Certain aneuploid people, such as those with trisomy 21, which causes Down syndrome in humans, are viable. Since aneuploidy frequently modifies gene dosage in ways that are harmful to the organism, it is unlikely to proliferate across populations.

2.5.1.5. Polyploidy

Whole genome duplication, also known as polyploidy, is the outcome of meiotic nondisjunction, which creates multiple copies of the entire genome. In addition to being common in plants, animals have also experienced polyploidy. Two rounds of whole genome duplication (2R event) occurred in the vertebrate lineage that gave rise to humans. Approximately 100 mya have also experienced it in hemascomycete yeasts. Following a complete genome duplication, there is a brief period of increased nucleotide substitution, widespread gene loss, and regulatory network rewiring. Gene dosage effects are a major factor as well. As a result, the majority of duplicates disappear quickly, but a sizeable portion do so. It's interesting to note that genes involved in regulation are kept more frequently. Moreover, adaptive innovation has resulted from the preservation of regulatory genes, particularly the Hox genes.

At the transcriptional level of duplicated genes, rapid evolution and functional divergence have been observed, typically due to point mutations in short transcription factor binding motifs. Furthermore, another factor that contributes to the survival and quick adaptation/neofunctionalization of duplicate genes is the rapid evolution of protein phosphorylation motifs, which are typically embedded within rapidly evolving intrinsically disordered regions. Therefore, it appears that genome evolution and gene regulation are related, at least in the post-translational sense. Since offspring of polyploid species have a different number of chromosomes than their parent species, they are frequently unable to interbreed with non-polyploid organisms, making polyploidy another well-known source of speciation. Since each gene's relative dosage should be the same, whole genome duplications are believed to be less harmful than aneuploidy.

2.5.2. Limitations of Developmental Genetics

Although developmental genetics is a powerful field for understanding how genes control development, researchers must overcome a number of inherent limitations in order to conduct their work. These restrictions result from the conceptual and technical difficulties that come with researching the complex processes of development.

2.5.2.1. Technical Limitations

1. **Gene Redundancy:** Many organisms possess redundant genetic pathways and compensatory mechanisms, where multiple genes can compensate for the loss or alteration of a single gene. This redundancy complicates the interpretation of phenotypic outcomes from genetic manipulations, as compensatory mechanisms can mask or alter the expected effects of gene mutations.
2. **Temporal and Spatial Dynamics:** Developmental processes are highly dynamic and regulated in a temporal and spatial manner. Genes may have different functions or regulatory roles at different stages of development or in different tissues. Thus, accurately dissecting the roles of specific genes throughout development requires precise temporal and spatial control over gene expression and manipulation, which can be technically challenging.
3. **Gene-Environment Interactions:** Developmental outcomes are influenced not only by genetic factors but also by environmental

cues and epigenetic modifications. Understanding how genes interact with environmental factors to produce phenotypic diversity requires sophisticated experimental designs and analyzes.

2.5.2.2. Conceptual and Interpretative Limitations

1. **Complexity of Gene Networks:** Developmental processes are governed by complex gene regulatory networks involving multiple genes and signaling pathways. Deciphering these networks and determining the hierarchy of gene interactions can be challenging, especially when considering feedback loops and non-linear interactions.
2. **Genetic Variability and Genetic Background Effects:** Genetic variability within populations and differences in genetic backgrounds among experimental organisms can influence developmental outcomes. Variations in genetic backgrounds can lead to phenotypic variability, making it difficult to generalize findings from one genetic context to another.
3. **Ethical and Practical Constraints:** In studies involving animal models or human embryonic tissues, ethical considerations and practical constraints (such as limited sample availability) can restrict the scope and design of experiments, limiting the ability to perform comprehensive developmental genetics studies.

CASE STUDY

CHROMOSOMES

Cells package their DNA not only to protect it, but also to regulate which genes are accessed and when. Cellular genes are therefore similar to valuable files stored in a file cabinet — but in this case, the cabinet's drawers are constantly opening and closing; various files are continually being located, pulled, and copied; and the original files are always returned to the correct location.

Of course, just as file drawers help conserve space in an office, DNA packaging helps conserve space in cells. Packaging is the reason why the approximately two meters of human DNA can fit into a cell that is only a few micrometers wide. But how, exactly, is DNA compacted to fit within eukaryotic and prokaryotic cells? And what mechanisms do cells use to access this highly compacted genetic material?

What are Chromosomes?

Cellular DNA is never bare and unaccompanied by other proteins. Rather, it always forms a complex with various protein partners that help package it into such a tiny space. This DNA-protein complex is called **chromatin,** wherein the mass of protein and nucleic acid is nearly equal. Within cells, chromatin usually folds into characteristic formations called **chromosomes**. Each chromosome contains a single double-stranded piece of DNA along with the aforementioned packaging proteins.

Eukaryotes typically possess multiple pairs of linear chromosomes, all of which are contained in the cellular nucleus, and these chromosomes have characteristic and changeable forms. During cell division, for example, they become more tightly packed, and their condensed form can be visualized with a light microscope. This condensed form is approximately 10,000 times shorter than the linear DNA strand would be if it was devoid of proteins and pulled taut. However, when eukaryotic cells are not dividing — a stage called **interphase** — the chromatin within their chromosomes is less tightly packed. This looser configuration is important because it permits transcription to take place (Figures 1 and 2).

In contrast to eukaryotes, the DNA in prokaryotic cells is generally present in a single circular chromosome that is located in the cytoplasm. (Recall that prokaryotic cells do not possess a nucleus.) Prokaryotic chromosomes are less condensed than their eukaryotic counterparts and don't have easily identified features when viewed under a light microscope. During interphase, the cell's DNA is not condensed and is loosely distributed. A stain for heterochromatin (which indicates the position of chromosomes) shows this broad distribution of chromatin in a mouse cell (upper left). The same stain also shows the organized, aligned structure of the chromosomes during mitosis. Scale bars = 10 microns.

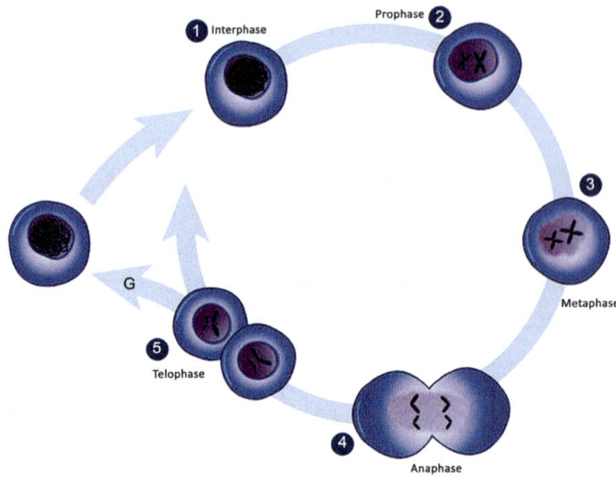

Figure 1. *Chromatin condensation changes during the cell cycle. During interphase (1), chromatin is in its least condensed state and appears loosely distributed throughout the nucleus. Chromatin condensation begins during prophase (2) and chromosomes become visible. Chromosomes remain condensed throughout the various stages of mitosis (2–5).*

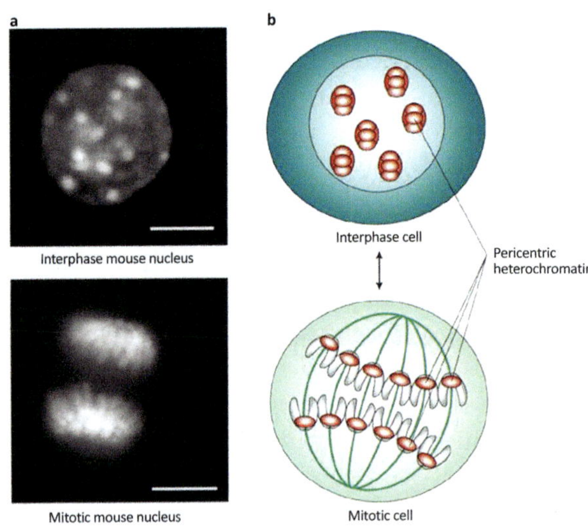

Figure 2. *The appearance of DNA during interphase versus mitosis.*

How are Eukaryotic Chromosomes Structured?

Eukaryotic chromosomes consist of repeated units of chromatin called **nucleosomes**, which were discovered by chemically digesting cellular nuclei and stripping away as much of the outer protein packaging from the DNA as possible. The chromatin that resisted digestion had the appearance of "beads on a string" in electron micrographs

— with the "beads" being nucleosomes positioned at intervals along the length of the DNA molecule (Figure 3).

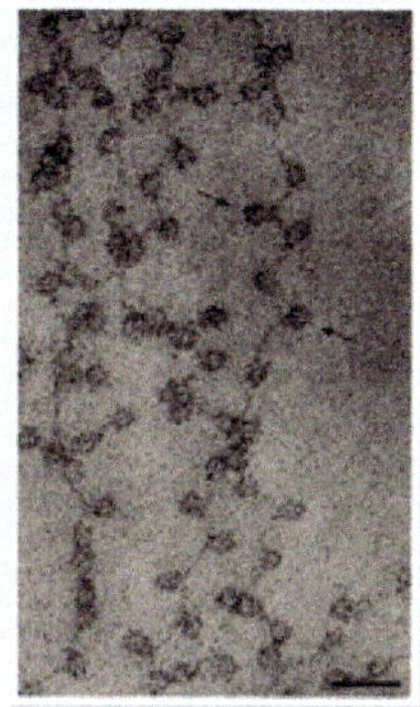

Figure 3. *The length of the DNA molecule.*

Nucleosomes are made up of double-stranded DNA that has complexed with small proteins called **histones**. The core particle of each nucleosome consists of eight histone molecules, two each of four different histone types: H2A, H2B, H3, and H4. The structure of histones has been strongly conserved across evolution, suggesting that their DNA packaging function is crucially important to all eukaryotic cells (Figure 4).

Histones carry positive charges and bind negatively charged DNA in a specific conformation. In particular, a segment of the DNA double helix wraps around each histone core particle a little less than twice. The exact length of the DNA segment associated with each histone core varies from species to species, but most such segments are approximately 150 base pairs in length. Furthermore, each histone molecule within the core particle has one end that sticks out from the particle. These ends are called **N-terminal tails**, and they play an important role in higher-order chromatin structure and gene expression.

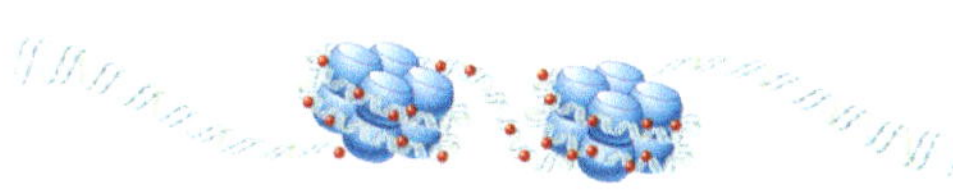

Figure 4. *The nucleosome structure within chromatin.*

Each nucleosome contains eight histone proteins (blue), and DNA wraps around these histone structures to achieve a more condensed coiled form.

Why Is Complex Packing Critical for Eukaryotic Chromosomes?

Although nucleosomes may look like extended "beads on a string" under an electron microscope, they appear differently in living cells. In such cells, nucleosomes stack up against one another in organized arrays with multiple levels of packing. The first level of packing is thought to produce a fiber about 30 nanometers (nm) wide. These 30 nm fibers then form a series of loops, which fold back on themselves for additional compacting (Figure 5).

The multiple levels of packing that exist within eukaryotic chromosomes not only permit a large amount of DNA to occupy a very small space, but they also serve several functional roles. For example, the looping of nucleosome-containing fibers brings specific regions of chromatin together, thereby influencing gene expression. In fact, the organized packing of DNA is malleable and appears to be highly regulated in cells.

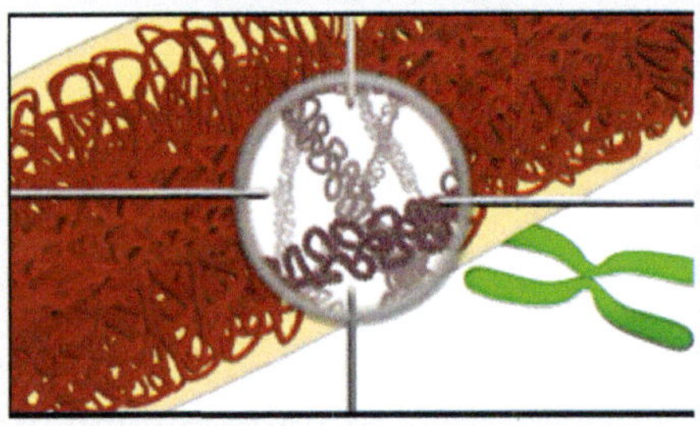

Figure 5. *To better fit within the cell, long pieces of double-stranded DNA are tightly packed into structures called chromosomes.*

Chromatin packing also offers an additional mechanism for controlling gene expression. Specifically, cells can control access to their DNA by modifying the structure of their chromatin. Highly compacted chromatin simply isn't accessible to the enzymes involved in DNA transcription, replication, or repair. Thus, regions of chromatin where active transcription is taking place (called **euchromatin**) are less condensed than regions where transcription is inactive or is being actively inhibited or repressed (called **heterochromatin**) (Figure 6).

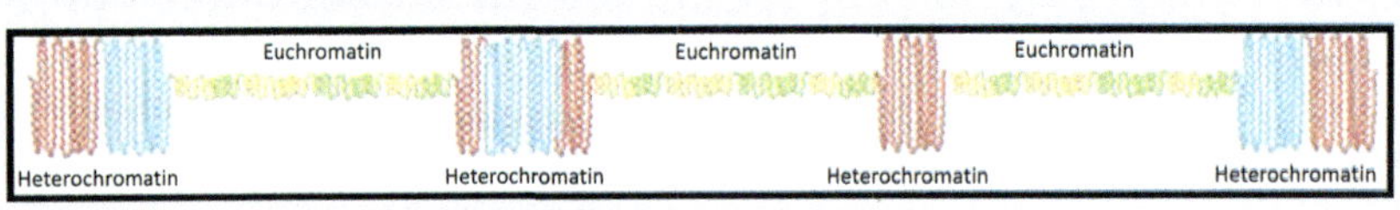

Figure 6. *The structure of chromatin in interphase.*

Heterochromatin is more condensed than euchromatin. Typically, the more condensed chromatin is, the less accessible it is by transcription factors and polymerases.

The dynamic nature of chromatin is regulated by enzymes. For example, chromatin can be loosened by changing the position of the DNA strands within a nucleosome. This loosening occurs because of chromatin remodeling enzymes, which function to slide nucleosomes along the DNA strand so that other enzymes can access the strand. This process is closely regulated and allows specific genes to be accessed in response to metabolic signals within the cell. Another way cells control gene expression is by modifying their histones with small chemical groups, such as methyl and acetyl groups in the N-terminal tails that extend from the core particle. Different enzymes catalyze each kind of N-terminal modification. Scientists occasionally refer to the complex pattern of histone modification in cells as a "histone code." Some of these modifications increase gene expression, whereas others decrease it.

How are Chromosomes Organized in the Nucleus?

In electron micrographs, eukaryotic interphase chromatin appears much like a plate of spaghetti — in other words, there is no obvious pattern of organization. In recent years, however, investigators have begun using fluorescent probes for each of the different interphase chromosomes. In doing so, they have discovered that these chromosomes have functional and decidedly nonrandom arrangements.

One of the first things these scientists noted was that uncondensed chromosomes occupy characteristic regions of the nucleus, which they termed **chromosome territories**. The spatial localization of these territories is thought to be important for gene expression. In fact, with the advent of gene-specific probes, researchers are beginning to understand how the arrangement of chromosome territories can bring particular genes closer together. A second major observation related to chromosome territories is that the position of chromosomes relative to one another differs from cell to cell. Such differences reflect variation in gene expression patterns.

Conclusion

The prokaryotic genome typically exists in the form of a circular chromosome located in the cytoplasm. In eukaryotes, however, genetic material is housed in the nucleus and tightly packaged into linear chromosomes. Chromosomes are made up of a DNA-protein complex called chromatin that is organized into subunits called nucleosomes. The way in which eukaryotes compact and arrange their chromatin not only allows a large amount of DNA to fit in a small space, but it also helps regulate gene expression.

CLASS ACTIVITY

In this class activity, students will explore genetic traits through a survey and analysis. Each student will receive a questionnaire listing observable genetic traits such as eye color, hair texture, and tongue rolling ability. They will collect data from their classmates, recording each participant's traits. After collecting data, students will compile and analyze the results to identify patterns of inheritance. Through group discussion and analysis on a whiteboard or flip chart, students will interpret their findings, discussing dominant and recessive traits, as well as the role of genetics in determining physical characteristics. This activity promotes understanding of genetic principles in a hands-on and interactive way.

SUMMARY

- Developmental genetics explores how genes regulate the growth, differentiation, and patterning of cells and tissues during the development of organisms.
- Developmental mutants are genetic variants of organisms that exhibit abnormal or altered developmental patterns compared to wild-type individuals.
- Apoptosis is an orderly process, distinct from the one that disposes of cells injured or killed by trauma, infection, or other environmental insults.
- Sex chromosomes are chromosomes that carry the genes that determine the sex of an individual. The human sex chromosomes are a typical pair of mammal allosomes.
- The sex chromosomes of the fruit fly *Drosophila melanogaster'* have played a particularly important role in our understanding of heredity.
- Maternal and zygotic genes play distinct roles in the early development of many organisms, particularly during embryogenesis.
- Dysregulation of pathways can lead to diseases such as cancer, metabolic disorders, and developmental abnormalities. Targeting specific nodes within pathways with drugs or gene therapies holds promise for treating these conditions by restoring normal pathway function.
- Genetic mosaicism refers to the phenomenon where an organism contains cells with different genotypes, typically resulting from genetic mutations occurring after fertilization.

REVIEW QUESTIONS

1. What are the key differences between genetic determination and environmental influence in developmental genetics?
2. Describe the process of morphogenesis and its importance in developmental biology.
3. Explain the role of homeotic genes in determining body patterning during development.
4. How do mutations in HOX genes affect embryonic development and body segmentation?
5. Discuss the concept of cell differentiation and its regulation during embryonic development.
6. What are the major stages of embryonic development and what genetic mechanisms control each stage?
7. How do signaling pathways like Notch, Wnt, and Hedgehog contribute to pattern formation in embryonic development?
8. Compare and contrast the roles of transcription factors and signaling molecules in developmental genetics.

9. Describe how epigenetic mechanisms influence gene expression during development.
10. Discuss the significance of genetic regulatory networks in coordinating gene expression patterns during development.

REFERENCES

1. Cline, T. W., & Meyer, B. J. (1996). Vive la difference: Males vs. females in flies vs. worms. *Annual Review of Genetics, 30*, 637–702.
2. Crews, D., et al. (1995). The role of estrogen in turtle sex determination and the effect of PCBs. *Environmental Health Perspectives, 103*, 73–77.
3. Gilbert, S. F. (2006). *Developmental Biology*. Sinauer.
4. Koopman, P., et al. (1991). Male development of chromosomally female mice transgenic for Sry. *Nature, 351*, 117–121. https://doi.org/10.1038/3511117a0.
5. McLaren, A. (1991). The making of male mice. *Nature, 351*, 96. https://doi.org/10.1038/351096a0.
6. Mouse Genetics Core. (2011). *Transgene Design* [Archived]. Washington University.
7. Nanda, I., et al. (1999). 300 million years of conserved synteny between chicken Z and human chromosome 9. *Nature Genetics, 21*, 258–259. https://doi.org/10.1038/6769.
8. Page, D. C., De la Chapelle, A., & Weissenbach, J. (1985). Chromosome Y-specific DNA in related human XX males. *Nature, 315*, 224–226. https://doi.org/10.1038/315224a0.
9. Parisi, M., et al. (2004). A survey of ovary-, testis-, and some-biased gene expression in *Drosophila melanogaster* adults. *Genome Biology, 5*, R40. https://doi.org/10.1186/gb-2004-5-6-r40.
10. Pieau, C., et al. (1994). Temperature sensitivity of sexual differentiation of gonads in the European pond turtle. *Journal of Experimental Zoology, 270*, 86–93.
11. Saccone, G., Pane, A., & Polito, L. C. (2002). Sex determination in flies, fruit flies, and butterflies. *Genetica, 116*, 15–23.
12. Smith, C. A., & Sinclair, A. H. (2004). Sex determination: Insights from the chicken. *BioEssays, 26*, 120–132.
13. Traut, W., Sahara, K., & Marec, F. (2007). Sex chromosomes and sex determination in Lepidoptera. *Sexual Development, 1*, 332–346. https://doi.org/10.1159/000111765.

CHAPTER 3

Approaches to Development: Experimental Embryology and Its Molecular Basis

LEARNING OBJECTIVES

After studying this chapter, you will be able to:

- Define normal development.
- Describe developmental commitment.
- Explain quantitative experimental embryology.
- Elaborate cell removal as an essential tool in experimental embryology.
- Discuss the tissue embedding as an essential tool in experimental embryology.

KEY TERMS FROM THIS CHAPTER

Chimerism	Embryology
Embryonic regulation	Experimental embryology
Fate map	Fertility
Hemisphere	Molecular level
Neuroepithelium	Oogenesis
Orthotopic graft	Perturbation
Transcriptome	Ventral region

INTRODUCTION

Experimental embryology is a field dedicated to understanding the intricate processes that govern the development of organisms from fertilization to maturity. This discipline involves the study of how cells, tissues, and organs form and differentiate, often through experimental manipulation and observation. It provides vital insights into developmental mechanisms that are essential for normal growth and can also shed light on congenital anomalies.

One of the primary approaches in experimental embryology is the manipulation of embryos at various stages of development. Techniques such as microinjection, gene knockout, and transplantation experiments allow scientists to alter specific genes or cellular environments to observe the effects on development. By doing so, researchers can identify critical genes and pathways that regulate development, helping to map out the complex interactions that guide embryogenesis.

At the molecular level, the basis of development is rooted in the intricate regulation of gene expression and signaling pathways. Genes are switched on and off in precise sequences, and signaling molecules such as growth factors and hormones guide cells to divide, differentiate, and migrate. The spatial and temporal expression of these genes is controlled by a network of regulatory elements and transcription factors, which together ensure that cells develop into the right types and form the correct structures at the appropriate times.

3.1. NORMAL DEVELOPMENT

The term "normal development" refers to the developmental trajectory that a typical embryo takes under controlled laboratory settings, free from experimental perturbations. It takes a solid understanding of normal development to comprehend how experimental manipulations affect the subject. Several common descriptive terms are used to characterize embryos. An animal's anterior or cranial end is referred to as its front end. The posterior, or caudal, end is the term for the back end. Dorsal is the upper surface and ventral is the lower. Transverse microscope sections are those that are taken across the animal's long axis. Longitudinal axes are those that run parallel to them. If a vertical longitudinal section is to one side of the midline, it is referred to as parasagittal, and if it is in the midline, it is known as sagittal. Frontal or coronal refers to a horizontal longitudinal section that divides the dorsal and ventral sides.

Under a dissecting microscope, external features can be used to identify each of the published tables of stage series for the model species used in laboratory work. These tables describe the developmental course as a series of standard stages. Because the development of embryos is predictable, one can be certain that an embryo will reach stage 20 a specific number of hours after it has reached stage 10 at a given time. The normal physiological temperature is where development happens for mammals and birds, but the rate of development varies for free-living embryos like zebrafish and Xenopus. Because these tables are available, researchers can use embryos of the same stage, regardless of the lab's temperature on any given day, to standardize their procedures.

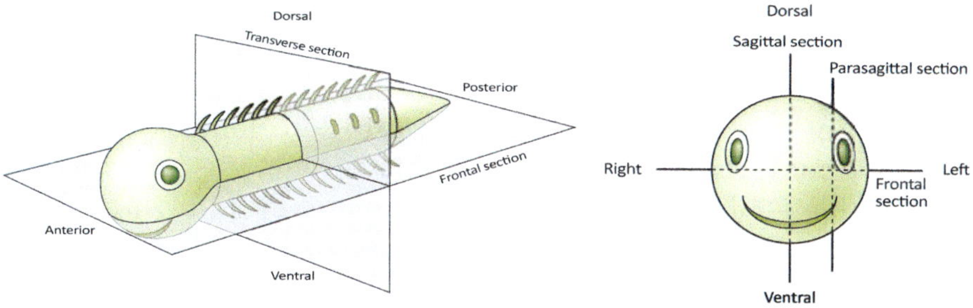

Figure 3.1. *Axes and planes of section used for describing embryos.*

Source: Essential Developmental Biology, Third Edition. Jonathan M.W. Slack. © 2013 John Wiley & Sons, Ltd. Published 2013 by John Wiley & Sons, Ltd.

When characteristics of development result from elements present in the egg that have accumulated during oogenesis, they are referred to as maternal features. If they result from newly synthesized components by the embryo itself after fertilization, they are referred to as zygotic. Figure 3.1 illustrates the axes and planes of section used for describing embryos.

3.1.1. The Fate Map

A "fate map" is a schematic representation of the embryo's development that indicates where the various regions will go, how they will change shape, and what structures they will become (Figure 3.2). Because of morphogenetic movements and growth, the fate map will vary from stage to stage. Consequently, a series of fate maps will show the path taken by each component from the fertilized egg to the adult. The degree of random cell mixing that takes place during development determines how accurate a fate map is. The fate map can be accurate down to the cellular level if there isn't any, as in the case of the nematode Caenorhabditis elegans. Due to some local mixing of similar cells in most embryo types, the fate maps are not quite this accurate. However, the fate map is a key idea in embryology, and understanding it is necessary to interpret almost all experiments involving early developmental decisions.

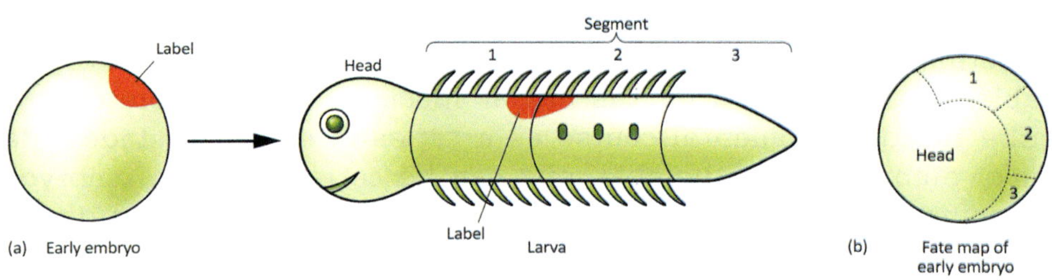

Figure 3.2. *Fate mapping. (a) A label placed at a particular position in the early embryo ends up in a reproducible position on the animal. (b) A possible fate map of the early embryo.*

Source: Essential Developmental Biology, Third Edition. Jonathan M.W. Slack. © 2013 John Wiley & Sons, Ltd. Published 2013 by John Wiley & Sons, Ltd.

In order to create fate maps, individual cells or areas of embryos are labeled, and the location and form of the labeled patch are then determined at a later stage of development. The labeling techniques that are employed include either grafting labeled tissue to replace an identical portion in the host embryo or applying an extracellular label to a patch of cells or injecting an intracellular label into a single cell. In the experiment depicted in Figure 3.3, green fluorescent protein (GFP) mRNA was injected into two nearby early blastomeres of a Xenopus embryo, and the labeled domains were located at the neurula stage. The fate map for a specific stage can be created by combining the outcomes of numerous separate labeling experiments. It is important to remember that nothing about developmental commitment is indicated by a fate map. Every component of the embryo has a destiny during development, but intercellular interactions are typically the means by which commitment to form specific structures or cell types is gained.

An embryo is referred to as regulative in older embryological literature if an isolated part forms more structures than expected from the fate map, and as a mosaic if experimentally isolated parts develop strictly according to the fate map. Actually, depending on the area of the embryo and the developmental stage in question, all types of embryos exhibit some characteristics of both mosaic and regulative behavior. Double gradient systems, such as the ADMP–Chordin system in Xenopus, which can adapt to changes in embryonic size, provide the basis for the explanation of regulative behavior. It should be noted that the term "mosaic" in this context has nothing to do with genetic mosaics, which are organisms made up of cells with different genotypes, nor does it refer to embryonic regulation in the same way as it does to gene regulation or physiological regulation.

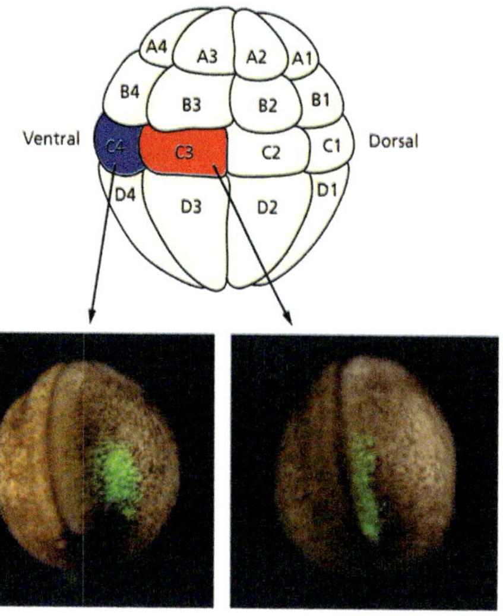

Figure 3.3. *A fate mapping experiment in Xenopus. GFP mRNA was injected either into the C3 or C4 blastomere at the 32-cell stage.*

Source: Essential Developmental Biology, Third Edition. Jonathan M.W. Slack. © 2013 John Wiley & Sons, Ltd. Published 2013 by John Wiley & Sons, Ltd.

3.1.2. Clonal Analysis

A type of fate mapping known as clonal analysis involves labeling a single cell and subsequently identifying the location and cell types of its progeny. One cell may be injected with a lineage label to complete the labeling process. For organisms that do not grow significantly, like sea urchins, zebrafish, or early stage Xenopus, this is a simple method with large cells.

It is preferable to introduce a genetic label that will persist without dilution into a single cell of organisms that do grow, such as mouse or chick embryos. This could be a result of a genetic recombination event that produces an identifiable marker or insertion by a replication-incompetent retrovirus. Determining whether a cell is committed to form a specific structure or cell type at the time of labeling is the most valuable use of clonal analysis (Figure 3.4).

Its descendants will only be able to populate structure A and nothing else if it is committed to forming a specific structure named A. As a result, if a clone labels both structures A and B, the cell cannot have been committed to developing into structure A or structure B at that point. Occasionally, it is discovered that if a label is applied later, it will only populate either A or B, whereas an early application will span A and B. This could be as a result of the cells becoming committed during the interval between the two labels. That could, however, just be the case because the later-induced clones are smaller and therefore less likely to occupy multiple structures. Clonal analysis can therefore demonstrate a lack of commitment but not its presence.

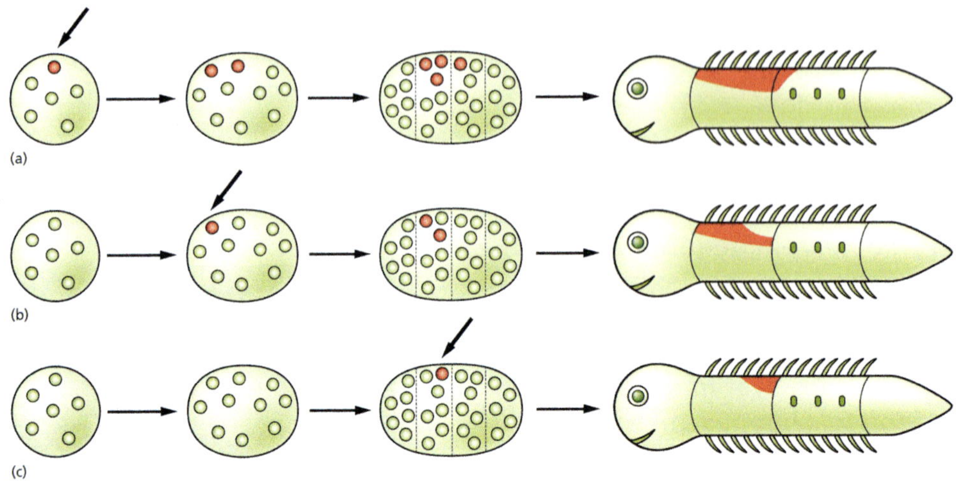

Figure 3.4. *Clonal analysis. No clonal restriction means no determination, but the converse is not necessarily true. (a) Single cell labeled before determination; progeny span later boundary. (b) Cell labeled before determination but progeny fail to span boundary because clone is too small. (c) Cell labeled after determination and cannot cross boundary.*

Source: Essential Developmental Biology, Third Edition. Jonathan M.W. Slack. © 2013 John Wiley & Sons, Ltd. Published 2013 by John Wiley & Sons, Ltd.

Clonal analysis has been widely applied, especially in the study of vertebrate hindbrain patterning and Drosophila segmentation. A section of an embryo whose limits are clonal restrictions is sometimes referred to as a compartment. No cells may enter or exit a compartment once it has been created. Stated differently, every cell in a compartment is descended from the original cells. A compartment is typically associated with an outwardly visible structure, like a segment or an organ rudiment. It is kept in place either by physical barriers to cell migration, like basement membranes, or by the cells' differential adhesion to one another, which keeps the cells inside the compartment from mixing with the outside. The boundaries of the anterior and posterior compartments of Drosophila imaginal discs are one example of a situation where the clonal restriction boundary does not match an obvious anatomical boundary.

3.2. DEVELOPMENTAL COMMITMENT

Previously uncommitted cells commit to form specific body parts or cell types as development moves forward. We now believe that commitment is directly visualizable through *in situ* hybridization, which allows us to observe the expression of the relevant genes. Commitment is encoded as a combination of transcription factors found in the cell. However, historically, studies using embryonic models were conducted to examine commitment. Two operational definitions that resulted from this—often referred to as specification and determination—are still relevant today.

If a cell or tissue explant can develop into a specific structure on its own after being isolated from the embryo, it is said to be specified to become that structure. By conducting numerous experiments and combining the results, it is feasible to create a specification map of the embryo. This demonstrates the functions that the cells are programmed to perform at that specific developmental stage. A region's definition does not always correspond with how it will develop normally. For instance, when cultivated in isolation, the potential neural plate of a *Xenopus blastula* will differentiate into epidermis rather than neuroepithelium. The neuroepithelium requires an inductive signal from the mesoderm in order to form.

Figure 3.5 depicts the tests for fate, specification, and determination.

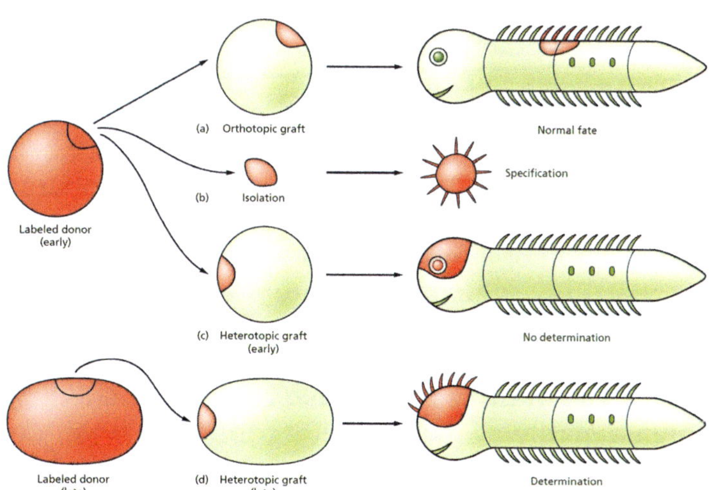

Figure 3.5. *Tests for fate, specification, and determination.*

Source: Essential Developmental Biology, Third Edition. Jonathan M.W. Slack. © 2013 John Wiley & Sons, Ltd. Published 2013 by John Wiley & Sons, Ltd.

(a) The labeled region will normally contribute to the spiny dorsal part of the animal (fate). (b) When isolated this tissue still forms dorsal spines (specification). (c) When grafted at an early stage to another region it differentiates according to the new position (not determined). (d) When grafted at a later stage to another region it differentiates according to its original position (determined).

While a designated area of tissue will also grow on its own in isolation, it will be committed in an irreversible way to the variety of environments found in the embryo. Put another way, it won't stop developing on its own if it is transferred to another area of the embryo. Transplanting a section of tissue from one location to another is a common method used in embryological experiments to determine whether the tissue will develop according to its new location or its previous one. In order to distinguish between the tissues of the graft and the host, grafts are typically labeled using one of the techniques mentioned. They can be carried out with a great deal of precision with practice. A graft of GFP-labeled tissue is seen in Figure 3.6 in the posterior neural fold region of an embryonic Xenopus closed neurula. One common technique for fate mapping is an orthotopic graft, which involves transferring an embryo to the exact location of another embryo. Heterotopic grafting, which is a test for determination, is a graft to a different location within the host. The tissue is defined as determined if the graft has no effect on the developmental pathway. Even though it might have been specified at the time of grafting, it was not determined if a heterotopic graft grows in accordance with its new position. A sequence of these grafts carried out at various phases typically demonstrates the point at which the tissue becomes determined. For instance, a Xenopus embryo's potential neural plate is not identified at the blastula stage since it will develop into either mesoderm or epidermis if grafted to another location within the embryo. During gastrulation, the formation of the neural plate is inevitable because grafts to other areas of the embryo will always differentiate into neuroepithelium after this point. Determination, from a molecular perspective, indicates that the cells are no longer competent or responsive to the signals that initially activated the appropriate combination of transcription factors. This could be the result of the cells losing signaling machinery components such as receptors, or it could be the result of factors other than the ones that initially turned on the transcription factor combination maintaining it.

Every embryo has a hierarchy of commitment linked to progressive physical subdivision during development. The ectoderm, mesoderm, and endoderm, for instance, are the three germ layers that form first. Each germ layer is then further divided, for instance, the ectoderm into the neuroepithelium, neural crest, and epidermis. Subregions of the neuroepithelium, such as individual rhombomeres of the hindbrain, will eventually be divided into smaller subregions. This implies that any given cell region in the embryo will go through multiple states of commitment, each of which is determined by a unique set of transcription factors. At any level of this hierarchy, the embryological techniques for defining specification and determination can be used. Figure 3.7 illustrates the hierarchy of regional specification in development.

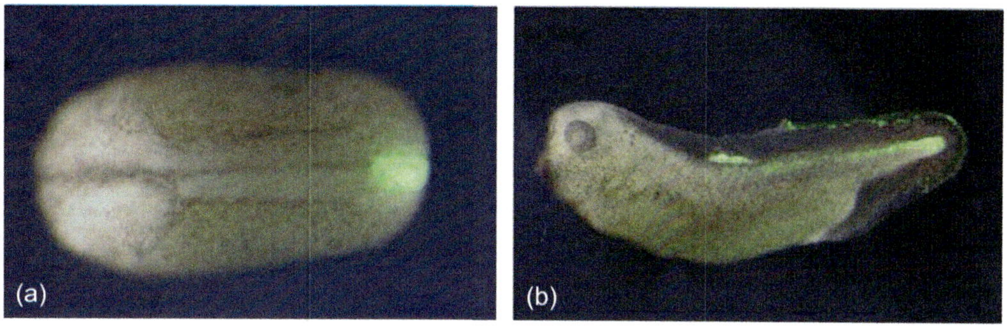

Figure 3.6. *(a) Graft of transgenic GFP-labeled tissue to posterior neural fold of a Xenopus neurula. (b) The graft later contributes to spinal cord and neural crest.*

Source: Essential Developmental Biology, Third Edition. Jonathan M.W. Slack. © 2013 John Wiley & Sons, Ltd. Published 2013 by John Wiley & Sons, Ltd.

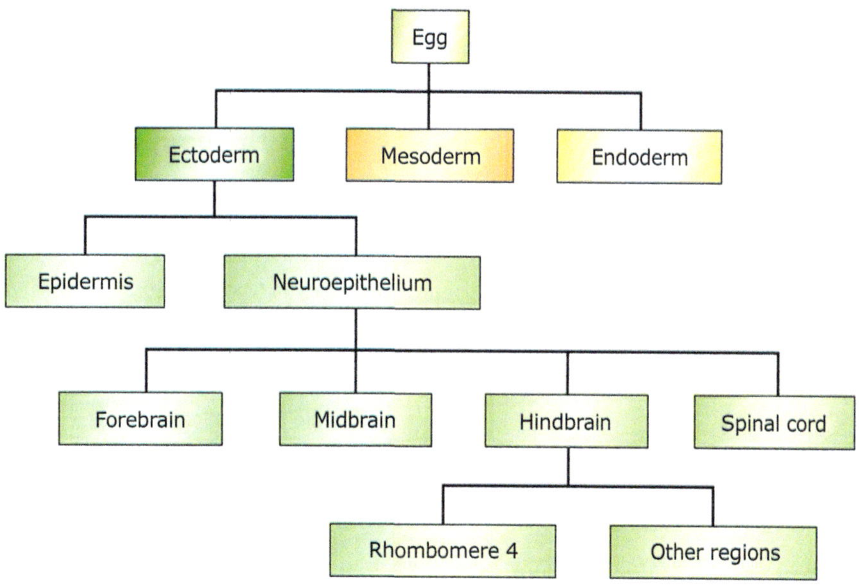

Figure 3.7. *A hierarchy of regional specification in development. Cells that become rhombomere 4 of the hindbrain first need to "decide" to be ectoderm, then to be neuroepithelium, then to be hindbrain, and finally to be rhombomere 4.*

Source: Essential Developmental Biology, Third Edition. Jonathan M.W. Slack. © 2013 John Wiley & Sons, Ltd. Published 2013 by John Wiley & Sons, Ltd.

Since the initial stages of the hierarchy of commitment don't align with specific body parts, they are frequently denoted by position (e.g., anterior/posterior and dorsal/ventral. For students who are just starting to study developmental biology, this can be extremely confusing. Because states of commitment can be changed by experimental manipulation, one may come across a statement like "overexpression of X makes dorsal cell population Y ventral." This indicates that the embryo's dorsal cells have been forced to adopt a state of commitment that is identical to what is typically seen in the ventral region. Therefore, it's critical to understand clearly from publications whether terms with positional connotations, such as "dorsal," actually refer to a position or to a normal developmental state of commitment associated with that position.

Sometimes, the range of potential cell types or structures that a specific cell population can develop into is referred to as its potency. This is comparable to competence, but it may also refer to developmental pathways that can be triggered *in vitro* by external environments that are not typically present in the embryo. These days, it's common to refer to tissue-specific stem cells, which can form cell types specific to their own tissue, as multipotent, and embryonic stem cells, which can form any type of cell in the body, as pluripotent.

3.2.1. Cytoplasmic Determinants

A cytoplasmic determinant is a substance or substances present in a portion of an egg or blastomere that ensure the cells that inherit them during cleavage will assume a specific state of commitment (Figure 3.8). The cell division is asymmetrical by nature, resulting in two distinct developmental pathways for the daughters. A determinant-containing cytoplasm can be grafted to another area of the egg, and the new cells that contain it will then form the proper structure.

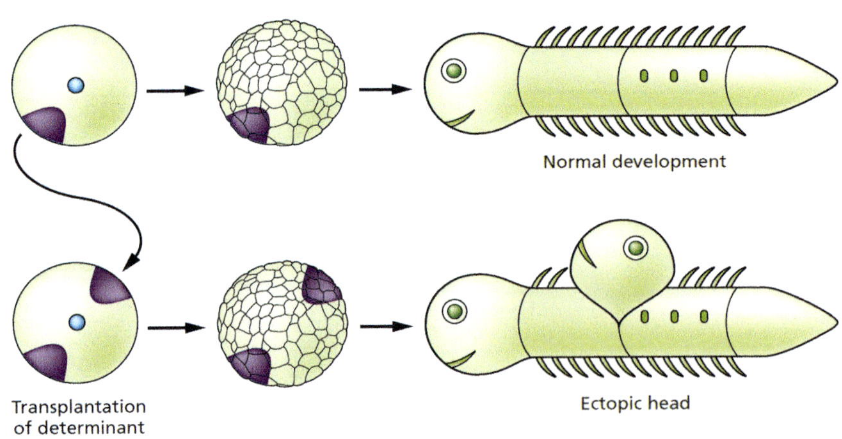

Figure 3.8. *Operation of a cytoplasmic determinant coding for the head of the embryo. In normal development it ensures that the head forms from the cells that inherit it. If the determinant is transplanted elsewhere than it causes formation of an ectopic head.*

Source: Essential Developmental Biology, Third Edition. Jonathan M.W. Slack. © 2013 John Wiley & Sons, Ltd. Published 2013 by John Wiley & Sons, Ltd.

Since cytoplasmic determinants are frequently in charge of establishing the first two or three clearly defined regions in the embryo, they are extremely significant for the very earliest stages of embryonic development. Interactions between these initial domains result in the pattern's subsequent complexity. Sometimes mRNAs localized to a specific area of the cell in conjunction with the cytoskeleton serve as determinants; two examples of these mRNAs are the bicoid and nanos mRNAs found in the Drosophila egg. They might be proteins as well. In the initial phases of C. elegans development, a complex that is localized in the anterior and regulates the fate of the first two blastomeres. This complex consists of PAR3, PAR6, and aPKC. In other animal species, neurogenesis and epithelial polarization involve a similar system.

3.2.2. Induction

The majority of regional differences in development are caused by extracellular cues known as inducing factors. In other contexts, many of them are referred to as growth factors, cytokines, or hormones. Competence is the capacity to react to an inductive signal; it necessitates the presence of particular receptors as well as a functional signal transduction pathway connected to transcription factor regulation.

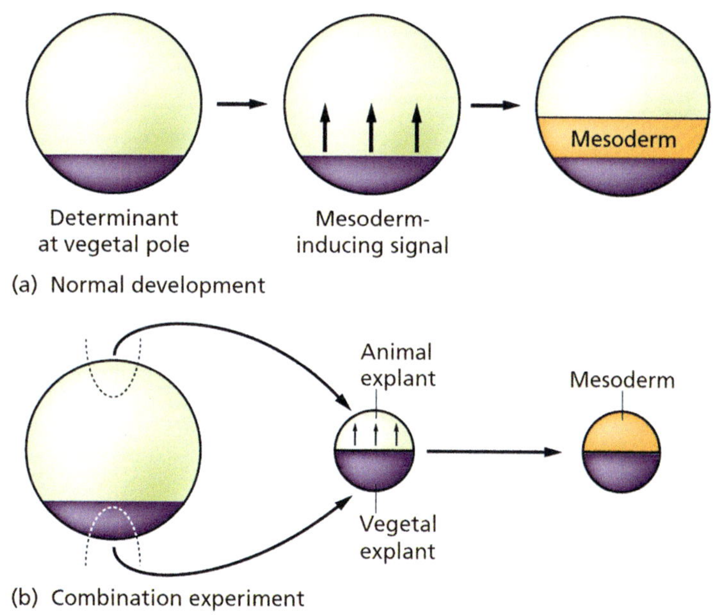

Figure 3.9. *Mesoderm induction in Xenopus. (a) As it occurs in normal development. (b) As it occurs in an animal–vegetal combination experiment.*

Source: Essential Developmental Biology, Third Edition. Jonathan M.W. Slack. © 2013 John Wiley & Sons, Ltd. Published 2013 by John Wiley & Sons, Ltd.

For instance, in the Xenopus embryo, signaling molecules called Nodals released from the vegetal region cause the mesoderm to be induced from the animal hemisphere tissue (Figure 3.9). The signals cause the expression of several transcription factors, including the T box protein Brachyury, to increase. These transcription factors define the mesodermal state. Both the remaining animal hemisphere and the entire animal hemisphere by itself turn into ectoderm. It has been possible to demonstrate that this interaction takes place during the blastula stages by using fragments from embryos of various stages. This interaction can happen between tiny pieces of the blastula cultured together. Because the responding tissue has a choice—either mesoderm or ectoderm—and because the interaction increases the complexity of the embryo during normal development, this kind of interaction is known as an instructive induction.

There are two distinct forms of informative induction, and their effects on regional specificity are a little different. It's possible that the signaling center, which is the origin of the signal substance's concentration gradient, is located at one end of a cell sheet. A number of territories form in response to the gradient because the surrounding tissue's competence embodies distinct threshold responses to varying concentrations. In a situation like this, where there are multiple favorable outcomes, it is now customary to refer to the signal substance as a morphogen. The gradients of sonic hedgehog protein in the neural tube and limb bud, active bone morphogenetic protein (BMP) in the early Xenopus embryo, or decapentaplegic protein (Dpp) in the Drosophila imaginal discs are well-known examples of morphogen gradients.

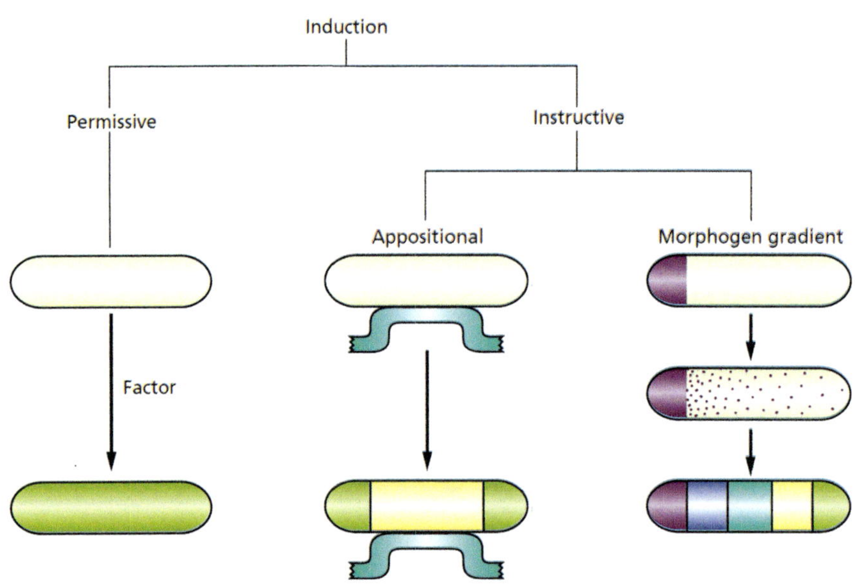

Figure 3.10. *Types of induction: permissive and instructive. Instructive inductions may be appositional or morphogen type.*

Source: Essential Developmental Biology, Third Edition. Jonathan M.W. Slack. © 2013 John Wiley & Sons, Ltd. Published 2013 by John Wiley & Sons, Ltd.

An alternative scenario is that the responding cells are located in one cell sheet and the signaling centers in another. A single threshold response in the areas of the responding tissue that are directly next to the signaling centers causes the right structures to be induced when they are brought together. This occurs, for instance, when the underlying tissues influence the induction of nasal, lens, and otic placodes from the head epidermis of vertebrate neurulae. The epidermis differentiates as normal epidermis when the signal is absent, but forms a placode when it is present. We refer to this as appositional induction. Even though the same substance might act as a morphogen in another context, the inducing factor would not be considered a morphogen if the responding tissue only produced one threshold response. Figure 3.10 shows the types of induction.

Permissive inductive interaction is a different type of inductive interaction. In this case, the signal is required for the responding tissue to successfully self-differentiate, but it has no control over the chosen developmental pathway. During the later stages of development, permissive interactions are crucial. For instance, when the ureteric bud sends out a permissive signal, the mesenchyme forms tubules during the development of the kidney. It just stops developing and doesn't form any other tissue when the signal is absent. The primary distinction between permissive and instructive inductions is that the former cause a division of the competent tissue, whereas the latter do not.

The prevailing belief is that inductive signals permeate the extracellular environment. Nonetheless, a large number of the variables in question are growth factors that are firmly bound by glycosaminoglycans found outside of cells. Alternative diffusion pathways have been explored for these and other reasons; one such pathway is transport *via* the tiny intercellular structures known as cytonemes, which are present in a variety of developing tissues. Even in situations where inducing factors do traverse the intercellular space, their spatial distribution may be influenced by endo- and exocytosis processes within the cell. Since multiple mechanisms for the transport of inducing factors have been identified, the Drosophila imaginal discs provide the best understanding of this process (Figure 3.11).

3.2.3. Lateral Inhibition

Lateral inhibition is a significant subclass of cell communication that is typically taken into consideration independently of induction. The Notch–Delta system's behavior, which functions in various scenarios where individual cells from a uniform population follow one developmental pathway while the surrounding cells follow another, is the best example of this. Examples are the development of endocrine cells in the gut's epithelia or neurogenesis in the early neural plate. In theory, lateral inhibition systems function by establishing several tiny signaling centers that impede the development of the signal in the cells around them. A field of committed cell type B cells that are impulsively moving toward commitment to form cell type A is shown in Figure 3.12. The earliest cells to differentiate into type A produce two substances: an "activator" that encourages the development of type A cells and a "inhibitor" that counteracts the activator's effects to prevent inhibited cells from progressing toward type A and instead cause them to revert to type B cells. The system produces patterns because the inhibitor has a long range and can move freely through diffusion, while the activator is short range and may only be active through intercellular contacts. This indicates that in close proximity to the activator source, the activator will outweigh the inhibitor and ensure the establishment and upkeep of cell type A, which is responsible for producing both substances. The inhibitor will outweigh the activator in the surrounding area, suppressing the formation of cell type A. The inhibitor's action will become ineffective beyond a certain point, allowing type A cells to continue forming signaling centers. This will eventually lead to the formation of numerous signaling centers, roughly uniformly distributed throughout the cell field. The specifics of the mechanism determine how consistent the final pattern is.

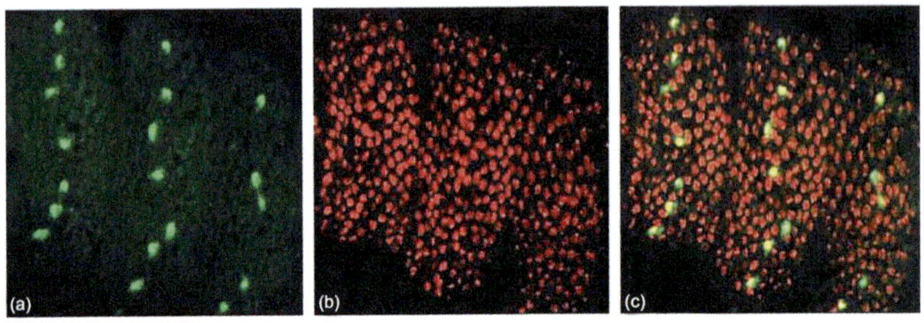

Figure 3.11. *Sensory progenitors arise as individual cells in Drosophila imaginal discs following a lateral inhibition process. (a) Early sensory cells visualized by immunostaining for SuH protein (green). (b) A general nuclear stain (propidium iodide). (c) A merged view.*

Source: Gho et al. Development 1996; 122: 1673–1682, with permission from Company of Biologists Ltd.

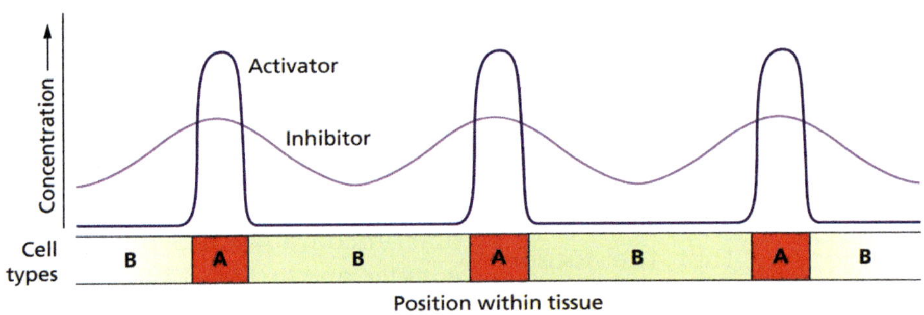

Figure 3.12. *Lateral inhibition. Cell type A produce both the activator and the inhibitor. Where the activator prevails cell type A is stabilized, where the inhibitor prevails cell type A is suppressed.*

Source: Essential Developmental Biology, Third Edition. Jonathan M.W. Slack. © 2013 John Wiley & Sons, Ltd. Published 2013 by John Wiley & Sons, Ltd.

3.2.4. Stochasticity in Development

Creating patterns out of homogeneity appears to be a common aspect of development. When a lateral inhibition system activates, this happens. When a cytoplasmic determinant localizes to one end of the cell instead of the other, it happens. It also happens when different cell types, even those grown *in vitro* in a controlled environment, seem to diverge from a single type of progenitor. These procedures all entail the breaking of symmetry. Positive feedback causes very tiny, naturally occurring variations in the concentration of certain substances or the activity of particular genes to become amplified at certain locations and repressed in nearby regions as a result of the amplification. The limited quantity of regulatory molecules within a cell is ultimately the cause of these fluctuations. Each cell may contain several hundred copies of a specific transcription factor, and in order to locate their binding sites in the DNA, researchers must search through a vast amount of DNA sequence data. Every gene is found in pairs, and each copy of a regulatory sequence has either a transcription factor attached to it at that specific moment, or it does not. A population of seemingly identical cells will actually differ in terms of the instantaneous occupation of regulatory sites due to the lengthy residence times of the individual molecules, which are measured in minutes. Such fluctuations can readily be magnified to become macroscopic and irreversible compositional differences because positive feedback systems exist.

3.2.5. The "Epigenetic Landscape"

The remarkable concept of developmental commitment was first presented by the embryologist and geneticist C. H. Waddington during World War II. The "epigenetic landscape" is this (Figure 3.13). His diagrams are once again being copied often and have recently seen a resurgence in popularity. A typical diagram shows a landscape where valleys emerge from branches as they descend a hillside, simulating how water sculpts a natural landscape. A ball positioned at the summit will tumble down the gullies. Every bifurcation represents a symmetry breaking event because it experiences minor perturbations at each point where it will move left or right. It will eventually settle into one of the valleys with the lowest tier. Each bifurcation in the ball represents a developmental choice, the valleys states of developmental commitment, and the ball itself is meant to represent a cell in an embryo. Naturally, a large number of balls would represent the entire embryo, and some balls would take every conceivable path across the terrain to produce the full range of stable cell types. As of right now, the diagram encapsulates a few key concepts that the molecular era has confirmed. It does, however, also have certain issues. One issue is that the diagram solely depicts cell states; it does not in any way depict the embryo's structure. This implies that there is no representation of the interactions between the various parts of the embryo, which are crucial due to the inductive signals that are shared. The majority of developmental decisions are actually deterministic processes driven by inductive signals rather than symmetry-breaking events.

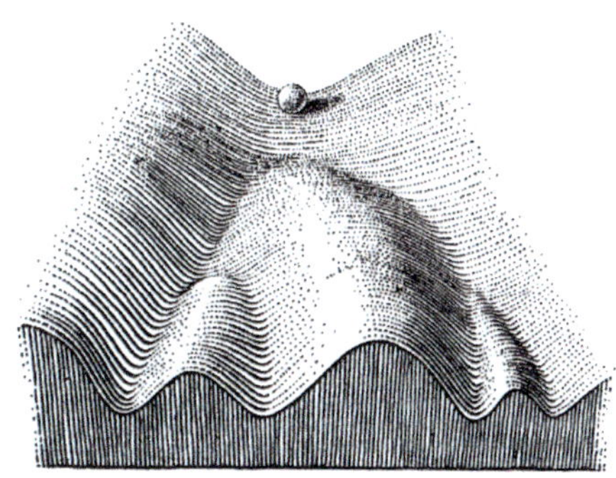

Figure 3.13. *Waddington's "epigenetic landscape." The ball represents a cell, and the valleys represent different developmental choices that it can make.*

Source: Slack (2002), with permission from Nature Publishing Group.

Exactly determining what the surface represents is a somewhat more abstract problem. Actually, it is intended to be a multidimensional state space containing a "potential energy surface" that represents all of the concentrations of various substances and gene products present in cells. It's still unknown if such surfaces can actually be computed or exist at all. Waddington coined the term "epigenetics" to refer to the process of regulating gene expression, realizing that this process is essentially synonymous with development. Today, though, the term has a different meaning. These days, the term "epigenetic" primarily refers to the comprehension of gene expression in terms of chromatin structure as opposed to DNA alone. This acknowledges the role that DNA methylation plays in controlling the expression of genes, as well as the various chemical changes that affect histones and other chromosomal proteins and collectively either promote or inhibit transcription. Although these rare instances of inheritance unrelated to DNA sequences are not very significant for animal development, the term is also used to refer to them.

3.3. QUANTITATIVE EXPERIMENTAL EMBRYOLOGY

Controlling the processes of development is a multi-scale problem by nature. One could think about how changes in cell state are caused over time by the activity of genes and their regulatory networks. Such alterations in cell state have an effect on how cells behave dynamically when they move, alter in size or shape, or interact with surrounding cells and the extracellular matrix. These intercellular interactions affect the structure, rate of expansion, and/or physical characteristics of tissues. Afterwards, tissues work together to create embryos and entire organisms through a mix of chemical and mechanical signaling interactions. Each of these biological organization levels simultaneously experiences multi-scale feedback interactions, such as when gene regulatory networks react to cell-to-cell interactions and modifications in their mechanical surroundings, or when tissue morphogenesis reacts to inductive interactions and multi-tissue mechanics. One method of investigating such multi-scale interactions in developmental biology is experimental embryology; specifically, this method can be used to examine how changes at higher organizational levels affect those at lower levels through downward causation. Because it has a lengthy history in the field and predates the addition of molecular biology techniques to the developmental biologist's toolkit, experimental embryology is regarded as a classical approach. But we contend that it is at the forefront of developmental biology because of its use in examining important and timely questions like scaling, self-organization, developmental timing, and the mechanics of morphogenesis. Moreover, quantitative experimental embryology becomes a modern classic with the growing use of quantitative read-outs of such manipulations.

In general, there are three types of manipulation in experimental embryology, which we cover in the three sections of this review. First, the insertion of cells into an undeveloped area of the embryo or into an *in vitro* tissue conjugate. The second method involves ablating specific cells or tissues to investigate the mechanisms governing pattern regulation and regeneration. Lastly, by using tissue confinement and *in vitro* cell culture to change the mechanical environment of cells. The study of neural crest development has benefited greatly from the application of each of these methods, both in terms of the early specification and the later migration of the cell. The Roberto Mayor laboratory has been able to use experimental embryology techniques to ascertain the rules by which cells, cellular collectives, and tissue interact with one another during neural crest development by utilizing the unique advantages of the *Xenopus laevis* embryo. Crucially, these results are invariably connected to *in vivo* investigations to ascertain whether the recently discovered neural crest development mechanisms apply to cells in their native habitat. We intend to examine a variety of recent instances of experimental embryology's use, using this method as additional motivation to encourage a return to it as a fundamental tool for developmental biologists. Our goal is to give the interested reader a selection of quantitative experimental embryology applications that convey the breadth of its application, rather than reviewing all of these examples as this would be outside the purview of a single review. Table 3.1 explains the experimental embryology techniques.

Table 3.1. *Experimental Embryology Techniques and Examples of Relevant Biological Questions That They Can Help to Elucidate*

Experimental Embryology Techniques	**Questions**
Adding cells	Scaling
Homotypic grafts	Cell competition
Heterotypic grafts	Cell autonomous vs. non cell autonomous processes
Embryonic aggregates	Fate mapping
	Inductive reprogramming
Removing cells	Regeneration
Single cell removal	Scaling
Genetically targeted ablation	Mechanical regulation
Tissue dissection	Multi tissue coupling
	Competence
Confining cells	Force generation
Agarose gels	Intrinsic vs. extrinsic mechanical signals
Matrigel	Mechanical versus biochemical signals
Biochemically & Mechanically defined Hydrogels	Force adaptation

Source: J. Dev. Biol. 2022, 10(4), 44; https://doi.org/10.3390/jdb10040044.

Table 3.2 summarizes the examples of modern quantitative adaptations of experimental embryology techniques.

Table 3.2. *Examples of Modern Quantitative Adaptations of Experimental Embryology Techniques*

Paper	**Experimental Embryology Method**	**Modern/Quantitative Addition**
Solovieva et al., 2022	Grafting	Single cell RNA-sequencing and live imaging
Guibentif et al., 2021	Chimera production (blastula aggregation)	Single cell RNA-sequencing

Yamaguchi et al., 2016	Single cell ablations	Used laser ablation.
Tata et al., 2013, Aztekin et al., 2019	Cell population ablation	Genetically-targeted ablation. (Tet-On diptheria toxin; Nitroreductase).
Almuedo-Castillo, et al., 2018; Ishimatsu et al., 2018; Huang & Umulis, 2019	Tissue removal	Used classical methods to remove cells (capillary; hairloop; needle). Combined with imaging and mathematical modeling to predict how the system scales and other perturbations, such as protein-anchoring, to test the model's predictions.
Caldarelli et al., 2021	Tissue removal (embryo bisection)	Used laser ablation. Coupled with high-resolution imaging to quantify the mechanical forces of tissue movement.
Savin et al., 2011	Tissue removal	Used classical tissue dissection followed by mathematical modeling of tissue properties.
Reinhardt et al., 2005	Tissue ablation	Used laser ablation.
McLaren & Steventon 2021; Ozelci et al., 2022	Tissue ablation	Used laser ablation. Coupled with high resolution live imaging.
Zhou et al., 2015; Huebner et al., 2022	Embedding	Quantification of force generation (displacement of fluorescent micro-beads and tissue buckling)
Ranga et al., 2016	Embedding	Robotics and automatic image analysis
Elosegui-Artola et al., 2022	Embedding	Computer simulations to study how fingering behaviour depends on the mechanics of the substrate

Source: J. Dev. Biol. 2022, 10(4), 44; https://doi.org/10.3390/jdb10040044.

3.3.1. Cell Addition as an Essential Tool in Experimental Embryology

For many years, transplantation (grafting) experiments have allowed experimental embryologists to add cells to embryos. These manipulations have produced significant findings, such as the idea of organizer regions within the embryo and the development of techniques for fate mapping. We will discuss two important instances of research that have contributed to our comprehension of regulative development: the first is the role of inter-embryo cell combinations (chimaera formation) in illuminating the extraordinary capacity for regulation of mammalian embryos, and the second is the particular case of inductive reprogramming. The primary distinction between these two instances is that the cells being added in the first are frequently equivalent (in terms of stage and potency), whereas the cells being added in the second are clearly different and have the ability to trigger a response in the host tissue. We will contend that the two kinds of experimental manipulations of embryology that were described paralleled real-world occurrences in which cells were added to the early embryo and talk about unanswered issues in this area. Given that these events occur spontaneously in the natural world, modern molecular approaches hold great promise for understanding the underlying mechanisms that allow regulation of development when new cells are added. These discoveries could have significant implications for medicine and fertility.

3.3.1.1. Chimaeras, Homotypic Grafts and Size Regulation

A chimaera, according to Greek mythology, was a creature that had the head of a lion, the body of a goat, and the tail of a dragon fused together. The term has gained greater medical significance in today's lexicon and is now used to describe any living thing with at least two genetically distinct cells. Chimaeras are naturally occurring in humans and other animals, though many remain completely undetected during the animal's lifetime because they don't seem to have any obvious negative effects. The medical literature has described the phenotypic effects of chimerism that can occur in a small subset of human cases, including the first recorded case of human chimerism in which a patient presented with mixed blood type. Additional effects that have been reported include patchy skin, intersex phenotypes, and non-matching eye colors. These can all be caused by the fusion of two genetically different cell types within a single person. Occasionally, these phenotypes can have detrimental effects, such as psychological effects or infertility.

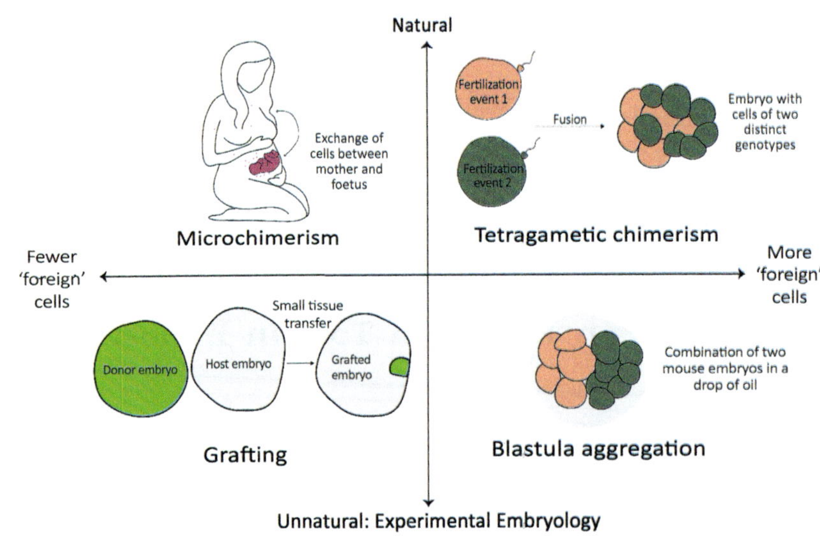

Figure 3.14. *A spectrum of chimeras. This diagram conceptualizes examples of chimeras discussed across two axes: degree of naturality (i.e., naturally occurring or produced by experimental manipulation), and the proportion of foreign cells relative to total cell number. Naturally occurring chimeras include micro chimeras (top left) and tetragametic chimeras (top right). Experimentally produced chimeras include grafted embryos (bottom left) and blastula aggregation experiments (bottom right).*

Source: https://www.mdpi.com/jdb/jdb-10-00044/article_deploy/html/images/jdb-10-00044-g001-550.jpg

Naturally occurring chimaeras can be categorized as micro chimaeras or tetragametic chimaeras, which have different natural origins during gestation, based on the proportion of "foreign" cells in an individual. A micro chimaera is a person whose body contains a small number of distinct genetically made cells that are produced during gestation by the transfer of a small number of cells between the mother and the fetus (or vice versa), or in some cases between siblings. On the other hand, a tetragametic chimaera is made up of a significant amount of cells from both genotypes and is created when the zygotes and early embryos fuse in utero after two separate fertilization events. There has been prior description of this incident as involving "a disappearing twin." Although not within the purview of this review, the specifics of natural chimaera formation are included here because of the striking similarities to experimental embryological manipulations. The ability for these cell exchanges between individuals during gestation is made possible by viviparity; in reproductive modes in which the embryo develops outside of the parent under a protective covering, opportunities for these exchanges will be rare, if they exist at all.

These medical examples closely resemble the kinds of scenarios in which experimental embryologists have added cells to embryos, even though they are not products of the field. The examples of adding cells to embryos that we will discuss are conceptualized along two axes: the number of foreign cells added and the degree of naturalness. This is shown in Figure 3.14. The human medicine examples that were discussed are on the "natural" end of the spectrum. However, we will now look at the analogous examples found in experimental embryology.

3.3.1.2. Blastula Aggregation

Between 1961 and 1962, two separate scientists fused two murine embryos at the 8-cell stage. They discovered that the resultant aggregate, known as chimaera, could develop normally and produce live offspring. Other groups have since expanded on this finding to produce much larger embryonic aggregates; for instance, combining five to nine embryos can result in the production of giant blastocysts that undergo normal development. These aggregation experiments were and are still remarkable in their success, as they are now widely used in mouse research to produce transgenic or mutant offspring. A mechanism of size regulation is necessary for the chimeric embryo to develop to term and generate offspring of the proper size. This mechanism must be able to detect and react to changes in the embryo's size. It should be noted that the proper assignment of cell fate is a necessary additional component, but it won't be covered in detail here. Lewis & Rossant (1982) demonstrated that the embryo's response to size perturbation is accomplished through adjusting the length of the cell cycle. The aggregates showed no changes in apoptosis or the populations of dividing cells, but the dividing cells in these embryos exhibited different dynamics from control embryos. Crucially, the double-sized embryos did not exhibit the burst of proliferation that the control embryos did at 6 and 8 hours after coitus. Therefore, by adjusting the proliferative dynamics of its cells, the size of the murine embryo is regulated upon the addition of new cells. These findings partially explain how a cell senses whether it is in a larger or smaller embryo, but they do not explain the downstream mechanism of size regulation once a perturbation is detected.

Potential molecular inputs that might be responsible for this sensing step have been reported more recently. Applying saturating concentrations of the mitogen Platelet-Derived Growth Factor (PDGF) to cultured embryonic stem cells can affect how the cells divide. While PDGF levels *in vivo* will progressively decrease as cell populations grow, high PDGF application in culture or in transgenic embryos can artificially stimulate cell division independent of the embryo's cells' previous division state. But unlike the spinal cord progenitors examined in the aggregation experiments above, these experiments were carried out with them considerably later in the developmental process. What this cue might be in relation to the embryo at the morula stage is still unknown.

While the remarkable ability of mammalian embryos to regulate themselves is demonstrated by experimental embryology, during the normal development of the mouse, thirty-five percent of the embryonic cells are eliminated. Thus, given the natural cell loss and consequently plastic cell numbers in early development, the observation of regulative and size-robust mammalian development may not be surprising. Research on cell competition has demonstrated the significance of mTOR and cMyc-related molecular pathways in these processes. Normally, cells exhibiting a loser phenotype (e.g., 4N cells) are engineered to exhibit increased activity on the mTOR pathway; this rescues them from their typical apoptotic state and keeps them within the population. These cells exhibit improved competitive abilities and do not become extinct from the population when injected into an early mouse embryo to create a chimera. In this context, it was demonstrated that p53 controls the mTOR pathway, which is involved in some aspects of cell competition. Removing p53 activity can result in the creation of "super-competitor" cells. Comparably, competitiveness and cMyc protein levels are correlated, and overexpression of the protein can drive the eradication of otherwise healthy cells from the population. All of these studies have shed light on how less fit cells can be removed from the embryo during the early stages of development, with the consequence that the embryo needs to be able to regulate its size at that point. As previously stated, the method by which the remaining cells in the population determine the size of the population remains unclear in this system. This is essential for efficient size regulation, and we contend that it is a crucial question in the area. Combining cutting-edge molecular techniques with experimental embryology is a viable strategy to answer this query.

Recently, single-cell transcriptome analysis of Brachyury (T) mutant chimeric embryos provides a promising example of this. A powerful method for comprehending the inclinations and behaviors of individual cells is the chimera created when cells with different genotypes are combined in the early mouse embryo. This system, in particular, enables researchers to investigate the relative significance of the non-autonomous and cell-autonomous effects of these mutations. The ability to assay the entire transcriptome in individual cells has been made possible by the development of single-cell RNA sequencing technologies in recent years. This allows for the examination of both the heterogeneity that occurs within these populations as well as the differences between wildtype and mutant cell states. Guibentif and associates. (2021) demonstrated the existence of two distinct populations of somitic progenitor cells, each with a high and low T state, in the mouse embryo by combining single-cell RNA sequencing with experimental embryology—the creation of T mutant chimeric embryos. This prompted them to suggest that T function has no bearing on the development of the mouse embryo's most anterior somites.

In summary, the initial reports of mouse chimeras, which were released in the early 1960s, have had a significant impact. An illustration of experimental embryology is the creation of a chimera, which is the addition of cells to the early embryo. A few cells can be injected into the blastocoel on a small scale, or up to nine early embryos can aggregate on a large scale. Although the creation of chimeras in mice and other vertebrate species is a remarkable achievement, it can be explained in light of more recent findings, such as the significant natural exclusion of cells with fitness defects from the early embryo. It has been successful to create chimeras across species. Changes in cell division have been demonstrated to impact size regulation in these systems, but the upstream perception of population size in the early embryo is still unclear. Significantly, chimeras are naturally occurring in humans, indicating that the mechanisms of size regulation we discuss may apply to human health and fertility in addition to experimental systems.

3.3.1.3. Homotypic Grafting Experiments

A homotypic graft is defined as the transplantation of a cell population from a donor embryo to the corresponding location in a host embryo; if the host cells have a different genotype than the donor cells, this process will ultimately produce a chimeric embryo. If the donor tissue is labeled so that it can be identified from the host tissue, homotypic grafts offer a potent tool for fate mapping. When this was first developed, it was common practice to use closely related species as the donor and host that could be identified. For instance, quail-chick chimeras were widely used in early experimental embryology because quail cells could be identified by immunostaining or cytology. Chick nuclei have condensates throughout the nucleoplasm, whereas quail nuclei have noticeable heterochromatic condensates connected to the nucleolus. A comparable method has been applied in amphibian studies, where the distinction in pigmentation between species facilitates the identification of donor and host tissue in grafts. Transgenic lines, which have been developed in many model organisms more recently, have the added benefit of only involving cells from a single species, eliminating any potential unintended effects brought on by species differences. This allows for the clear distinction of donor and host tissue under fluorescent imaging. The other benefit of this method is that, in contrast to post-fixation staining (e.g., quail-chick phantoms. Solovieva et al., describe a particularly amazing use of the GFP transgenic chicken embryos for fate mapping. (2022). Using single cell grafts, the Hensen's node fate map was reexamined in order to pinpoint the precise location of self-renewing stem cells, which were found to be in the posterior part of the node. Because of the remarkable regulative nature of vertebrate embryos, adding cells to them is an ideal method for fate mapping; these grafts rarely disrupt normal development and integrate well with the host. During this process, the host tissue may be entirely removed in certain instances or retained in others. Several species, such as Xenopus and chicken, have detailed development fate maps thanks in part to homotypic grafting and dye injections. For instance, a significant amount of research has used experimental embryology to map the fate of the primitive streak stage embryo in chickens, revealing the arrangement of various progenitor cells within and surrounding the primitive streak.

Transgenic lines have recently been employed in Xenopus, where injection of fluorescent dextran has been widely used to fate map pre-gastrulation, allowing analysis to be extended to post-metamorphosis stages (by which time fluorescent dextran has been diluted out). Such fate maps offer fundamental insight into which regions of the embryo will contribute to which tissues, and they open the door to more intricate experiments that look into the interactions between these various populations during development. As such, their value to our understanding of development should not be understated.

3.3.1.4. Inductive Reprogramming

Spemann & Mangold (1924) carried out one of the most well-known studies in experimental embryology, which identified the amphibian dorsal blastopore lip as the embryonic organizer. Similar to this, Saunders' polarizing region grafts in the chick limb bud are well known for their profound effect on limb patterning, causing the digits in the AP axis to duplicate in mirror image. We have since learned, through extensive research, about the significance of the signals originating from these signaling centers in the development of the vertebrate primary body axis and limbs as a result of these groundbreaking embryology experiments. The importance of these examples of adding cells to embryos should not be understated: though they are examples of highly localized signals that elicit particularly dramatic responses in host tissue upon grafting, they have pointed us toward fundamental principles of development—i. e. Combining heterotypic grafts with tissue labeling allows responses in host tissue to be separated from the autonomous production of embryonic structures from donor tissue, shedding light on how non-equivalent tissues interact to produce structures in the early embryo. Using differently pigmented Triton (newt) species, it was possible to demonstrate that the ectopic organizer in Spemann's organizer grafts causes host tissue to respond, resulting in the organizer's incorporation into the secondary body axis. The "reprogramming" of the typical ventral fates that cells would adopt in a control embryo is the result of this response. When competent staged embryos' blastocoels were transplanted with Einsteck grafts, a critical study investigating the neural induction aspects of organizer function in Xenopus embryos revealed region-specific differences in the cell types produced from distinct regions of the archenteron roof. In addition to lateral regions producing cell types known to be derived from the neural crest, such as pigment cells, more medial regions were able to induce cell types corresponding to the central nervous system. The first gene to identify early neural crest tissue, snail2 (formerly known as slug), was cloned, allowing for the repetition and analysis of these tests at earlier phases after grafting. This made it possible to examine the sequence of inductive interactions that result in the specification of the neural crest during the stages of development known as gastrulation and neurulation in much greater detail. The ability to fully characterize the gene-regulatory network that reacts to the established inductive interactions of early neural crest development has been made possible by the development of single cell sequencing and chromatin accessibility assays (Williams et al., 2019). Although the addition of cells has greatly aided in our understanding of the mechanisms governing the specification of cell fate during normal development, we will now discuss another set of embryological techniques that have allowed researchers to investigate the regulative ability of pattern formation.

3.4. CELL REMOVAL AS AN ESSENTIAL TOOL IN EXPERIMENTAL EMBRYOLOGY

A vital tool in the arsenal of an experimental embryologist is the ability to extract cells from a developing tissue. The term "cell removal" refers to the elimination of cells at any level of organization, ranging from individual cells or clusters of cells to entire populations or tissues. Furthermore, our definition includes the total removal of cells from their environment as well as their ablation inside a tissue. It is significant to note that in cases where cells are removed from a tissue, the presence of dead cells may influence the body's reaction in a different way than when the cells are removed. There are several ways to remove cells, including laser-mediated ablation and surgical manipulation. Cell removal can provide information about whether a particular phenotype requires the removed entity in all combinations. In situations where the system is resilient to cell loss, cell removal can be utilized to further explore the regulatory mechanisms. Because of this, it's a great place to start when analyzing the specifics of how cells and tissues behave in a particular system. Here, we highlight a variety of studies that effectively use the removal of cells or tissues to start to understand some of the most important unsolved issues in developmental biology. We discuss the various length scales at which this can be studied in the embryo, ranging from the cellular to the tissue and multi-tissue levels.

Cell removal in early embryos was a common technique used in classical experimental embryology to demonstrate that while regulation did occur and normal structures formed in some species, it did not occur in others. The ideas of mosaic and regulative development benefited from this. Although helpful in characterizing the outcome of modifying a particular developmental process, most systems are neither completely regulative nor completely mosaic. A modern interpretation of the age-old practice of cell removal serves as an example of this. Early annelid embryo cells are frequently linked to mosaicism because they frequently have stereotypical fates. For instance, the larval eyes of Capitella telata are produced by two of the four main micromeres. About 75% of embryos that undergo micromere ablation lose the corresponding eye. The remaining embryos, however, successfully develop into two eyes. The percentage of embryos lacking an eye rises when both an eye and non-eye forming micromere are removed. This suggests that the differentiation of larval eye cells has some regulatory potential and that the early stages of development have not yet fully determined the ability to form eyes.

Additionally, cell ablation has proven to be a very effective method for examining the behaviors of cells in tissues during late development and adulthood. Many tissues possess some capacity for regeneration in response to injury or normal cell turnover. To eliminate tissue-resident stem cells, genetically targeted ablation is frequently employed. For instance, a Tet-On transgenic line resistant to diphtheria toxin can be used to eliminate the basal stem cells that express CK5 in the mouse airway epithelium. Secretory cells dedifferentiate upon ablation of basal stem cells and aid in the population's regeneration of stem cells. Notably, the degree of dedifferentiation and the secretory cells' differentiation state are inversely correlated. Studies on the removal of cells can shed light on behavioral changes in addition to differentiation state changes. After being amputated, the Xenopus tadpole tail can regenerate at specific developmental stages. Nevertheless, there is no indication of trans differentiation between cell types or the emergence of a population with a multipotent signature alongside this regeneration. Rather, it seems that a distinct cell population controls the growth of progenitors specific to a given tissue to precisely coordinate regeneration. Moreover, tail regeneration is decreased upon genetic ablation of these regeneration-organizing cells via the nitro-reductase system. Therefore, the migration and signaling of the cells that organize regeneration as well as the proliferation of the other progenitors are the primary cell behaviors that propel the regulation of this system.

How patterns scale with changes in tissue size is a longstanding and intricate question in developmental biology, the specifics of which were recently reviewed. From an evolutionary standpoint, where tissue size varies but patterning mechanism is conserved across species, this is significant. The excision of cells from the early zebrafish embryo is a prime illustration of the patterning's resilience. After gastrulation and the removal of thirty percent of the zebrafish blastula, an embryo with a regular pattern is formed. This is due to the mesoderm scales' specification in reaction to size reduction, which results in germ layers with the proper proportions. The authors predicted and showed that the scaling of the pattern is caused by the high diffusivity of the Nodal-repressor Lefty by combining mathematical modeling and *in vivo* protein anchoring. Under high Nodal levels, Lefty is expressed at the blastoderm margin, but it diffuses more quickly than Nodal. Beyond the margin, the concentration of Lefty is sufficient to impede Nodal, thereby restricting mesendoderm specification. A high concentration and diffusivity of Lefty are necessary in an embryo with a reduced blastoderm in order for Lefty to build up more quickly and limit the mesoderm to fewer cells. Other research has investigated somite scaling and dorso-ventral Bone Morphogenetic Protein (BMP) gradient scaling by combining a blastoderm reduction technique with mathematical modeling and quantitative imaging (Huang et al., 2019).

The highly regulative nature of primitive streak formation in the early chick embryo has long been understood through the use of the chick epiblast in classical embryological experiments. Recent research in quails has shown that mechanical signals scale robustly after tissue removal by combining high resolution imaging with laser-mediated embryo bisection (Caldarelli et al., 2021; biorxiv). It is commonly known that extensive tissue flows are involved in the formation of the primitive streak in the chick embryo. The development of the primitive streak is driven by tissue contractility in the posterior of the embryo, according to quantification of these movements in both chicks and quail. To limit the localization of the streak, tension from the embryonic margin must counterbalance this contraction. Following bisection, there is less embryonic margin, which causes the contractile domain to shrink and the tension within the tissue to rise. In the half-embryo, this promotes scaling of the primitive streak and related gene expression. These studies highlight the significance of integrating contemporary quantification methods with cell removal techniques to comprehend the details of tissue patterning and signaling feedback.

The proper development of surrounding tissues can be influenced by the way a single tissue develops at the level of multiple tissues. It is possible to dissect these interactions by separating the tissues. It has been demonstrated that the attached dorsal mesentery tissue is necessary for the gut tube's looping morphogenesis. Gut loops are lost as a result of both partial removal of the mesentery in ovo and complete separation of the two tissues ex ovo. The mesentery tissue contracts, indicating that the gut was under tension, while the gut relaxes, indicating that it was under compression. Through this experiment, the authors were able to differentiate between various theories regarding the cause of gut looping and ultimately show that differential tissue growth is the driving force behind it. Understanding how tissues communicate with one another can also be accomplished by removing entire tissues. Given their high degree of adaptation to tissue loss, plants make an ideal model system for this. A new primordia formed close to the removal site of a tomato leaf primordia that was laser ablationed in one study, suggesting that forming leaf primordia have an inhibitory effect on surrounding tissue to ensure minimum shading between leaves. Eliminating previously recognized primordia also demonstrated how signals from multiple distinct established primordia are necessary to determine the precise location of a future primordium (Reinhardt et al., 2005).

Removing subsets of cells can also affect multiple tissues. For the other axial tissues to properly elongate during the formation of the zebrafish tail, the notochord must mature. During normal development, new cells are added from a progenitor pool in the tailbud and the anterior notochord cells expand through vacuolation. The notochord influences the elongation of the neighboring somites without influencing the formation of new somites, as demonstrated by ablation of notochord cells in both the anterior and posterior, followed by quantification of tissue changes from live imaging data. In general, this influences the tail's elongation. The notochord and somites are mechanically coupled through the extracellular matrix, just like the gut and mesentery above. This demonstrates how multi-tissue mechanical coupling, which occurs when cells are removed from one tissue, can impact the development of neighboring tissues.

3.5. TISSUE EMBEDDING AS AN ESSENTIAL TOOL IN EXPERIMENTAL EMBRYOLOGY

Compressing embryos between two coverslips is one of the mechanical manipulations of embryos that can be done with an experimental embryology toolbox without affecting the number of cells. Immersion of embryos, explants, or organoids in 3D substrates with well-defined properties is an emerging method for regulating the mechanical inputs that impact them. These novel methods enable the independent control of the substrate's chemical and physical properties as well as the measurement of the force applied by growing tissues. Embedding can assist in the analysis of the interplay between mechanical and biochemical signals, the study of complex phenotypes influencing force generation, and the distinction between intrinsic and extrinsic morphogenetical cues. We will briefly review some important examples from recent research into understanding the role of mechanical forces in guiding developmental processes in order to shed light on developments in this emerging quantitative experimental embryology approach.

3.5.1. Force Generation and Tissue Mechanics during Development

In growing tissues, the amount of force produced and passive mechanical characteristics like stiffness have an equal impact on morphogenesis. This makes it difficult to distinguish between situations in which force production is impacted or in which the passive mechanical properties of the tissue are altered by particular cellular or molecular processes.

Actomyosin contractility is crucially regulated by Rho kinase (Rok), which is also involved in a number of morphogenetic processes. Surprisingly, chemical inhibition of Rok did not result in a phenotype that was visible in developing frog embryos, despite the fact that it significantly decreased the stiffness of the dorsal tissue. Three reasonable explanations were put forth by them: (1) Rok has no role in the convergence and extension (CE) process; (2) the effects of Rok inhibition on force generation are offset by other molecular and cellular processes; or (3) Rok inhibition lowers force generation but is offset by a corresponding decrease in tissue stiffness. An embedding-based technique was created to evaluate force generation during the CE of Xenopus transplants in order to differentiate between these three possibilities. Through quantitative image analysis, they measured the displacement of fluorescent micro-beads dissolved in the substrate, which allowed them to calculate the bulk force produced by explants embedded in agarose gels with known viscoelastic properties. When compared to controls, unconstraint dorsal explants treated with a RoK inhibitor fail to elongate and produce noticeably less force under the mechanical load imposed by the agarose gel. Nevertheless, they do not exhibit significant differences in CE. This demonstrates how RoK inhibition lessens force production and tissue stiffness, allowing the tissue to elongate in the absence of resistance. Last but not least, they demonstrate using the same method that various explants with the notochord surgically removed or mock-operated have the same capacities to generate force, indicating that, in contrast to what has been observed in zebrafish embryos at later stages, the notochord is not necessary for tissue elongation at this stage of development.

A related study conducted more recently utilized agarose embedding to show that the Armadillo Repeat gene, a member of the catenin family required for force generation, was deleted in Velo-Cardio-Facial syndrome (ARVCF). The shorter, bent bodies of Arvcf Knock Down (KD) embryos indicate a defective CE because they exhibit defective anteroposterior elongation. Dorsal tissues from control and Arvcf KD embryos were explanted in order to examine the existence of CE defects. Remarkably, there were no discernible variations in elongation between Arvcf KD dorsal explants and controls. They postulated that Arvcf, or the amount of force required to overcome the embryo's surrounding tissues' resistance, could be significant during CE. By embedding the explants in soft agarose, they were able to replicate the mechanical load imposed by the other tissues and test this theory. Under these circumstances, the Arvcf KD dorsal explants fail to elongate, buckle as seen in the embryo, and produce noticeably less force than controls. In contrast, the control explants elongate effectively. Lastly, they estimated the stiffness of the embedded tissue by looking at its buckling. The Arvcf KD dorsal explants exhibited a significant increase in tissue compliance, as demonstrated by the calculation of buckling.

When taken as a whole, these studies demonstrate how embedding makes it possible to investigate the effects of complex phenotypes that influence force generation and tissue mechanical properties at the same time.

3.5.2. Intrinsic and Extrinsic Mechanical Cues in Development

Intrinsic mechanical cues can be produced by developing tissues to direct differentiation or morphogenesis. Tissue invagination can be rescued in myosin mutant drosophila embryos, indicating that extrinsic mechanical signals can replace these intrinsic signals. Nevertheless, developing organisms can also make use of these external mechanical triggers. This is most likely especially true for plants, as development doesn't happen in an egg that isn't exposed to mechanical or environmental stimuli.

Differential cell elongation in the hypocotyl of dicotyledonous plants produces a hook that shields the meristems from the soil during germination. Baral and associates. (2021) observed that when seedlings germinate on the agar surface, katanin mutants in Arabidopsis, a protein that inhibits branching and encourages microtubule bundling, are unable to produce the apical hook. Nevertheless, they saved the development of the hypocotyl hook when they sprouted the seedling in sterile soil, a more natural environment. The seedlings were germinated inside a soft agarose substrate to enable them to differentiate between the mechanical and biochemical constraints placed on the soil. The hypocotyl hook's formation was also saved by this circumstance, demonstrating that it can only be initiated by outside mechanical cues. Subsequent research demonstrated how mechanical cues from the soil can cause microtubule reorganization and, ultimately, tissue folding independently of katanin by combining pharmacological and genetic perturbations. In this way, they demonstrated that both an extrinsic mechanically dependent pathway and an intrinsic katanin-dependent pathway could activate the hypocotyl hook. Both routes most likely play a part in the hook's strong formation.

Agarose gels can be used in a manner similar to this to investigate the need for outside influences in animal development. Hiramatsu & associates (2013) cultivated mouse embryos at an early stage in agarose gels with varying stiffness. They discovered that an adequately stiff external environment is necessary for the conceptus to elongate properly. Complementary experiments demonstrated that the Distal Visceral Endoderm (DVE) is stimulated to differentiate by the external mechanical environment, indicating the characteristics of the uterine tissues. A recent review has examined the impact that changes in the mechanical constraint imposed by extra-embryonic tissues have on the evolution and development of gastrulation morphogenesis. This research agenda is expected to take exciting new turns as a result of developments in mammalian embryo imaging and ex utero culture (McDole et al., 2018).

3.5.3. Control of Mechanical and Biochemical Parameters

The process of morphogenesis and differentiation is significantly impacted by extrinsic mechanical and biochemical cues originating from neighboring tissues or the extracellular matrix (ECM). Distinguishing between the impact of distinct mechanical and biochemical parameters from outside sources on various biological processes is regrettably difficult. Newer hydrogel generations, however, make it possible to create chemically defined three-dimensional substrates with independently modifiable biochemical and physical characteristics.

An exemplary illustration of this methodology can be located in Ranga et al., (2016). Ranga and companions examined the effects of environmental factors on the development of neural tube organoids using a PEG-based hydrogel whose stiffness, biochemical composition, and degradability could all be combinatorically adjusted orthogonally. They investigated the effects of various biochemical and mechanical variations on other aspects of development, like patterning, proliferation, or the acquisition of cell polarity, by using automatic image analysis. For example, they demonstrate that an intermediate stiffness, a non-degradable matrix, and, to a lesser extent, laminin, all independently improve dorsoventral patterning. Additionally, they discovered evidence that, in this instance, extrinsic forces produced by the stiffer substrates cannot override intrinsic mechanical signals produced by the cells through the use of cytoskeleton contractility activators and inhibitors.

A more current instance is that of Elosegui-Artola et al. The viscoelastic properties of alginate hydrogels conjugated with integrin-binding sites (Arg-Gly-Asp peptide) can be adjusted independently of the stiffness, pore size, and adhesive ligands (2022). They demonstrate how the relative mechanical parameters of the substrate and tissue affect the formation of fingering morphologies of human breast epithelia and intestinal organoids using these gels in conjunction with computer simulations. Hydrogels have also been developed that allow the independent modification of both gel stiffness and extracellular matrix composition, allowing researchers to examine their respective effects on cells grown on different substrates.

Recent research on cell migration during embryonic development has examined the function of mechanical cues in directing cell migration. Through the process of cultivating primary explants of the cranial neural crest from Xenopus embryos on substrates varying in stiffness, it was demonstrated that a higher substrate stiffness led to an increase in chemotaxis efficiency. Crucially, as neural crest migration begins, the underlying mesodermal tissue becomes more elastic in appearance, and this increasing stiffness of the substrate was visible *in vivo*. A pan-placodal domain that covers the neural crest divides into separate cranial placodes at the beginning of neural crest migration and engages in a chase-and-run interaction to aid in guiding neural crest migration. Recent research has also demonstrated the involvement of mechanical forces in this interaction. Specifically, the responding placodal population is subjected to a stiffness gradient imposed by the neural crest, which works in molecular harmony with chemotaxis. An intriguing new area of study that takes into account the potential interactions between electrical signals and mechanical cues is the understanding of how various mechanisms work together to drive the directional migration of cells. The core of this strategy is the simplicity with which tissue and cell explants can be produced via experimental embryology.

CASE STUDY

THE ROLE OF RETINOIC ACID IN LIMB REGENERATION IN AXOLOTLS

Background:

Axolotls (*Ambystoma mexicanum*), also known as Mexican walking fish, are a unique species renowned for their extraordinary ability to regenerate limbs, spinal cords, and even parts of their hearts. This capability makes them an ideal model for studying regenerative biology and experimental embryology. Understanding the mechanisms behind their regenerative abilities can have significant implications for regenerative medicine and developmental biology.

Methodology

§ **Experimental Setup**

- **Subjects:** Twenty adult axolotls were selected for the experiment.
- **Groups:** The subjects were divided into two groups of ten.
- **Control Group:** Received no treatment.
- **Experimental Group:** Treated with retinoic acid.
- **Procedure:** Both groups underwent limb amputation at the same location on the right forelimb. The experimental group received a topical application of retinoic acid on the amputated limb.

§ **Observations and Data Collection**

- **Regeneration Monitoring:** Over the course of six weeks, limb regeneration was monitored and recorded.
- **Data Points:** Key data points included the rate of limb regrowth, limb morphology, and cellular activity in the regeneration blastema (the mass of cells capable of growth and regeneration at the amputation site).
- **Histological Analysis:** Tissue samples from regenerating limbs were taken for histological analysis to assess cellular differentiation and tissue organization.

Analysis

- **Comparative Analysis:** Regeneration rates and morphological features of the regenerating limbs were compared between the control and experimental groups.

- **Cellular Activity:** Cellular activity in the blastema, including cell proliferation and differentiation, was analyzed to identify differences influenced by retinoic acid.

Results

§ **Regeneration Rate**

The experimental group, treated with retinoic acid, exhibited a significantly faster regeneration rate compared to the control group.

Full limb regeneration in the experimental group occurred within four weeks, while the control group required six weeks for comparable regeneration.

§ **Limb Morphology**

The limbs of the experimental group showed enhanced structural complexity, with more defined digits and overall morphology closely resembling a fully developed limb.

The control group exhibited less complexity and some asymmetry in regenerated limbs.

Cellular Analysis

Histological analysis revealed increased cellular proliferation and differentiation in the experimental group's blastema.

Retinoic acid-treated limbs showed enhanced cartilage and bone formation, indicating a more rapid progression through the stages of limb regeneration.

Discussion

The results of this study suggest that retinoic acid plays a crucial role in accelerating and enhancing the regenerative process in axolotls. The findings indicate that retinoic acid not only increases the rate of cellular proliferation but also promotes more complex tissue differentiation and structural development in regenerating limbs. This has significant implications for regenerative medicine, particularly in the context of improving limb regeneration therapies in humans.

The study highlights the potential of using retinoic acid and similar compounds to enhance tissue regeneration and repair, providing a promising avenue for developing treatments for limb loss and other injuries requiring regenerative solutions. Future research could explore the molecular pathways through which retinoic acid exerts its effects, potentially leading to targeted therapies that mimic the regenerative capabilities observed in axolotls.

Conclusion

This case study demonstrates the utility of experimental embryology in understanding the mechanisms of limb regeneration and the potential applications for regenerative medicine. The influence of retinoic acid on axolotl limb regeneration underscores the importance of exploring biological processes in model organisms to uncover new strategies for tissue repair and regeneration in humans.

CLASS ACTIVITY

Students research a specific aspect of embryonic development and prepare a brief report on its significance and current research trends including embryonic development stages, morphogenesis, and the effects of environmental factors on embryogenesis.

SUMMARY

- Experimental embryology is a field dedicated to understanding the intricate processes that govern the development of organisms from fertilization to maturity.
- Normal development means the course of development that a typical embryo follows in standard laboratory conditions when it is free from experimental disturbance.
- A fate map is a diagram that shows what will become of each region of the embryo in the course of normal development: where it will move, how it will change shape, and what structures it will turn into.
- Clonal analysis is a form of fate mapping in which a single cell is labeled and the position and cell types of its progeny identified at a later stage.
- A cell or tissue explant is said to be specified to become a particular structure if it will develop autonomously into that structure after isolation from the embryo.
- A determined region of tissue will also develop autonomously in isolation but differs in that its commitment is irreversible with respect to the range of environments present in the embryo.
- A cytoplasmic determinant is a substance or substances, located in part of an egg or blastomere, which guarantees the assumption of a particular state of commitment by the cells that inherit it during cleavage.
- Experimental embryologists have added cells to embryos for many years in the form of transplantation (grafting) experiments, with these types of manipulations yielding influential results including the conception of organizer regions within the embryo, as well as providing methods for fate mapping.
- A homotypic graft may be described as the transplantation of a cell population from a donor embryo to the equivalent location in a host embryo—ultimately resulting in the production of a chimeric embryo if the host cells differ from donor cells in genotype.
- Morphogenesis in developing tissues is simultaneously affected by the amount of force generated and the passive mechanical parameters, such as stiffness.

REVIEW QUESTIONS

1. What do you understand by normal development in experimental embryology?
2. What is induction?
3. Explain about quantitative experimental embryology.
4. Distinguish force generation and tissue mechanics during development.
5. Explain intrinsic and extrinsic mechanical cues in development.

REFERENCES

1. Almuedo-Castillo, M., Bläßle, A., Mörsdorf, D., Marcon, L., Soh, G. H., Rogers, K. W., Schier, A. F., & Müller, P. (2018). Scale-invariant patterning by size-dependent inhibition of Nodal signaling. *Nature Cell Biology, 20*, 1032–1042.
2. Ashe, H. L., & Briscoe, J. (2006). The interpretation of morphogen gradients. *Development, 133*, 385–394.
3. Aztekin, C., Hiscock, T. W., Marioni, J. C., Gurdon, J. B., Simons, B. D., & Jullien, J. (2019). Identification of a regeneration-organizing cell in the Xenopus tail. *Science, 364*, 653–658.
4. Baral, A., Aryal, B., Jonsson, K., Morris, E., Demes, E., Takatani, S., Verger, S., Xu, T., Bennett, M., & Hamant, O., et al. (2021). External mechanical cues reveal a Katanin-independent mechanism behind auxin-mediated tissue bending in plants. *Developmental Cell, 56*, 67–80.e3.
5. Barriga, E. H., & Mayor, R. (2019). Adjustable viscoelasticity allows for efficient collective cell migration. *Seminars in Cell & Developmental Biology, 93*, 55–68.
6. Barriga, E. H., Franze, K., Charras, G., & Mayor, R. (2018). Tissue stiffening coordinates morphogenesis by triggering collective cell migration *in vivo*. *Nature, 554*, 523–527.
7. Becchio, B. (2006). *Encyclopedia of World Religions.* Foreign Media Group.
8. Bhattacharya, S., Zhang, Q., & Andersen, M. E. (2011). A deterministic map of Waddington's epigenetic landscape for cell fate specification. *BMC Systems Biology, 5*, 85.
9. Bonstein, L., Elias, S., & Frank, D. (1998). Paraxial-fated mesoderm is required for neural crest induction in Xenopus embryos. *Developmental Biology, 193*, 156–168.
10. Bowerman, B. (1995). Determinants of blastomere identity in the early *C. elegans* embryo. *BioEssays, 17*, 405–414.
11. Bowling, S., di Gregorio, A., Sancho, M., Pozzi, S., Aarts, M., Signore, M., Schneider, M. D., Martinez-Barbera, J. P., Gil, J., & Rodríguez, T. A. (2018). P53 and mTOR signaling determine fitness selection through cell competition during early mouse embryonic development. *Nature Communications, 9*, 1–12.
12. Bożyk, K., Gilecka, K., Humiecka, M., Szpila, M., Suwińska, A., & Tarkowski, A. K. (2017). Mouse↔rat aggregation chimaeras can develop to adulthood. *Developmental Biology, 427*, 106–120.
13. Brown, J. M., & Storey, K. G. (2000). A region of the vertebrate neural plate in which neighboring cells can adopt neural or epidermal fates. *Current Biology, 10*, 869–872.
14. Caldarelli, P., Chamolly, A., Alegria-Prévot, O., Gros, J., & Corson, F. (2021). Self-organized tissue mechanics underlie embryonic regulation. *bioRxiv*.
15. Čapek, D., & Müller, P. (2019). Positional information and tissue scaling during development and regeneration. *Development, 146*, dev177709.

16. Cepko, C., Ryder, E. F., Austin, C. P., Walsh, C., & Fekete, D. M. (1995). Lineage analysis using retroviral vectors. *Methods in Enzymology, 254*, 387–419.

17. Chuai, M., Serrano Nájera, G., Serra, M., Mahadavan, L., & Weijer, C. J. (2021). Reconstruction of distinct vertebrate gastrulation modes via modulation of key cell behaviors in the chick embryo. *bioRxiv*.

18. Collinet, C., & Lecuit, T. (2021). Programmed and self-organized flow of information during morphogenesis. *Nature Reviews Molecular Cell Biology, 22*, 245–265.

19. Dale, L., & Slack, J. M. W. (1987). Fate map for the 32-cell stage of *Xenopus laevis*. *Development, 99*, 527–551.

20. Driesch, H. (1982). Entwicklungsmechanische Studien III–VI. *Zoologische Wissenschaftliche, 55*, 1–62.

21. Dunsford, I., Bowley, C. C., Hutchinson, A. M., Thompson, J. S., Sanger, R., & Race, R. R. (1953). A human blood-group chimera. *British Medical Journal, 2*, 81.

22. Elosegui-Artola, A., Gupta, A., Najibi, A. J., Seo, B. R., Garry, R., Darnell, M., Gu, W., Zhou, Q., Weitz, D. A., Mahadevan, L., et al. (2022). Matrix viscoelasticity controls spatio-temporal tissue organization. *bioRxiv*.

23. Elowitz, M. B., Levine, A. J., Siggia, E. D., & Swain, P. S. (2002). Stochastic gene expression in a single cell. *Science, 297*, 1183–1186.

24. Essex, L. J., Mayor, R., & Sargent, M. G. (1993). Expression of *Xenopus* snail in mesoderm and prospective neural fold ectoderm. *Developmental Dynamics, 198*, 108–122.

25. Ferreira, F., Moreira, S., & Barriga, E. H. (2021). Stretch-induced endogenous electric fields drive neural crest-directed collective cell migration *in vivo*. *bioRxiv*.

26. Garcia-Bellido, A., Lawrence, P. A., & Morata, G. (1979). Compartments in animal development. *Scientific American, 241*, 90–98; 102–111.

27. Goldberg, A. D., Allis, D., & Bernstein, E. (2007). Epigenetics: A landscape takes shape. *Cell, 128*, 635–638.

28. Goldstein, B., & Macara, I. G. (2007). The PAR proteins: Fundamental players in animal cell polarization. *Developmental Cell, 13*, 609–622.

29. Gray, S. D., & Dale, J. K. (2010). Notch signaling regulates the contribution of progenitor cells from the chick Hensen's node to the floor plate and notochord. *Development, 137*, 561–568.

30. Gross, J. B., Hanken, J., Oglesby, E., & Marsh-Armstrong, N. (2006). Use of a ROSA26 transgenic line for long-term *Xenopus* fate-mapping studies. *Journal of Anatomy, 209*, 401–413.

31. Guibentif, C., Griffiths, J. A., Imaz-Rosshandler, I., Ghazanfar, S., Nichols, J., Wilson, V., Göttgens, B., & Marioni, J. C. (2021). Diverse routes toward early somite's in the mouse embryo. *Developmental Cell, 56*, 141–153.

32. Gurdon, J. B. (1987). Embryonic induction—Molecular prospects. *Development, 99*, 285–306.

33. Hartenstein, V., Technau, G. M., & Campos-Ortega, J. A. (1985). Fate mapping in wild-type *Drosophila melanogaster* 3. A fate map of the blastoderm. *Wilhelm Roux's Archives of Developmental Biology, 194*, 213–216.

34. Hatada, Y., & Stern, C. D. (1994). A fate map of the epiblast of the early chick embryo. *Development, 120*, 2879–2889.

35. Hejnol, A., Martindale, M. Q., & Henry, J. Q. (2007). High-resolution fate map of the snail *Crepidula fornicata*: The origins of ciliary bands, nervous system, and muscular elements. *Developmental Biology, 305*, 63–76.

36. Henrique, D., & Schweisguth, F. (2003). Cell polarity: The ups and downs of the PAR6/aPKC complex. *Current Topics in Genetics and Development, 13*, 341–350.

37. Hiramatsu, R., Matsuoka, T., Kimura-Yoshida, C., Han, S. W., Mochida, K., Adachi, T., Takayama, S., & Matsuo, I. (2013). External mechanical cues trigger the establishment of the anterior-posterior axis in early mouse embryos. *Developmental Cell, 27*, 131–144.

38. Huang, Y., & Umulis, D. M. (2019). Scale invariance of BMP signaling gradients in zebrafish. *Scientific Reports, 9*, 1–9.

39. Huebner, R. J., Weng, S., Lee, C., Sarıkaya, S., Papoulas, O., Cox, R. M., Marcotte, E. M., & Wallingford, J. B. (2022). ARVCF catenin controls force production during vertebrate convergent extension. *Developmental Cell, 57*, 1119–1131.e5.

40. Iimura, T., Yang, X., Weijer, C. J., & Pourquié, O. (2007). Dual mode of paraxial mesoderm formation during chick gastrulation. *Proceedings of the National Academy of Sciences USA, 104*, 2744–2749.

41. Ishimatsu, K., Hiscock, T. W., Collins, Z. M., Sari, D. W. K., Lischer, K., Richmond, D. L., Bessho, Y., Matsui, T., & Megason, S. G. (2018). Size-reduced embryos reveal a gradient scaling-based mechanism for zebrafish somite formation. *Development, 145*, dev161257.

42. Jessell, T. M., & Melton, D. A. (1992). Diffusible factors in vertebrate embryonic induction. *Cell, 68*, 257–270.

43. Kessler, D. S., & Melton, D. A. (1994). Vertebrate embryonic induction–Mesodermal and neural patterning. *Science, 266*, 596–604.

44. Kimmel, C. B., & Warga, R. M. (1986). Tissue-specific cell lineages originate in the gastrula of the zebrafish. *Science, 231*, 365–368.

45. Kimmel, C. B., Warga, R. M., & Schilling, T. F. (1990). Origin and organization of the zebrafish fate map. *Development, 108*, 581–594.

46. Kondo, S., & Miura, T. (2010). Reaction-diffusion model as a framework for understanding biological pattern formation. *Science, 329*, 1616–1620.

47. Labouesse, C., Tan, B. X., Agley, C. C., Hofer, M., Winkel, A. K., Stirparo, G. G., Stuart, H. T., Verstreken, C. M., Mulas, C., Mansfield, W., et al. (2021). StemBond hydrogels control the mechanical microenvironment for pluripotent stem cells. *Nature Communications, 12*, 1–17.

48. le Douarin, N. (1980). Migration and differentiation of neural crest cells. *Current Topics in Developmental Biology, 16*, 31–85.

49. Lewis, J. (2008). From signals to patterns: Space, time, and mathematics in developmental biology. *Science, 322*, 399–403.

50. Lewis, J., Slack, J. M. W., & Wolpert, L. (1977). Thresholds in development. *Journal of Theoretical Biology, 65*, 579–590.

51. Lewis, N. E., & Rossant, J. (1982). Mechanism of size regulation in mouse embryo aggregates. *Development, 72*, 169–181.

52. Li, T., Wang, L., Zhang, X., Jiang, L., Li, Y., Mao, J., Cui, T., Li, W., Wang, L., & Zhou, Q. (2018). Generation of rat-mouse chimeras by introducing single cells of rat inner cell masses into mouse blastocysts. *Science, 1673*, 8527.

53. Madan, K. (2020). Natural human chimeras: A review. *European Journal of Medical Genetics, 63*, 103971.

54. Marchant, L., Linker, C., Ruiz, P., Guerrero, N., & Mayor, R. (1998). The inductive properties of mesoderm suggest that the neural crest cells are specified by a BMP gradient. *Developmental Biology, 198*, 319–329.

55. McAdams, H. H., & Arkin, A. (1999). It's a noisy business! Genetic regulation at the nanomolar scale. *Trends in Genetics, 15*, 65–69.

56. McDole, K., Guignard, L., Amat, F., Berger, A., Malandain, G., Royer, L. A., Turaga, S. C., Branson, K., & Keller, P. J. (2018). In toto imaging and reconstruction of post-implantation mouse development at the single-cell level. *Cell, 175*, 859–876.

57. McGrew, M. J., Sherman, A., Lillico, S. G., Ellard, F. M., Radcliffe, P. A., Gilhooley, H. J., Mitrophanous, K. A., Cambray, N., Wilson, V., & Sang, H. (2008). Localized axial progenitor cell populations in the avian tail bud are not committed to a posterior Hox identity. *Development, 135*, 2289–2299.

58. McLaren, S. B. P., & Steventon, B. J. (2021). Anterior expansion and posterior addition to the notochord mechanically coordinate zebrafish embryo axis elongation. *Development, 148*, dev199459.

59. Meinhardt, H. (1982). *Models of Biological Pattern Formation*. New York: Academic Press.

60. Mintz, B. (1962). Experimental study of the developing mammalian egg: Removal of the zona pellucida. *Science, 138*, 594–595.

61. Montero, S. P., Bowling, S., Pérez-Carrasco, R., & Rodríguez, T. A. (2022). Levels of p53 expression determine the competitive ability of embryonic stem cells during the onset of differentiation. *bioRxiv*. https://doi.org/10.1101/2022.06.15.496485.

62. Nance, J. (2005). PAR proteins and the establishment of cell polarity during *C. elegans* development. *BioEssays, 27*, 126–135.

63. Nichols, J., Lima, A., & Rodríguez, T. A. (2022). Cell competition and the regulative nature of early mammalian development. *Cell Stem Cell, 29*, 1018–1030.

64. Nishida, H., & Sawada, K. (2001). macho-1 encodes a localized mRNA in ascidian eggs that specifies muscle fate during embryogenesis. *Nature, 409*, 724–729.

65. Okada, T. S. (2004). From embryonic induction to cell lineages: Revisiting old problems for modern study. *International Journal of Developmental Biology, 48*, 739–742.

66. Ozelci, E., Mailand, E., Ruegg, M., Sakar, M. S., & Oates, A. C. (2022). Deconstructing body axis morphogenesis in zebrafish embryos using robot-assisted tissue micromanipulation. *bioRxiv.* https://doi.org/10.1101/2022.08.12.503771.

67. Patten, I., Kulesa, P., Shen, M. M., Fraser, S., & Placzek, M. (2003). Distinct modes of floor plate induction in the chick embryo. *Development, 130*, 4809–4821.

68. Petters, R. M., & Mettus, R. V. (1984). Survival rate to term of chimeric morulae produced by aggregation of five to nine embryos in the mouse, *Mus musculus. Theriogenology, 22*, 167–174.

69. Pouille, P. A., Ahmadi, P., Brunet, A C., & Farge, E. (2009). Mechanical signals trigger myosin II redistribution and mesoderm invagination in *Drosophila* embryos. *Science Signaling, 2*, ra16.

70. Psychoyos, D., & Stern, C. D. (1996). Fates and migratory routes of primitive streak cells in the chick embryo. *Development, 122*, 1523–1534.

71. Ranga, A., Girgin, M., Meinhardt, A., Eberle, D., Caiazzo, M., Tanaka, E. M., & Lutolf, M. P. (2016). Neural tube morphogenesis in synthetic 3D microenvironments. *Proceedings of the National Academy of Sciences USA, 113*, E6831–E6839.

72. Raven, C. P., & Kloos, J. (1945). Induction by medial and lateral pieces of the archenteron roof with special reference to the determination of the neural crest. *Acta Neerlandica Morphologica, 5*, 348–362.

73. Reinhardt, D., Frenz, M., Mandel, T., & Kuhlemeier, C. (2005). Microsurgical and laser ablation analysis of leaf positioning and dorsoventral patterning in tomato. *Development, 132*, 15–26.

74. Rogers, K. W., & Schier, A. F. (2011). Morphogen gradients: From generation to interpretation. *Annual Review of Cell and Developmental Biology, 27*, 377–407.

75. Ruffins, S. W., & Ettensohn, C. A. (1996). A fate map of the vegetal plate of the sea urchin (*Lytechinus variegatus*) mesenchyme blastula. *Development, 122*, 253–263.

76. Sancho, M., Di-Gregorio, A., George, N., Pozzi, S., Sánchez, J. M., Pernaute, B., & Rodríguez, T. A. (2013). Competitive interactions eliminate unfit embryonic stem cells at the onset of differentiation. *Developmental Cell, 26*, 19–30.

77. Saunders, J. W. (1968). Ectodermal-mesenchymal interactions in the origins of limb symmetry. *Epithelial-Mesenchymal Interactions, 28*, 78–97.

78. Savin, T., Kurpios, N. A., Shyer, A. E., Florescu, P., Liang, H., Mahadevan, L., & Tabin, C. J. (2011). On the growth and form of the gut. *Nature, 476*, 57–62.

79. Sawada, K., & Aoyama, H. (2002). Fate maps of the primitive streak in chick and quail embryo: Ingression timing of progenitor cells of each rostro-caudal axial level of somite's. *International Journal of Developmental Biology, 43*, 809–815.

80. Schoenwolf, G. C., & Sheard, P. (1990). Fate mapping the avian epiblast with focal injections of a fluorescent-histochemical marker: Ectodermal derivatives. *Journal of Experimental Zoology, 255*, 323–339.

81. Selleck, M. A. J., & Stern, C. D. (1992). Commitment of mesoderm cells in Hensen's node of the chick embryo to notochord and somite. *Development, 114*, 403–415.

82. Selleck, M. A., & Bronner-Fraser, M. (1995). Origins of the avian neural crest: The role of neural plate-epidermal interactions. *Development, 121*, 525–538.

83. Selleck, M. A., & Stern, C. D. (1991). Fate mapping and cell lineage analysis of Hensen's node in the chick embryo. *Development, 112*, 615–626.

84. Serrano Nájera, G., & Weijer, C. J. (2020). Cellular processes driving gastrulation in the avian embryo. *Mechanisms of Development, 163*, 103624.

85. Shellard, A., & Mayor, R. (2021). Collective durotaxis along a self-generated stiffness gradient *in vivo*. *Nature, 600*, 690–694.

86. Slack, J. M. W. (1991). *From Egg to Embryo: Regional Specification in Early Development* (2nd ed.). Cambridge: Cambridge University Press.

87. Slack, J. M. W. (2002). Timeline–Conrad Hal Waddington: the last renaissance biologist? *Nature Reviews Genetics, 3*, 889–895.

88. Solovieva, T., Lu, H. C., Moverley, A., Plachta, N., & Stern, C. D. (2022). The embryonic node behaves as an instructive stem cell niche for axial elongation. *Proceedings of the National Academy of Sciences USA, 119*, e2108935119.

89. Spemann, H., & Mangold, H. (1924). Über Induktion von Embryonalanlagen durch Implantation artfremder Organisatoren. *Archiv für Mikroskopische Anatomie und Entwicklungsgeschichte, 100*, 599–638.

90. Spratt, N. T., Jr. (1942). Location of organ-specific regions and their relationship to the development of the primitive streak in the early chick blastoderm. *Journal of Experimental Zoology, 89*, 69–101.

91. Spratt, N. T., Jr., & Haas, H. (1960). Integrative mechanisms in the development of the early chick blastoderm. I. Regulative potentiality of separated parts. *Journal of Experimental Zoology, 145*, 97–137.

92. Steventon, B., Araya, C., Linker, C., Kuriyama, S., & Mayor, R. (2009). Differential requirements of BMP and Wnt signaling during gastrulation and neurulation define two steps in neural crest induction. *Development, 136*, 771–779.

93. Steventon, B., Busby, L., & Arias, A. M. (2021). Establishment of the vertebrate body plan: Rethinking gastrulation through stem cell models of early embryogenesis. *Developmental Cell, 56*, 2405–2418.

94. Steventon, B., Mayor, R., & Streit, A. (2016). Directional cell movements downstream of Gbx2 and Otx2 control the assembly of sensory placodes. *Biology Open, 5*, 1620–1624.

95. Strome, S., & Lehmann, R. (2007). Germ versus soma decisions: Lessons from flies and worms. *Science, 316*, 392–393.

96. Tabata, T., & Takei, Y. (2004). Morphogens, their identification and regulation. *Development, 131*, 703–712.

97. Tadros, W., & Lipshitz, H. D. (2009). The maternal-to-zygotic transition: A play in two acts. *Development, 136*, 3033–3042.

98. Tarkowski, A. K. (1961). Mouse chimaeras developed from fused eggs. *Nature, 190*, 857–860.

99. Tata, P. R., Mou, H., Pardo-Saganta, A., Zhao, R., Prabhu, M., Law, B. M., Vinarsky, V., Cho, J. L., Breton, S., Sahay, A., et al. (2013). Dedifferentiation of committed epithelial cells into stem cells *in vivo*. *Nature, 503*, 218–223.

100. Teillet, M. A., Ziller, C., & le Douarin, N. M. (2008). Quail–chick chimeras. *Molecular Embryology, 461*, 337–350.

101. Theveneau, E., Steventon, B., Scarpa, E., Garcia, S., Trepat, X., Streit, A., & Mayor, R. (2013). Chase-and-run between adjacent cell populations promotes directional collective migration. *Nature Cell Biology, 15*, 763–772.

102. Tlili, S., Yin, J., Rupprecht, J. F., Mendieta-Serrano, M. A., Weissbart, G., Verma, N., Teng, X., Toyama, Y., Prost, J., & Saunders, T. E. (2019). Shaping the zebrafish myotome by intratissue friction and active stress. *Proceedings of the National Academy of Sciences USA, 116*, 25430–25439.

103. van Heyningen, P., Calver, A. R., & Richardson, W. D. (2001). Control of progenitor cell number by mitogen supply and demand. *Current Biology, 11*, 232–241.

104. Williams, R. M., Candido-Ferreira, I., Repapi, E., Gavriouchkina, D., Senanayake, U., Ling, I. T. C., Telenius, J., Taylor, S., Hughes, J., & Sauka-Spengler, T. (2019). Reconstruction of the global neural crest gene regulatory network *in vivo*. *Developmental Cell, 51*, 255–276.

105. Yamaguchi, E., Dannenberg, L. C., Amiel, A. R., & Seaver, E. C. (2016). Regulative capacity for eye formation by first quartet micromeres of the polychaete *Capitella teleta*. *Developmental Biology, 410*, 119–130.

106. Zhang, B., Li, H., Hu, Z., Jiang, H., Stablewski, A. B., Marzullo, B. J., Yergeau, D. A., & Feng, J. (2021). Generation of mouse–Human chimeric embryos. *Nature Protocols, 16*, 3954–3980.

107. Zhou, J., Kim, H. Y., & Davidson, L. A. (2009). Actomyosin stiffens the vertebrate embryo during crucial stages of elongation and neural tube closure. *Development, 136*, 677–688.

108. Zhou, J., Pal, S., Maiti, S., & Davidson, L. A. (2015). Force production and mechanical accommodation during convergent extension. *Development, 142*, 692–701.

109. Zhu, A. J., & Scott, M. P. (2004). Incredible journey: How do developmental signals travel through tissue? *Genes & Development, 18*, 2985–2997.

CHAPTER

4

Approaches to Development: Cell and Molecular Biology Techniques

LEARNING OBJECTIVES

After studying this chapter, you will be able to:

- Understand about microscopy.
- Study the gene expression by molecular biology methods.
- Study the gene expression by in situ methods.
- Learn about microinjection.
- Describe the cell-labeling methods.

KEY TERMS FROM THIS CHAPTER

Cell biology	Chick embryos	Compound microscope
Confocal microscopy	Embryos	Fluorescence microscope
Immunoprecipitation	Immunostaining	Microarrays
Microinjection	Microscopy	Molecular biology
Multiphoton microscope	Nomarski optics	Optical system
Radioactivity	Xenopus embryos	

INTRODUCTION

All living organisms are made of cells. From the tiniest bacteria to enormous redwood trees and blue whales, they are the smallest unit of life. The earliest major divisions of life into the three kingdoms of Archaea (ancient bacteria), Eubacteria (modern bacteria), and Eukaryota (everything else, including humans) were based on differences in cell structure and internal mechanisms. Thus, any understanding of life itself requires an understanding of cells.

The study of cells and their functions, including the subcellular mechanisms that maintain their viability and their interactions with one another, is known as cell biology. While cell biology studies how these molecules are used by the cell to survive, reproduce, and perform regular cell functions, molecular biology focuses primarily on the molecules of life, which are primarily proteins and nucleic acids.

The field of molecular biology studies how genes are transcriptionally produced into RNA, how that RNA is translated into proteins, and how those proteins affect cellular function. Molecular biologists have been creating techniques to recognize, separate, and work with the molecules that make up cells, such as proteins, RNA, and DNA, since about 1960.

In this section, we'll look at a few methods that come from the fields of cell and molecular biology but are especially applicable to the study of development.

4.1. MICROSCOPY

The study of cell structure, architecture, primary cellular and subcellular functions, and intercellular interactions is the focus of the biological field known as cell biology. Numerous methodologies have been employed to examine the intricacies of the cell's structure and operation. Microscopy is one imaging technique that has shown to be very helpful in studying many aspects of the architecture of cells, tissues, and different organelles that are present in them. Microscopy techniques facilitate the easy study of minute structural details of the cell, cellular processes, and other characteristics. These techniques yield information that enhances our comprehension of the diverse cellular processes occurring within a living organism. Microscopy and biochemical techniques have greatly advanced our understanding of cellular structures and functions.

The structural characteristics of the cells and the activities occurring within them are studied using a variety of methodologies. The most popular method for examining the intricate structural makeup of the cells, tissues, and organelles found in both plants and animals is microscopy. A microscope is a tool used to observe small objects that are invisible to the unaided eye. The words "micro," which means small, and "scope," which means to look at, are the roots of the word microscope. A microscope greatly enlarges objects so that specimens are clearly visible. The ability to see cells and microorganisms through the use of microscopy is advantageous because, when magnified, the images appear much larger.

Due to their small size, embryos necessitate the use of microscopes during study. Using a dissecting microscope, experiments frequently require manual manipulations. This is a binocular microscope with a ×10 to ×50 magnification range and a good working distance between the specimen and the objective lens (Figure 4.1). Dissecting microscopes give an accurate three-dimensional image that helps with performing manipulations like microsurgery and microinjection and allows the observer to perceive depth. Dissecting microscopes don't invert the image like most compound microscopes do. Incident lighting is used with opaque specimens, like chick embryos or Xenopus. In other words, the specimen receives light from the light source that is shone down onto it. A fiber optic light guide is used to provide a cold but bright illumination for living specimens because it is crucial that they not be overheated by a strong incident light beam. A transmitted light base is typically used when the specimens are transparent, such as mouse, sea urchin, or zebrafish embryos.

When examining sections or entire specimens that are small enough to be transparent, like zebrafish, Drosophila, or Caenorhabditis elegans embryos, the compound microscope (Figure 4.2) is utilized. Specimens that fall outside of sections are called whole-mounts. The compound microscope's magnification range is approximately ×40 to ×1000. The maximum magnification is limited by the wave nature of light, which makes it impossible to resolve points that are closer together than 0.2µm. In most compound microscopes, the image is inverted. This is a normal byproduct of the optical system and is not typically corrected because additional lenses would be needed to achieve this. In most cases, there are no issues with the inversion.

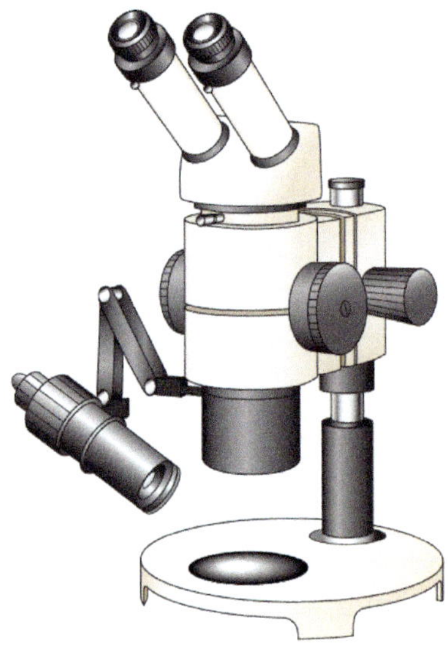

Figure 4.1. *Dissecting microscope.*

Source: Jonathan M.W. Slack. Essential Developmental Biology, 3rd Edition. Wiley-Blackwell. ISBN: 9781118387603.

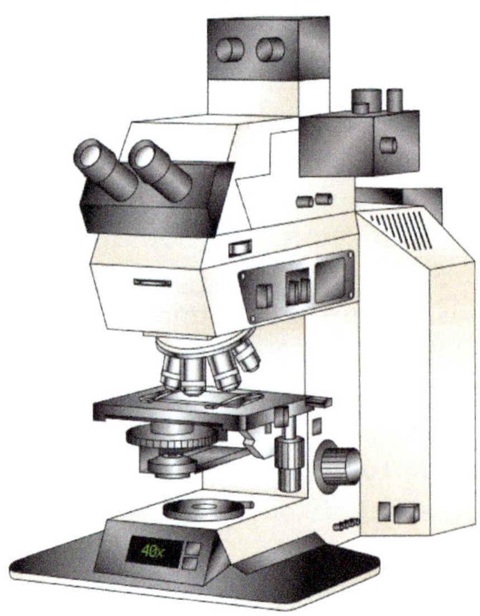

Figure 4.2. *Compound microscope.*

Source: Jonathan M.W. Slack. Essential Developmental Biology, 3rd Edition. Wiley-Blackwell. ISBN: 9781118387603.

4.1.1. Optical Techniques

The compound microscope can be used with a variety of optical techniques. Stained sections or wholemounts are examined using regular transmitted light (Figure 4.3a). Traditional histological stains or colored histochemical reaction products resulting from the identification of particular antigens or reporter proteins may be used as the staining agents. When viewed in transmitted light, wholemounts must be extremely thin because thick specimens often become too opaque to see detail.

Phase contrast microscopy is a technique that translates small variations in the specimen's refractive index into significant variations in light intensity (Figure 4.3b). It is the preferred technique for studying isolated gametes or live cells grown *in vitro* and is applied to small transparent embryos like those in the mammalian preimplantation stages.

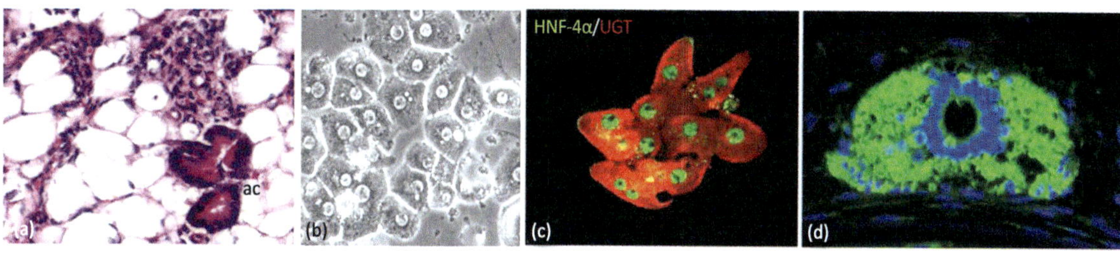

Figure 4.3. *Compound microscope images. (a) Section of a regenerating pancreas stained with hematoxylin and eosin. Two regenerating acini (ac) are clearly visible. (b) Phase contrast view of rat hepatocytes in culture. (c) Immunofluorescence of hepatocytes; green is the transcription factor HNF-4α, red is the enzyme UGT. (d) Section of a regenerated Xenopus tadpole spinal cord. Blue is DAPI, a fluorescent stain for DNA; and green is immunostain for neurofilament protein.*

Source: Jonathan M.W. Slack. Essential Developmental Biology, 3rd Edition. Wiley-Blackwell. ISBN: 9781118387603.

Differential interference contrast, sometimes referred to as Nomarski optics, causes slight variations in refractive index to appear to the observer as differences in height. Additionally, it offers a sharp resolution of a specific optical section within the specimen; thus, various optical sections become visible as one focuses the microscope up and down through the specimen. The entire cell lineage of the worm C. elegans was elucidated thanks to Nomarski optics, a technique that allows for the very clear visualization of individual cells within tiny transparent specimens. Preimplantation mammalian embryos as well as those of C. elegans, zebrafish, and Drosophila are all used with it.

There are several uses for fluorescence microscopy, but they are all based on identifying the presence of a fluorescent material, or fluorochrome, inside a sample (Figure 4.3c,d). These consist of the visualization of fluorochromes added into the specimen to label subsets of cells, fluorescent antibody staining, and fluorescent *in situ* hybridization (FISH). According to the principle of fluorescence, a fluorochrome absorbs light with a specific energy and emits light with a lower energy. A longer wavelength or a color that is more shifted toward the red end of the spectrum indicate lower energy. A specific fluorochrome's excitation spectrum will demonstrate how the fluorescence intensity changes with the excitation wavelength. It will also feature an emission spectrum (Figure 4.4) that illustrates the distribution of the intensity of the fluorescence emitted across the wavelength spectrum. The substance's properties can be seen in the emission and excitation spectra. The specimen is illuminated by the excitation beam through an attachment on a fluorescence microscope. This includes a strong lamp, most likely an LED (light emitting diode) type lamp these days, but it was originally a mercury arc lamp. After a dichroic mirror reflects the excitation beam down onto the specimen, a filter chooses a narrow excitation wavelength band that is appropriate for the fluorochrome being used. The dichroic mirror transmits wavelengths above and below a predetermined cutoff. Since the specimen's emission has a longer wavelength, it will pass through the dichroic mirror and be visible to the observer (Figure 4.4). One alternative filter set for each fluorochrome in use is typically included in a fluorescence microscope. With the right filter set, it should be possible to see each of the two or three fluorochromes separately in a specimen. Green fluorescent protein and other related fluorescent proteins are used in many applications, necessitating the examination of fluorescence in whole specimens. For this reason, fluorescent dissecting microscopes are now widely used.

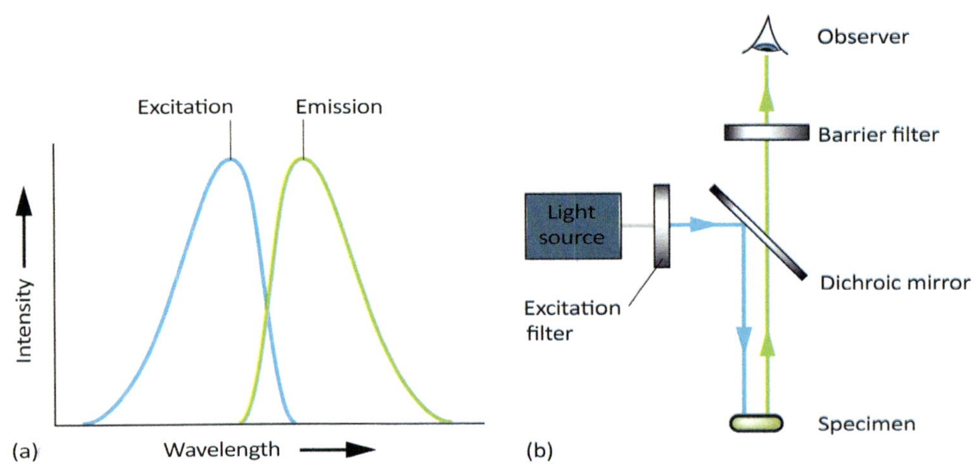

Figure 4.4. *Fluorescence microscopy. (a) Typical spectra for excitation and for emission of a fluorochrome. (b) Arrangement of components in a fluorescent microscope.*

Source: Jonathan M.W. Slack. Essential Developmental Biology, 3rd Edition. Wiley-Blackwell. ISBN: 9781118387603.

Some methods that are widely used in cell biology are not as useful when studying development. For dark-field microscopy to work, light must be applied at an extremely oblique angle so that only areas of the specimen that widely scatter light can be seen. The specimen is mostly dark except for these bright spots. It is primarily utilized in developmental biology to visualize radioactive *in situ* hybridizations, where the signal is an accumulation of silver grains coated in a photographic emulsion. However, advancements in nonradioactive detection have rendered this method outdated. The specimen is placed between crossed Polaroid filters for polarization microscopy. A specimen is said to exhibit birefringence if it contains elements, such as muscle fibers, that can change the light beam's plane of polarization and appear as bright patches. Again, due to advancements in other techniques for *in situ* detection of particular molecules, this technique is now outdated.

4.1.1.1. Confocal and Multiphoton Microscopes

A "pinhole" is used in confocal microscopy, an advanced microscopy technique, to remove out-of-focus light. Once out-of-focus light above and below the plane has been eliminated, the discrete optical sections can be gathered. The sample is illuminated by high power lasers. They aid in gathering light coming from the intended focal plane. Only light from the intended focal plane can enter the detector as the laser beam scans across the sample.

Because fluorescence from cells above and below the plane of focus would otherwise overwhelm the image, conventional fluorescence microscopy is only useful on sections or very thin wholemounts. However, the confocal scanning microscope makes it easy to see fluorescence from thicker wholemounts. Since this apparatus uses a laser for illumination, excitation is only possible using that laser's single wavelength. It illuminates a single point at a time as opposed to the entire specimen. One optical section at a time is scanned point by point, with each point's fluorescence recorded by a detector and used to reconstruct the section's image. The optical section has very high quality because only one point is viewed at a time. Points of light that are above or below the focused plane will scatter and add very little to the signal. A confocal microscope's image is inherently a digital image saved on a computer.

Compared to the confocal microscope, the multiphoton microscope can provide even better optical sectioning of fluorescent specimens. In this case, the energy of the light source is insufficient to excite the fluorochrome. Consequently, an excitation can only happen when two or more photons strike a single fluorochrome molecule at the same time, and this can only happen at the excitation beam's point of focus, where the photon density is very high.

4.1.1.2. Image Capture

Prior to the 1990s, charge coupled devices (CCDs) were used for all image capture; previously, photography was used to capture microscopic images. A CCD camera's detection chip is made up of a series of pixels that can be "filled up" with electrons. The amount of charge that each pixel accumulates over time is directly related to the amount of light that strikes it. After an exposure is made, a detector sequentially reads out the charge on each pixel, converting it into a digital image. Each pixel's intensity in an 8-bit image would be represented by a number between 0 and 255. Three colors—red, green, and blue—would each have one 8-bit number in a 24-bit color image. CCD cameras come in various varieties for a variety of uses. The most sensitive function is limited to counting single photons and is only available in black and white. It can also gather lengthy exposures from extremely faint specimens. Less sensitive ones allow for the real-time recording of movies and can gather images quickly enough to show on a video screen.

After the picture is captured, it is sent to a computer, where it is shown on the screen and can be processed in a number of ways. For example the contrast and brightness can be altered, the image can be smoothed or sharpened, densitometric measurements can be taken, or multiple fluorescent images obtained through different monochrome channels can be recombined so that several colors are visualized on the same image. The ability to increase an image's signal-to-noise ratio by averaging multiple exposures is just one of the many benefits of collecting images digitally. Because the signal is constant throughout all exposures, averaging won't change it; however, noise varies throughout all exposures and will consequently become low and consistent.

When studying the morphogenetic movements of cells, organ cultures, or embryos, time-lapse imaging is particularly helpful. Intervals are used to capture individual images, which are subsequently replayed at shorter intervals. For instance, a 4-hour process can be seen in 8 seconds if an image is captured once every minute for 4 hours and then replayed at 30 frames per second. Typically, the issue arises when the specimen shifts out of the field of view while being processed. Timelapse recording microscopes of today come with programmable stage controls that allow the microscope to image a sequence of specimens at predetermined intervals and then bring each one back to the light path one at a time. More advanced models will additionally make adjustments for slight shifts in focus or location. An environmental chamber, akin to a tissue culture incubator, is necessary for filming mammalian or avian embryos or organ cultures. This chamber must allow the embryos to be maintained at 37°C and 5% CO_2, or in any other suitable environment.

4.1.2. Histological Methods

The resolution at which the structure of a specimen can be resolved is limited, even with Nomarski or confocal microscopy, despite the fact that wholemounts are widely used in developmental biology. As a result, histological section preparation and examination are always necessary (Figure 4.3a, d) (Bancroft & Gamble, 2007). Fixing a specimen is the first step in getting it ready for any kind of microscopical analysis. Fixation is the process of killing it and strengthening its mechanical structure so that it can survive osmotic shocks or a specific level of handling. Fixatives function in different ways. Formalin, a mixture of formaldehyde gas, is the most widely utilized. When this interacts with the sulfhydryl or amino groups in macromolecules, some denaturation and cross-linking are produced. Glutaraldehyde is an extremely powerful intermolecular cross-linking agent because it contains two aldehyde groups. Organic solvents and acids are additional frequently used ingredients in fixatives. These serve as precipitating and denaturing agents (Kiernan, 2008).

The specimen needs to be embedded in a sturdy supporting material that will allow it to be cut into extremely thin sections after it has been fixed. The most typical substance is paraffin wax. The specimen must be dehydrated in order for wax to penetrate it, and this is accomplished by putting it through a succession of ethanol baths with progressively higher concentrations. This range of concentrations is required because the solvents' mixing forces can damage tissue when pure ethanol is transferred straight from water. The specimen is equilibrated in a solvent miscible with wax, such as xylene, after the water has been removed. It is then submerged in melted wax at a temperature of roughly 60°C and left until the wax has completely penetrated every part. The block that results from allowing the wax to solidify can be kept indefinitely at room temperature. The block is placed on a microtome, which repeatedly moves it across an extremely sharp knife, advancing the block by a few microns each time, to create sections. As a result, a ribbon-like collection of connected sections is created. After mounting them on microscope slides, they can be processed for *in situ* hybridization or immunostaining, among other things.

The orientation of sections is crucial in embryology. It is essential to have an entire set of connected, or serial, sections since it is frequently desirable to analyze the disposition of structures throughout the entire specimen. Because of this, paraffin wax works well as an embedding material because the sections cut into a ribbon by nature. Its drawback is that the specimen must be heated to 60°C for the embedding period and must be dehydrated using organic solvents. This may result in the specimen's proteins or nucleic acids being damaged, which could impair *in situ* hybridization or immunostaining.

Using frozen sections is a common practice for these techniques. In this case, the specimen is quickly frozen in a medium with a high sucrose concentration—it might not even need to be fixed. This is put on a cryostat—basically, a microtome running in a cooled chamber—after being cut into a block. Since frozen sections don't form a ribbon, collecting them is very challenging, and their quality is typically inferior to that of paraffin wax. However, if only a few representative sections are needed, this is a very helpful technique.

Because it is difficult to cut paraffin wax sections thinner than approximately 5 μm, it may not provide sections of sufficient quality for certain purposes. Alternative embedding materials derived from different plastics are available that can be sliced at 1μm or even smaller fractions of a micron for use with an electron microscope. However, these materials frequently cannot produce serial sections and are incompatible with *in situ* hybridization and immunostaining. Developmental biology hardly ever uses electron microscopy of sections because it offers a higher magnification than is necessary to distinguish between different cell types in tissues or embryos. However, the scanning electron microscope is frequently used to visualize cell arrangements during morphogenetic movements and can offer bright, three-dimensional views of wholemounts.

4.2. STUDY OF GENE EXPRESSION BY MOLECULAR BIOLOGY METHODS

It is crucial to understand the typical expression patterns of the genes being studied when studying development. We must ascertain which regions of the embryo, at what developmental stage, and to what extent they are active. For identifying expression patterns, there are two primary categories of methods: *in situ* methods, which provide accurate anatomical information but limited quantification, and biochemical methods, which provide a reasonably accurate quantitative measure but no anatomical information. There are different approaches to studying mRNA and protein in each scenario. When feasible, it is preferable to accomplish both because a protein's presence in a given location does not always indicate that it has been synthesized there—it could have been transported from another site of synthesis—and a gene's transcription does not ensure that it will subsequently be translated into a protein. The development of the laser capture technique in recent years has allowed for the molecular analysis of individual cells or cell regions obtained from microscope sections.

Although the specimen can be dissected using the biochemical methods to obtain a limited amount of regional information, these methods are fundamentally techniques for estimating the amount of a particular messenger RNA or protein in the entire specimen. By analyzing groups of embryos at various developmental stages, they are frequently used to obtain a stage series for gene expression (Figure 4.5).

This is merely a synopsis of the biochemical techniques. While ^{32}P is still occasionally used because it offers the highest level of sensitivity, nonradioactive labeling and detection techniques are now utilized in all cases instead of radioactivity.

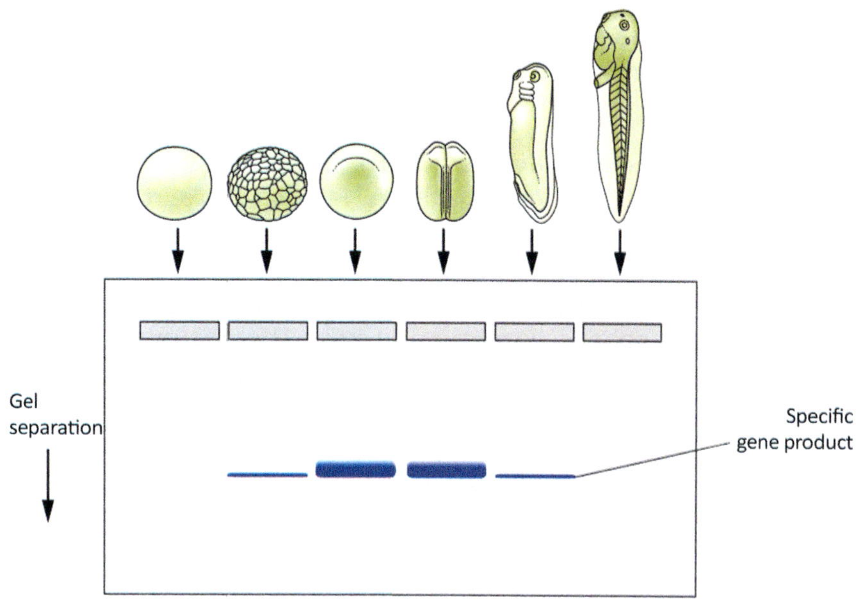

Figure 4.5. *A "stage series" for a particular gene product. This could be a specific mRNA or a protein extracted from whole embryos at the different stages and separated by a suitable technique.*

Source: Jonathan M.W. Slack. Essential Developmental Biology, 3rd Edition. Wiley-Blackwell. ISBN: 9781118387603.

4.2.1. Methods for Messenger RNA

RT-PCR, or reverse transcription polymerase chain reaction, is the accepted technique for identifying and quantifying mRNA. After removing the sample's total RNA, reverse transcriptase is used to convert it into complementary DNA (cDNA). Subsequently, two oligonucleotides are introduced, selected to align with specific distances between sequences in the target cDNA. Repeated cycles of synthesis, melting, and hybridization—which are carried out in a programmable heating block that modifies temperature so that each reaction step can take place in turn—amplify the DNA sequence between the primers. The reaction mixture is run on a gel and the DNA bands are visible through ethidium bromide staining once an appropriate number of cycles have passed. A band the size of the cDNA region between the two primers should be visible if the reaction was successful. In the event that no band is present, the sample did not contain any specific target cDNA. As the method only looks at the product concentration accumulated at the end of the reaction, after the exponential phase has ended and a plateau has been reached, it can only be described as semiquantitative. Furthermore, RT-PCR is highly susceptible to artifacts due to the exponential nature of the reaction, necessitating meticulous controls to guarantee specificity. To ensure that each sample contains a comparable amount of RNA, one of these is the loading control. Typically, this is a widely distributed messenger RNA (mRNA), like β-actin or glyceraldehyde-3-

phosphate dehydrogenase. In addition, a "no reverse transcription" control must be included in case contaminated DNA produces a signal. Three samples will be identified as genuine positives, negatives, and neutrals. Furthermore, using an excessive number of cycles is not advisable; more than 35 typically denotes trace amounts of mRNA with no biological significance.

Real-time polymerase chain reaction (PCR) technology can yield more precise quantitative measurements of particular mRNA levels. Here, the amplified product's formation rate is tracked in real time. The addition of the dye SYBR Green, which fluoresces brightly when bound to double-stranded DNA but not when bound to single-stranded DNA, is one of the many methods for accomplishing this that is frequently used. Thus, the rise in fluorescence corresponds with the accumulation of the amplified product. The double-stranded product surpasses a critical value, at which point the cycle number is recorded in an exponential fashion (Figure 4.6). This means that the starting concentration can be measured accurately in relation to a control mRNA because the early exponential phase of amplification is much more directly related to starting concentration than is the final plateau level. A ubiquitous mRNA is typically used as the control since it is utilized for loading controls in qualitative RT-PCR.

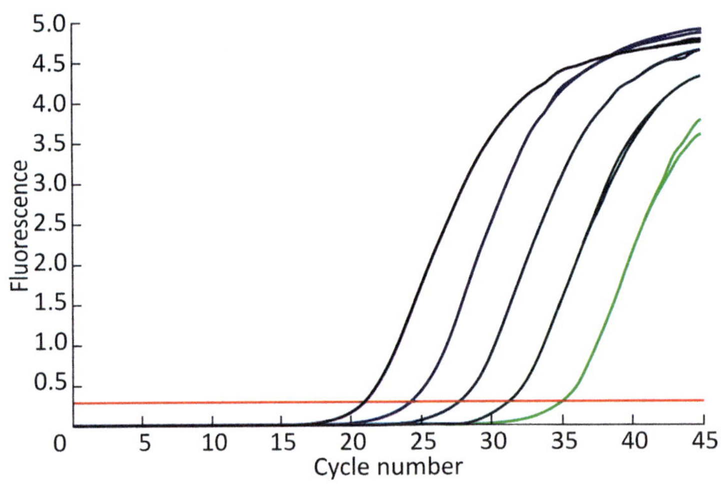

Figure 4.6. *Measurement of relative mRNA abundance by real time PCR. The instrument measures the cycle number at which the critical concentration of double stranded DNA is synthesized (red line).*

Source: Jonathan M.W. Slack. Essential Developmental Biology, 3rd Edition. Wiley-Blackwell. ISBN: 9781118387603.

You'll come across a few additional mRNA detection techniques in the literature. Prior to the widespread use of RT-PCR, the most sensitive technique for detecting particular mRNA was ribonuclease protection analysis, although it is now infrequently employed. The least sensitive and oldest method is northern blotting. Because it can display the quantity and size of mRNAs where multiple splice variants have arisen from the same gene, it is still occasionally used. Total mRNA is extracted from the sample and separated on a denaturing agarose gel using gel electrophoresis. Once a satisfactory separation is obtained, the gel's contents are "blotted" onto a membrane for hybridization. The particular probe, either ^{32}P or a chemical tag, is used to hybridize the membrane. An enhanced chemiluminescence reaction is used to detect the chemical tag, and either autoradiography or a phosphorimager is used to visualize it. The same techniques can be used to directly visualize radioactivity.

4.2.1.1. Microarrays

Microarrays allow the simultaneous analysis of many gene products. Although microarrays frequently only contain a subset of the entire genome, it is possible to examine the full inventory of cDNAs corresponding to the entire genome where the genome sequence is known. When a research project is just getting started, microarrays are usually used to find the genes that are most likely to be involved in a given system or process. They are sometimes called chips because they are manufactured with equipment from the computer industry. There are two primary varieties available: the Affymetrix oligonucleotide array and the cDNA array. cDNA arrays are made up of a regular pattern of closely spaced spots on a glass slide, each representing a distinct cDNA. The array is rectangular in shape. Occasionally, lengthy synthetic oligonucleotides are utilized in place of cDNAs. A greater number of short oligonucleotides are synthesized directly on the sides of the proprietary Affymetrix system. In this case, multiple oligonucleotides will stand in for different gene sequences. The array is utilized for nucleic acid hybridization in both situations. By removing RNA from the embryo or tissue sample and converting the mRNA back into cDNA, a probe is created.

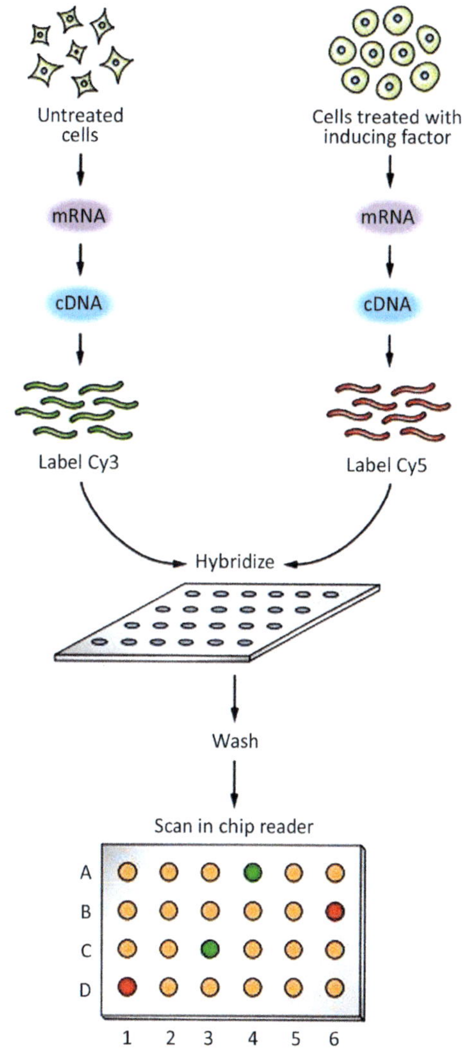

Figure 4.7. *Use of a microarray to compare gene expression in two cell populations. The results show that genes B6 and D1 have been upregulated, and genes A4 and C3 have been downregulated, by the inducing factor treatment.*

Source: Jonathan M.W. Slack. Essential Developmental Biology, 3rd Edition. Wiley-Blackwell. ISBN: 9781118387603.

Under the right circumstances, this is hybridized to the microarray after being labeled with a fluorescent dye. A chip reader, a fluorimeter that measures the fluorescence from the dye bonded to each spot, then scans the array. The amount of the complementary cDNA in the probe, and consequently the amount of that particular mRNA in the tissue sample, should be represented by the fluorescence intensity for a given cDNA on the array. Microarrays are widely used in developmental biology to compare, for instance, two embryonic stages or cells treated and untreated with an inducing factor (Figure 4.7). In this instance, there is a probe to be compared from each of the two distinct samples. These are marked with various fluorochromes, typically the red dye Cy5 and the green dye Cy3. Prior to hybridization, the probes are combined, and the ratio of green to red fluorescence is then determined. Due to the fact that both probes will bind, genes whose expression remains constant will produce a yellow signal; genes whose expression increases or decreases will produce a red or green signal, respectively. Both kinds of microarray analysis rely on highly advanced software that can match each spot's location to the cDNA's identity.

4.2.1.2. Deep Sequencing Methods

Expression analysis has been impacted by the remarkable ability of contemporary ("next generation") sequencing technology to produce DNA sequences, in addition to genomics. This has led to the development of a number of new methods that, if costs can be reduced, will probably see greater use in developmental biology. The capacity to quantify the transcriptome's whole composition in a cDNA sample is known as RNA-Seq. This is accomplished by sequencing a very large number of cDNAs, which is the equivalent of at least four times the genome's coverage. Following a comparison with the organism's known genome, each known exon is identified, and the frequency of each gene product is then tallied in relation to the others. This ought to provide a quantitative profile of the entire transcriptome, provided that both the initial RNA preparation and the reverse transcription reaction were conducted impartially.

Understanding chromatin dynamics is becoming more and more important for analyzing gene expression, as opposed to merely knowing the relative abundance of the mRNA product. These days, deep sequencing can be coupled with chromatin immunoprecipitation (ChIP) to provide a genome-wide view of the binding of particular proteins of interest, or with DNA methylation analysis to provide a genome-wide description of DNA methylation.

4.2.2. Methods for Protein

A collection of methods known as "proteomics" can be used to distinguish individual unknown proteins from a complex mixture. Similar to microarrays, they are typically utilized at the onset of an inquiry when determining which proteins undergo changes in expression level during a specific developmental event is necessary. Total protein extracts are separated using two-dimensional gel electrophoresis: a sodium dodecyl sulfate (SDS)-polyacrylamide slab gel separated by molecular weight in the second dimension, and an isoelectric focusing gel separated by isoelectric point in the first. This results in a pattern of spots, each of which stands for a different protein. Hopefully, there won't be many variations in the spot pattern when the preparations from the two samples under examination are compared. Mass spectrometry is then used to identify individual proteins. A mass spectrometer measures the material's time of flight to the detector after it has been accelerated in an electric field by ionizing and volatilizing it. This makes it possible to measure the molecular weight with extreme precision. Since protein molecular weight can be computed from the known amino acid composition of each polypeptide represented in the genome sequence, in cases where the genome has been sequenced, it may be possible to identify the polypeptides through their measurement. If not, tandem mass spectrometry can be used to sequence each individual protein. Trypsin breaks down the protein into a variety of peptides. The first cycle of mass spectrometry separates these, and the second cycle separates the fragments that each peptide is broken up into by ion bombardment. With the use of advanced software, the peptide's amino acid sequence can be determined from its distinctive fragmentation pattern.

4.2.2.1. Immunochemical Methods

The techniques used to quantify a specific protein in a sample typically require the use of an antibody that is specifically made against the protein. This antibody, which can be monoclonal or polyclonal, is produced by immunizing an animal with either a pure protein or a portion of a protein.

One technique to display the amount of a particular protein is a western blot. An acrylamide gel is used to extract and separate the sample's total proteins. Onto a membrane, the gel's contents are "blotted" or transferred. After that, the membrane is incubated with the particular antibody, which is designed to bind exclusively to the target protein's band and not to any other components that have been separated on the gel. A second antibody that is directed against the first antibody's constant region is then used to visualize the bound antibody. It is probable that the second antibody will be acquired commercially and altered for convenience of detection. Typically, this involves conjugating it with horseradish peroxidase (HRP), which possesses highly sensitive chemiluminescent substrates. A phosphorescent reaction product generated by a chemiluminescent substrate will decay spontaneously and release light during this process. Following the incubation period in the substrate mixture, the blot is either exposed to X-ray film or put in a phosphorimager. At the location where the antibody is bound, which corresponds to the location and amount of the target protein in the original sample, the phosphorescence of the reaction product is recorded.

An immunoprecipitation is a technique for isolating the protein that is recognized by a particular antibody; it is not a true precipitation. It is frequently used to determine whether a specific protein in an embryo or tissue sample is being synthesized for the first time. The steady state concentration displayed by a western blot may differ significantly from the new synthesis rate. For instance, the embryo might not be producing a certain protein anymore, but it might inherit a sizable maternal store from the egg. For this reason, all of the proteins produced during the labeling period become radioactive by incubating the live specimen in radioactive amino acids, typically ^{35}S-methionine or ^{35}S-cysteine. After being extracted, the whole proteins are incubated with the designated antibody. This will create an immune complex by binding to its target protein. By incubating the mixture with protein A attached to agarose beads, the immune complexes are separated. Bacterial protein A binds firmly to the constant region of antibodies of the IgG type. The immune complexes will be caught and rendered immobile on the beads. After washing the beads to get rid of any contaminating proteins, they are boiled in a sample buffer that is highly denaturing to release the target protein and bound antibody. This is run on a molecular weight-separating SDS protein gel, which is then dried. Phosphorimager or autoradiography are used to view the target protein's radioactive band. Nonradioactive immunoprecipitations are also frequently used to determine whether a specific protein is present in a sample. The precipitated protein is then typically detected by a western blot, chemiluminescent detection, and a particular antibody.

Using an antibody to a protein of interest—which could be a chromosomal protein like a histone bearing a specific chemical modification—chromatin is sheared into fragments approximately 500 nucleotides long in the process of chromatin immunoprecipitation (ChIP). The amount of the region of interest (for instance, the promoter region of a developmental control gene) that has been pulled down is then determined by RT-PCR analysis of the deproteinized sample's DNA. Microarray analysis and this method have been combined to allow for the simultaneous examination of thousands of loci (ChIP-Chip). The whole genome can now be analyzed thanks to its combination with deep sequencing (ChIP-Seq).

4.3. STUDY OF GENE EXPRESSION BY IN SITU METHODS

The purpose of *in situ* methods is to make the spatial domains of a specimen's gene expression visible. *In situ* procedures can be carried out on wholemounts if the specimens are small enough and transparent enough, which has the benefit of quick and easy three-dimensional visualization. Serial sections are typically used for embryological work; however, *in situ* methods can be applied to specimens that are too big or opaque.

4.3.1. In Situ Hybridization

The areas of a specimen where a particular mRNA is present can be determined by *in situ* hybridization (Harland, 1991). Since an antisense probe complementary to the mRNA to be detected is synthesized *in vitro*, the chemistry is the same as that of a northern blot. This is visualized after being hybridized with the specimen (Figures 4.8 and 4.9). The probe has an additional chemical group that can be identified with a specific, commercially available antibody. One of the nucleoside triphosphates needed for synthesis has the group attached to it. It is added during synthesis and has no effect on the probe's ability to hybridize. Digoxigenin, a plant sterol, and fluorescein are preferred groups for this use (Polak & McGee, 1999).

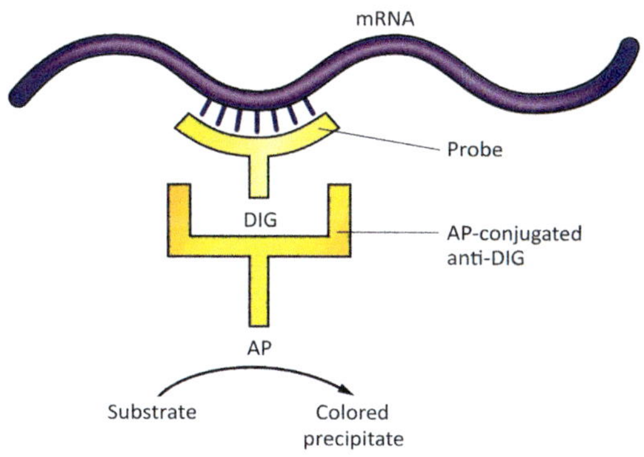

Figure 4.8. *In situ hybridization for detection of mRNA.*

Source: Jonathan M.W. Slack. Essential Developmental Biology, 3rd Edition. Wiley-Blackwell. ISBN: 9781118387603.

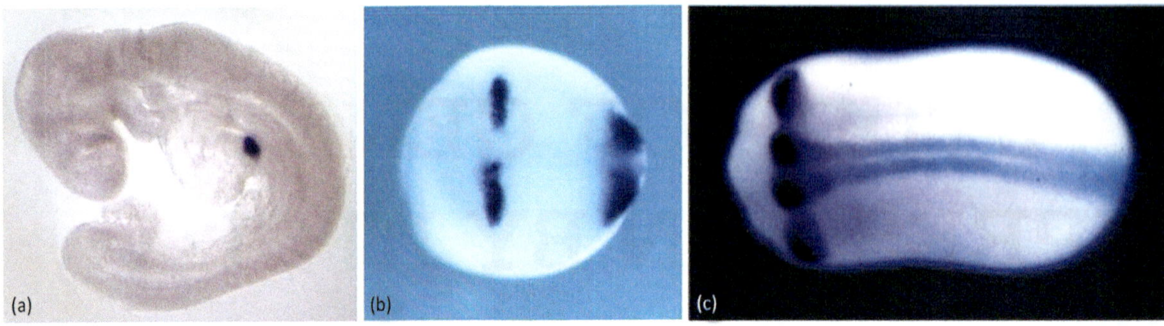

Figure 4.9. *Examples of wholemount in situ hybridization. (a) mRNA of the transcription factor C/EBP⊠ in the early liver bud of a 9.5 day mouse embryo. (b) two domains of expression of the same gene: mRNA of the inducing factor FGF3 in a xenopus neurula. (c) expression of two genes shown in different colors in a xenopus closed neurula. Dark blue shows mRNA for the transcription factor HoxA4; cyan shows mRNA for the transcription factor Xcad3.*

Source: Jonathan M.W. Slack. Essential Developmental Biology, 3rd Edition. Wiley-Blackwell. ISBN: 9781118387603.

In order for the large probe molecules to enter the cells during wholemount *in situ* hybridizations, the specimen will typically need to be permeabilized for a brief period of time using detergent or protease. The hybridization reaction is carried out overnight for wholemounts or sections, followed by a thorough washing of the specimen and the addition of an enzyme-linked antiprobe antibody (such as an anti-DIG antibody conjugated to alkaline phosphatase). By adding a suitable substrate mixture, which produces an intensely colored insoluble precipitate at the reaction site, one can determine the location of this. As a result, the color changes to indicate the location of the enzyme, the site of the probe's binding, and the location of the particular mRNA.

Some substrates are fluorescent and produce a signal that can be seen under the fluorescence microscope. The abbreviation "FISH" stands for fluorescence *in situ* hybridization. This is particularly useful for detecting regions of separation or overlap in the expression domains when multiple probes are used and visualized through various fluorescence channels. Several chemical tags must be applied to the probes in order to visualize more than one at once. These tags can be found using various antibody combinations and substrate mixtures. Although fluorescein is frequently used in probes, it should be noted that the quantity of fluorescein groups in the probe is insufficient to produce a fluorescent signal on its own. In this case, fluorescein visualization still necessitates antibody binding and amplification through an enzyme-substrate reaction.

Originally, radiolabeled probes were used to perform *in situ* hybridizations on sections. The main drawback was that they had to be developed at a specific time, which necessitated prior knowledge of the likely ideal exposure time. However, this had the benefit of high sensitivity. In contrast, a color reaction can be observed during its progression and terminated once the color reaches the intended intensity.

4.3.2. Immunostaining

In developmental biology, staining specimens with particular antibodies is also crucial (Figures 4.10 and 4.11). Almost any protein or carbohydrate molecule can be targeted with an antibody by injecting it into an animal along with an appropriate adjuvant. Regardless of its chemical makeup, the target of a particular antibody is known as its antigen. Epitopes are the specific regions of an antigen molecule that the antibody recognizes. Rabbits are frequently used to make polyclonal antibodies. There should be a high concentration of a particular antibody in the serum, known as a titer, following multiple immunizations. Even though the antibody is specific to the given antigen, it is made up of products from multiple B lymphocyte clones, each of which uses a different antibody gene. The serum will also include thousands of non-antibody proteins and thousands of additional antibodies that recognize different antigens in addition to the particular antibody. Because of the possibility of background or nonspecific staining caused by these additional ingredients, the antibody must typically be partially separated from the serum before usage. The portion of an antibody molecule that recognizes the antigen is called the variable region, and the portion that is specific to the immunoglobulin class and the animal species is called the constant region. A protein A column is a commonly used partial purification technique. This is the constant region of IgG-type antibody molecules being bound by a bacterial protein. As a result, all of the IgG, including the particular antibody, will be separated from the other serum components by the protein A column. If there is sufficient antigen available, a better approach would be to create an affinity column containing the antigen and isolate the particular antibody from the serum, eliminating both other antibodies and proteins (Polak, 2003).

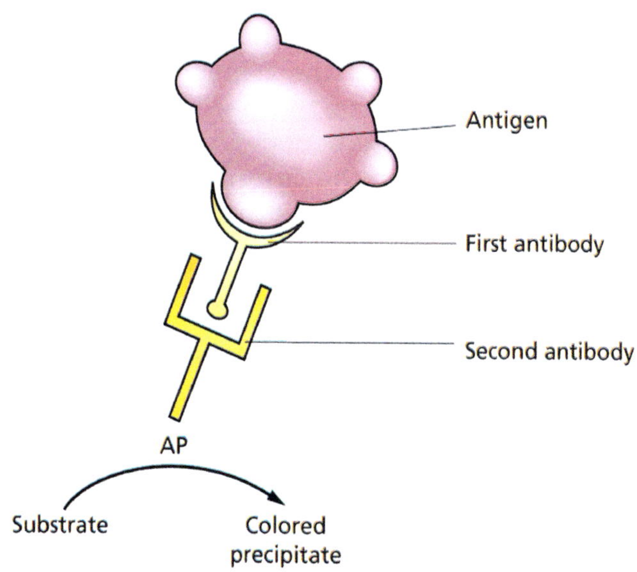

Figure 4.10. *Immunostaining for detection of protein.*

Source: Jonathan M.W. Slack. Essential Developmental Biology, 3rd Edition. Wiley-Blackwell. ISBN: 9781118387603.

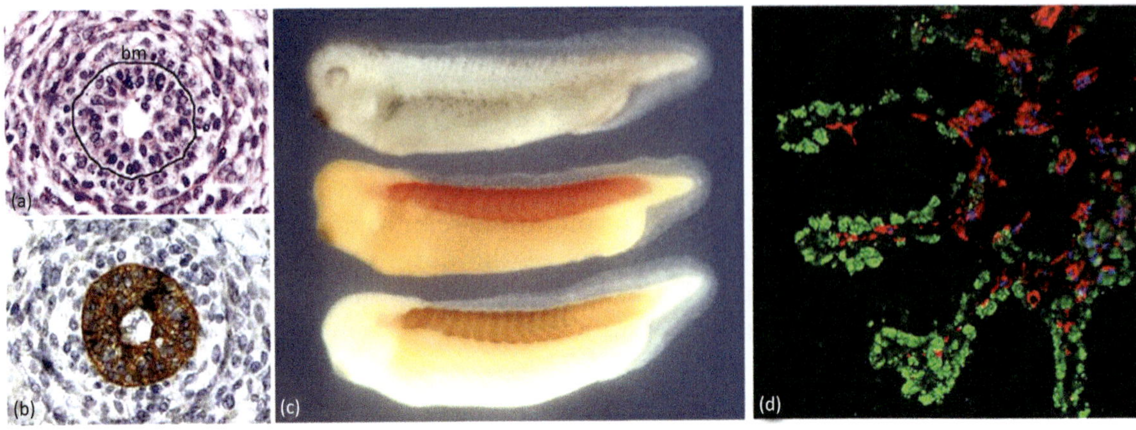

Figure 4.11. *Examples of immunostaining. (a, b) Section through a 13.5 day mouse embryo esophagus stained in hematoxylin–eosin (a) and immunostain for cytokeratin 8 with diaminobenzidine reaction (brown), plus hematoxylin as a nuclear stain (b); bm, position of the basement membrane. (c) Wholemount immunostaining (middle and below) of a muscle antigen in Xenopus tailbud-stage embryos. The antibody is the same in both but detection is with alkaline phosphatase and Fast Red (middle) and horseradish peroxidase with diaminobenzidine (below). (d) Wholemount stain of a mouse embryo pancreas organ culture. Three fluorescent antibodies were used: amylase is green, insulin red, and glucagon blue.*

Source: Jonathan M.W. Slack. Essential Developmental Biology, 3rd Edition. Wiley-Blackwell. ISBN: 9781118387603.

The typical process for creating monoclonal antibodies involves immunizing mice, then fusing their spleen cells with a human tumor cell line known as a myeloma. The hybrid cells, known as hybridomas, have the capacity to grow endlessly *in vitro* in addition to producing antibodies. To locate the necessary one, a number of multiwell plates with clones of the fused cells are screened. One specific monoclonal antibody can be obtained by cultivating and harvesting the culture medium of the isolated desired cell clone. The benefit of the monoclonal approach is that it can be used to immunize against completely uncharacterized tissue extracts; in fact, it does not require a pure antigen to begin with. Then, by using the same antibody to screen an expression library, the antibodies produced by individual hybridomas can be screened for intriguing expression patterns and molecularly cloned. However, producing monoclonal antibodies involves more effort and expertise than producing polyclonal ones, and the affinity of individual monoclonal antibodies for their target is frequently lower than that of polyclonals.

When it comes to *in situ* hybridization, sections or wholemounts can be immunostained. Sections offer a higher resolution by nature, whereas wholemounts only offer a single three-dimensional view of the antigen's location in the specimen. The specimen is first given a suitable amount of time to be incubated with the particular antibody in both cases. After that, it is cleaned and given another opportunity to be incubated with a different antibody. This antibody is marketed for use and targets the first antibody's species-specific constant region. Either an enzyme like alkaline phosphatase or horseradish peroxidase, or a fluorescent group like one of the dyes in the Alexa series, will be carried by it. The material can be directly inspected in the fluorescence microscope if a fluorescent secondary antibody was applied. The specimen is incubated in the substrate to allow the colored precipitate to develop, if an enzyme-linked second antibody was used. After that, it can be mounted and seen under transmitted light.

Due to the additional amplification step provided by the enzyme-substrate reaction, enzyme-linked methods are typically more sensitive than fluorescence. Moreover, the specimen can be dried and mounted using a nonaqueous medium after the precipitate has formed. Since nonaqueous media have a higher refractive index and make the specimen more nearly transparent, they are typically preferred over aqueous ones. It is impossible to dehydrate the specimen because the immune complex must remain intact for fluorescence immunostaining to work. For similar reasons, immune complexes eventually dissociate, suggesting that fluorescent specimens might not be permanent. Conversely, fluorescence techniques are easier to use because they involve fewer steps, and they offer the most effective means of examining multiple antigens in overlapping domains because each fluorochrome can be analyzed independently in a separate fluorescence microscope channel.

4.3.3. Reporter Genes

In many situations in developmental biology, it is practical to monitor a specific aspect of an organism's events using reporter genes, which encode a readily detectable product (Figure 4.12). Reporters are utilized, for instance, to denote the presence or activity of a transgene, to label a specific cell type for easy identification, or to show that a particular intracellular signaling pathway is active.

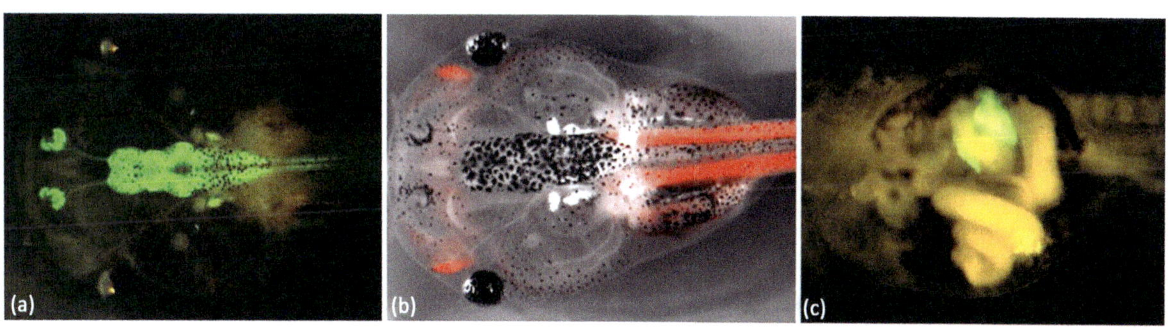

Figure 4.12. *Reporters. (a) A Xenopus tadpole carrying an N-tubulin reporter driving GFP in the central nervous system. (b) A muscle actin promoter driving tdTomato (a red fluorescent protein) in myotomal and jaw muscles. (c) An elastase promoter driving GFP, expressed only in the pancreas (two green buds).*

Source: Jonathan M.W. Slack. Essential Developmental Biology, 3rd Edition. Wiley-Blackwell. ISBN: 9781118387603.

Reporters are also widely utilized in the analysis of gene regulatory domains. Every brief segment originating from a potential regulatory area is linked to a reporter gene and inserted into embryos as a transgene. When a sequence is active, the reporter's expression is driven, allowing the activity's domain to be seen. Reporter expression is typically tracked using non-quantitative, *in situ* techniques with good spatial resolution. Although biochemical methods are quantitative, they don't provide spatial resolution, which makes them preferable for some purposes.

Overall, the most widely used reporter gene is the β-galactosidase-coding Escherichia coli lacZ gene. The lac operon's "Z" gene, which was the focus of traditional gene regulation research in the 1950s, is where the name originates. The large tetrameric enzyme β-galactosidase is able to hydrolyze a wide variety of β-galactosides, which are compounds that have a chemical group attached to the 1-carbon of galactose via a β linkage. Although β-galactosidase activity can be measured biochemically using a range of colorimetric or fluorescent substrates, this reporter is typically used *in situ* in developmental biology. 5-bromo-4-chloro-3-indolyl-βd-galactoside, or X-Gal for short, is the substrate used in this. The X component of the molecule instantly forms a green-blue insoluble precipitate when it is hydrolyzed apart from the galactose. Due to the reaction's sensitivity, extremely low expression levels can be found. Due to its high sensitivity and the lack of cross-reacting enzymes in most animal tissues, β-galactosidase is an extremely helpful reporter with low background staining. Since it functions after aldehyde fixation, it can be used in conjunction with other methods like immunostaining or conventional staining.

Keep in mind that the β-galactosidase protein is highly stable, so its presence could signal both current and previous gene activity. Thus, it is feasible to identify areas of an embryo where the β-galactosidase enzyme is present but the mRNA encoding it is not. This phenomenon, called perdurance, may be especially important for embryos that do not exhibit much growth. After the gene is turned off, the protein level in types that grow quickly will rapidly decrease through dilution.

When β-galactosidase is fused to another protein sequence, it usually stays active as a fusion protein. For some applications, such as the use of a construct called β-geo in mouse knockout technology, this is significant. Combining the activity of the neomycin resistance gene with the β-galactosidase enzyme, β-geo is a single molecule that carries both biological activities.

Since the late 1990s, green fluorescent protein (GFP) has been a basis for another significant class of reporter genes. This protein exhibits intense green fluorescent emission, as its name would imply. It comes from the Aequorea Victoria jellyfish. It frequently retains its activity when fused to other proteins, just like β-galactosidase. Several modifications have been made to the GFP coding sequence in order to change the fluorescence's color or intensity. As a result, a number of derivative proteins that emit blue, yellow, or red fluorescence are now available. Fluorescent proteins have a number of benefits, including the ability to be easily seen in living specimens and detected in fixed specimens, allowing for real-time analysis of the labeled cells. Some fluorescence is lost during fixation, but certain antibodies can still identify the molecules.

Firefly luciferase is another reporter that's frequently used. This enzyme is responsible for catalyzing the degradation of luciferin when ATP is present. Luminescence, the emission of light, is the result of the reaction. The emission spectrum has a green peak and extends to longer wavelengths that can readily pass through tissues. A luminometer is commonly used to measure the biochemical phosphorescence of cells or tissue explants. Using imaging devices with a highly cooled CCD camera that can detect single photons, *in situ* detection is also feasible. These can be used on small specimens but are typically used on larger specimens, like whole mice, to localize luciferase-labeled implants. A distinct luciferase with a blue emission is found in the sea pansy Renilla. It utilizes coelenterazine as a substrate. This is typically employed as a firefly luciferase normalization control.

4.4. MICROINJECTION

There are several reasons why it is frequently preferable to introduce a material into an embryo's single cell. The majority of Xenopus embryo overexpression experiments involve creating synthetic mRNA *in vitro* and injecting it into the fertilized egg. The ability to inject DNA into a fertilized egg's pronucleus is necessary to create transgenic mice. Similarly, to create transgenic Drosophila, DNA must be injected into the egg at the posterior end, just before the germ cells form. Inhibitors like particular antibodies or antisense oligonucleotides can also be injected into cells using microinjection techniques. When a specific cell lineage or graft needs to be identified by the presence of a visible substance known as a lineage label, microinjection is also necessary for a wide range of cell-labeling experiments (Carroll, 2008).

The size of the target cell determines the equipment needed for microinjection. But in order to control the injection visually, it will always be mounted on a microscope (Figure 4.13). A micromanipulator of some kind will be needed to hold the injection needle and limit the experimenter's manual movements to a size appropriate for the target cell. Glass tubing will be used to make the injection needle itself. By heating the glass to almost melting temperature and then applying the proper pull to extract it, this is drawn out into a fine-tipped injection needle using a needle-pulling machine. The injection material is introduced into the needle in one of two ways: either by capillarizing into the tip or by injecting it into the back end of the needle using a syringe. The needle is then linked via a flexible tube to an injection controller. A pressure device that applies abrupt pressure pulses to the needle and extracts a small volume from its sharp end could be the controller. Using an electric field applied across the needle to force appropriately charged molecules out of the tip, it might be an iontophoretic device. In either case, the micromanipulator is attached, the needle is filled, and the controller is connected. The researcher manipulates the controller to force a pulse of the material out of the tip while keeping an eye on the microscope. If the material is fluorescent—which a lineage label frequently is—the microscope will have a fluorescent attachment installed so that the injection's efficacy can be seen right away. It is possible to inject Xenopus embryos under a dissecting microscope; however, smaller specimens will need to be magnified using a compound microscope. While some setups use an upright microscope and others an inverted one, the microscope's optics are typically set up so that the image is not inverted, and its mechanism is such that the stage that holds the manipulators in place is fixed while the lenses move. These features are unique and not found on the typical upright compound microscope type.

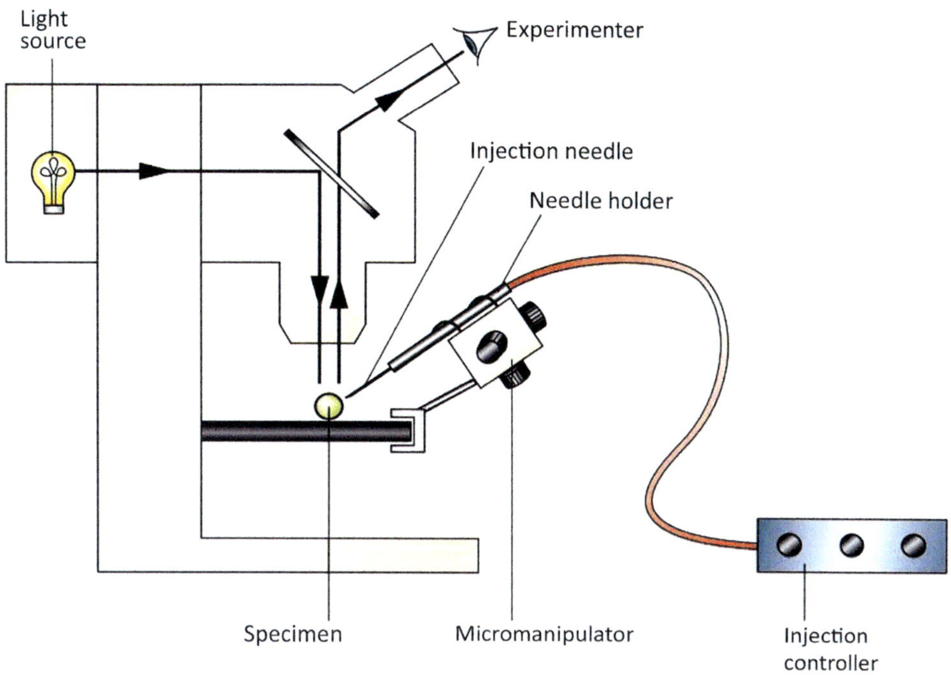

Figure 4.13. *Setup for microinjection under the fluorescence microscope.*

Source: Jonathan M.W. Slack. Essential Developmental Biology, 3rd Edition. Wiley-Blackwell. ISBN: 9781118387603.

4.5. CELL-LABELING METHODS

There are numerous applications for cell-labeling techniques. They are employed in fate mapping, which illustrates the typical course of development for various embryonic regions. They are employed in clonal analysis to mark individual cells. They are employed to mark cells that have been exposed to other materials like mRNA or antibodies. Whole embryos are labeled with them so that they can be used as donors for grafts that have been labeled. In developmental biology, labels are typically meant to be lineage labels, which identify only the offspring of the original labeled cells. However, in real life, this ideal is not always met. Because individual cells in all types of animal embryos can move around a great deal and it is impossible to keep track of them by observation alone, cell labeling is crucial.

4.5.1. Extracellular Labels

The first extracellular labeling technology was introduced for fate mapping in the early 1900s and is known as vital dyes. Nile blue and neutral red are the most often utilized colors. Because they don't have any harmful effects and can be absorbed by living cells in a reasonable amount of color, they are known as vital dyes. Using a small block of agar impregnated with the dye, they can be applied locally to identify particular areas of an embryo or to label entire specimens by soaking them in the medium (Figure 4.14b). Because they are affordable, easy to use, and quick, they are still occasionally used. They are not, however, true lineage designations. The dyes only give an approximate indication of the original labeling site because it is difficult to get the dye deep into a specimen and because they do spread and fade.

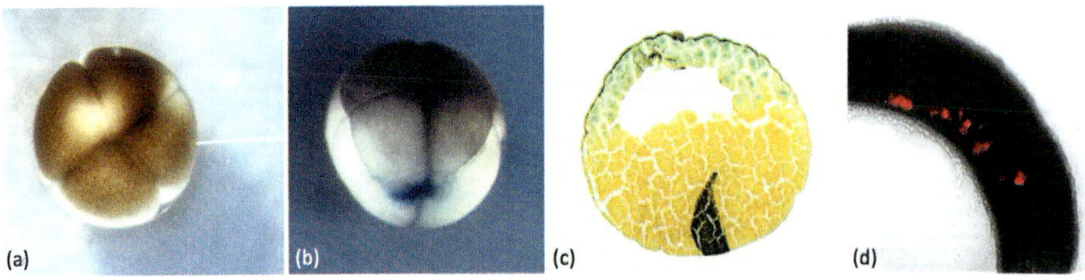

Figure 4.14. *Labeling techniques. (a) Microinjection of a four-cell stage Xenopus embryo. (b) A patch of the vital dye Nile blue applied to the dorsal side of a Xenopus embryo. (c) A single vegetal blastomere injected with horseradish peroxidase, which is then visualized on a section with the benzidine reaction. (d) A patch of DiI was applied to the early endoderm of a Xenopus embryo. The cells are later dispersed in the gut epithelium, indicating cellular intercalation movements.*

Source: Jonathan M.W. Slack. Essential Developmental Biology, 3rd Edition. Wiley-Blackwell. ISBN: 9781118387603.

In recent times, applications involving small patches of extracellular label have come to favor the carbocyanine dyes DiI and DiO. A microinjection device that is extracellular in nature is used to administer them. They dissolve in the lipid membranes of the labeled cells due to their extreme hydrophobicity, and the progeny retain them well. The fluorescence microscope is used to visualize them because they are highly fluorescent, with DiI producing red emission and DiO producing green emission. Organic solvents extract DiI and DiO from the specimen, but during histological processing, certain chemical derivatives are preserved and can be analyzed in paraffin wax sections.

4.5.2. Intracellular Labels

Fluorescent dextrans is the intracellular labeling material most frequently used. Dextran is a glucose polymer that is metabolically inert and freely soluble in water. Water-soluble molecules are unable to diffuse through the lipid plasma membrane to enter or exit cells, and while small molecules can pass through gap junctions to move from one cell to another, they are unable to transport molecules with molecular weights greater than or equal to approximately 1000. As a result, a dextran with a molecular weight of roughly 10,000 will stay inside the cell that it was injected into. Dextran is conjugated to a fluorochrome, such as rhodamine or fluorescein, to make it visible, and to the amino acid lysine, which has a free amino group that reacts easily with aldehyde fixatives, to make it fixable. FDA (fluorescein dextran amine) and RDA (rhodamine dextran amine) are two of this group's frequently used substances.

Horseradish peroxidase (HRP), an additional intracellular label that is frequently used, is likewise retained in cells following injection (Figure 4.14c). The most common substrate that peroxidases catalyze is diaminobenzidine, which is transformed into a brown, insoluble substance that is stable to paraffin wax histology. HRP can be seen in formalin-fixed wholemounts; however, if the specimen is too large for wholemount staining, frozen sections must be cut because the enzyme is not stable to paraffin wax histology.

4.5.3. Genetic Labels

Although substances like those have also been successfully used for brief periods of mammalian or chick development, they are most helpful in embryos that exhibit little to no growth during development, i.e., free-living species like Xenopus and zebrafish. The substances do not dilute in the absence of growth, and they can stay visibly distinct for several days. But as they grow, they quickly get diluted and disappear from view. In these conditions, the best labels are those that are integrated into the cell's genome so that they are replicated and undiluted during each cell cycle. These kinds of genetic labels were used extensively in early Drosophila and mouse experiments, though they all had significant drawbacks. For instance, the different electrophoretic mobility of the enzyme found in various mouse strains was necessary for the mouse glucose phosphate isomerase isozyme system to function. However, this had very low spatial resolution because it could only be visualized *in situ* through biochemical analysis of sections of dissected tissue. Although widely used, the Drosophila mutants with yellow and multiple wing hairs were only expressed in the cuticle of adult flies. They were unsuitable for analyzing larvae or the adult's internal tissues. These days, much better genetic labels that can be seen *in situ* and are expressed in all tissues and developmental stages are available. Transgenic mouse lines that express E, for instance. Human alkaline phosphatase (hPLAP) or coli β-galactosidase (encoded by the lacZ gene) in every tissue (Figure 4.15). Sensitive histochemical techniques can identify these, and hPLAP offers the benefit of having enzyme activity that remains stable when embedded in paraffin wax, allowing for histochemical visualization of sections. It is feasible to stain the specimen and identify which cells originated from each of the original components in any experiment involving a mixture of cells from labeled and unlabeled sources.

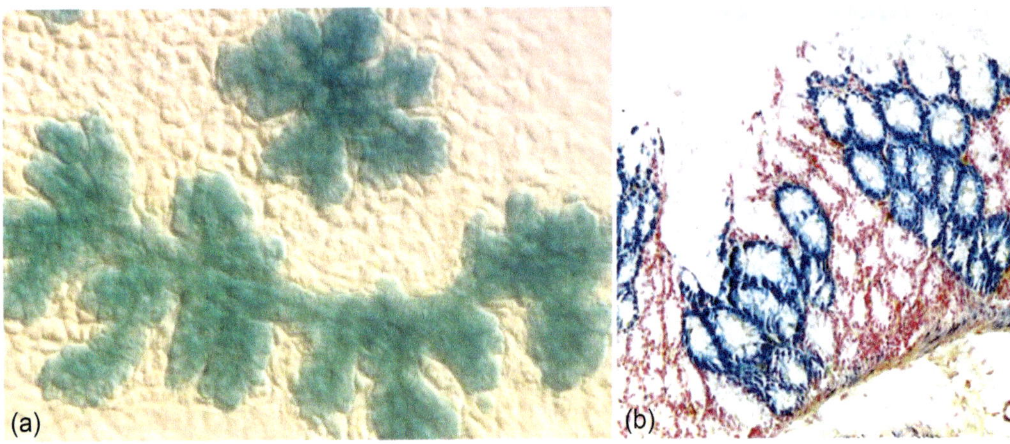

Figure 4.15. *Use of E. coli lacZ as a genetic label. Detection by XGal staining. (a) An organ culture of mouse embryo pancreas, similar to Figure 4.11d, but this one is formed by recombining the epithelium from a ROSA26 mouse, expressing E. coli β-galactosidase in all tissues, and the mesenchyme from a wild-type mouse. Only the epithelium is stained. (b) Intestinal crypts from an H253 mouse. This transgenic mouse strain has the lacZ gene on its X chromosome. Adult females inactivate the transgene in 50% of cells, leading to the mosaic appearance shown here.*

Source: Jonathan M.W. Slack. Essential Developmental Biology, 3rd Edition. Wiley-Blackwell. ISBN: 9781118387603.

A replication-incompetent retrovirus can also be used to introduce a genetic label. RNA viruses that contain the enzyme reverse transcriptase are known as retroviruses. When a cell is infected, the reverse transcriptase copies the viral genome into DNA, which the cell can subsequently incorporate into its chromosome and pass on to all of its offspring. Because a replication-incompetent virus lacks the genes required for virus particle assembly, its genome simply remains in the chromosome and is unable to replicate or lyze a cell. Replication-incompetent viruses are propagated in packaging cell lines, a type of cell line that has the missing viral genes and can fill in the functions the virus lacks. A lacZ gene that is driven by the viral long terminal repeat, a potent promoter sequence, is typically present in retroviruses used for cell labeling. Since clonal analysis is their primary application, infection is induced at low multiplicity, meaning that each infection focus represents the offspring of a single labeled cell.

ROLE MODEL

ANTONIE VAN LEEUWENHOEK

Antonie van Leeuwenhoek (1632–1723) was the most important microscopist of the Scientific Revolution. The Dutchman made over 500 microscopes, many with a magnification far superior to contemporary models. His discoveries include bacteria, protozoa, red blood cells, spermatozoa, and how minute insects and parasites reproduce.

Early Life

Unusually for one of the most prominent scientists of the 17th century, Antonie van Leeuwenhoek did not belong to the upper classes. He was born in Delft in the Netherlands into a low middle-class family on 24 October 1632. Antonie's family were brewers, and he did not have the opportunity to study at university or learn Latin or French, the two principal languages of scientific communication at the time. These handicaps were serious, especially regarding finance since most scientists had independent means to fund their often expensive research. Leeuwenhoek was obliged to fund his research into microscopy by working both as a draper (he had trained in this profession in Amsterdam) and as a minor municipal officer in Delft. Significantly, perhaps, for his future achievements, in his work inspecting cloth, he used magnifying glasses.

Leeuwenhoek's Microscopes

In the second decade of the 17th century, the microscope quickly followed the telescope as one of the most important instruments of the Scientific Revolution. A whole new miniature world was opened up to scientists. Anatomists, entomologists, and botanists were particularly keen to use this new invention to further their understanding of the natural world. Life under the microscope suddenly became unimaginably more intricate and complex.

Leeuwenhoek seems to have first made microscopes in the early 1670s. He may have become interested in the subject following the raging success of *Micrographia*, a richly illustrated book by Robert Hooke (1635–1703) on the wonders that could be observed on a microscopic slide. Certainly, Leeuwenhoek's investigations seem to follow the order of Hooke's book, suggesting he had his own copy. A microscope made by Leeuwenhoek is mentioned in a 1673 letter between two noted scientists, Reinier de

Graaf and Henry Oldenburg. Leeuwenhoek made over 500 microscopes in his career, grinding his own glass lenses. He invented a small and beautifully simple microscope that used, instead of the usual glass lens, a tiny glass bead fixed in position between two metal plates. Leeuwenhoek's bead microscope was capable of a highly impressive magnification of 270 times. Leeuwenhoek added a scale of measurement to his mini-microscope, and this allowed a more accurate comparison of views and specimens. In an age where sharing scientific data became paramount, he also devised a way for other microscopists to compare scale by using a base measurement such as the thickness of a strand of human hair or a grain of sand. In this way, microscopists elsewhere could better understand each other's work.

Leeuwenhoek's Discoveries

Far from being a mere technician, Leeuwenhoek put his microscopes to good use and made several important discoveries in this still-new field of inquiry. His "contributions include the discovery of red blood cells, of the circulation of blood through the capillaries, of the existence of protozoa, and of the nature of the male sperm cells". He also discovered bacteria, what he called "animalcules," seen in such things as samples taken from animal intestines. Another discovery was that minute insects, invisible to the naked eye, reproduced just like larger life-forms and not through some sort of spontaneous reaction as was then widely supposed.

Leeuwenhoek was particular since he was interested in viewing just about anything in his microscopes, as here summarized by the historian W. E. Burns:

He asked his neighbors for items to examine, like hair clippings or blood samples; he obtained from local businesses items such as infested grain and pieces of butchered animals; and he went to the dock to sample the wide range of goods available there, such as whale skins. Nor was Leeuwenhoek averse to putting portions of his own body under the lens, publishing the results of microscopic examination of tartar from his teeth in 1683...and detailing the analysis of a substance found between Leeuwenhoek's toes after he had left his stockings on for two weeks.

In just one example of Leeuwenhoek's unquenchable curiosity and willingness to use the people around him, he investigated what it was that caused the common ailment of gout. He knew a sufferer with the unfortunate symptom of a chalk-like substance excreted from two holes in his joints, one from the heel and the other from an elbow. Leeuwenhoek summarized what happened next in a letter:

I asked him to let me have some of the chalk which he willingly granted me...I observed the solid matter which to our eye resembles chalk, and I saw to my great astonishment that I was mistaken in my opinion, for it consisted of nothing but long, transparent little particles, many pointed at both ends. These particles which Leeuwenhoek observed, we now know, are deposits of sodium urate caused by an excess of uric acid in the blood.

In another letter, Leeuwenhoek gives the following description of how he discovered protozoa, as it happens, entirely by accident since the Dutchman was really trying to "discover the cause of the pungency of pepper upon our tongue":

I put about one third of an ounce of whole pepper in water, placing it in my study, with the sole design thereby that the pepper being rendered soft I might be enabled the better to observe what I proposed to myself. This pepper having lain three weeks in the water, to which I had twice added some snow water [the purest water available] (the other water being exhaled) I looked upon it the 24th of April, 1676, and discerned to my great wonder, an incredible number of very small animals of divers kinds.

Leeuwenhoek's Theory of Life

Leeuwenhoek's discovery of spermatozoa around 1677 led him to formulate a theory of the reproduction cycle. Scientists were still divided over exactly how new life was created, most pertinently in humans. Traditionally, the male was considered more important in reproduction, a view held by males to bolster their own ideas that they should be more important. However, by the 17th century, it had become clear that most new life came from a female-produced egg of one sort or another, and so the prevailing theory was 'ovism', that is, the essential characteristics of the new life form were all present in an egg which awaited fertilization. Leeuwenhoek proposed the opposite since he believed the spermatozoa and not the egg contained such characteristics as the life form's soul. He "developed a preformationist theory in which new creatures are elaborated in the male's semen (governed by the father's animal soul) before conception and subsequently grow in the mother's womb with the nutriment provided by the egg".

Leeuwenhoek's theory was not widely accepted, and he even got into trouble for discussing such intimate matters in public (some also wondered where and how he had got his specimens from). Nevertheless, Leeuwenhoek's discovery of spermatozoa and other previously unknowns was an example of how in the Scientific Revolution such discoveries were flames used to test and purify existing explanatory theories or destroy them or forge entirely new ones.

International Recognition

Leeuwenhoek was aware that to make his discoveries more widely known and useful to the scientific community he needed to capture them in drawings. Leeuwenhoek employed local draughtsman to create precise images of what he had seen in his microscopes. He sent almost all of his research findings, including the specially commissioned illustrations, to the Royal Society in London. The historian L. Jardine explains the microscopist's approach:

He made no pretense to expertise in anything apart from the manipulation of his microscopes; he presented his observations as raw data, in day-by-day diary form, 'that they might be better credited in England and elsewhere.' 'I beg you and the Gentlemen under whose eyes this happens to come to bear in mind that my observations and opinions are only the result of my own impulse and curiosity,' he explained, with studied humility.

Hundreds of these papers were then translated from the Dutch originals and published in the society's unofficial magazine *Philosophical Transactions* between 1673 and 1723. Many of Leeuwenhoek's letters to the society were subsequently published in collected volumes, too. In 1680, Leeuwenhoek was invited to become a fellow of the society. Leeuwenhoek was also a long-time contributor to the Dutch journal *De Boekzaal van Europe*.

Leeuwenhoek became so accomplished at preparing and observing specimens on microscopic slides that he was far ahead of most of his fellow scientists in what he made possible to see viewed under high magnification. Indeed, so high was the magnification and clarity that no other microscopists could replicate what he saw, and so some doubted that Leeuwenhoek had ever seen such things. There was certainly some mystery involved because Leeuwenhoek refused to reveal just how he lit up his specimens, a necessity to be able to see anything clearly under high magnification. Eventually, other scientists like Hooke caught up and proved that Leeuwenheok was indeed seeing things never before seen.

The Dutchman's work became so well known that he was able to offer his house in Delft as a sort of open museum where the public could visit and see prepared slides in the many microscopes he had set up. Monarchs visited, including James II of England (r. 1685–1688) and Peter the Great of Russia (1682–1721), as did eminent scientists such as Christiaan Huygens (1629–1695), the inventor of the pendulum clock.

Antonie van Leeuwenhoek died in Delft on 26 August 1723 at the ripe old age of 90. He had outlived two wives and had only one child. He had shown the way for other scientists to follow and had created a specialization in new fields of inquiry such as protozoology and bacteriology. The microscope, always difficult to use even in accomplished hands, fell out of favour in the 18th century but returned in the next century, particularly in the field of medicine, when the technical deficiencies of lenses, lights, and slide preparation were greatly improved. Leeuwenhoek did much, then, to ensure this instrument became a vital tool in modern science.

CLASS ACTIVITY

Many living things are microscopic, which means they are too small to be seen with the human eye. To see them you must view them under a microscope. Study different microscopic images and have a go at labeling them!

SUMMARY

- Cell biology is the study of cells and how they function, from the subcellular processes which keep them functioning, to the way that cells interact with other cells. Whilst molecular biology concentrates largely on the molecules of life (largely the nucleic acids and proteins), cell biology concerns itself with how these molecules are used by the cell to survive, reproduce and carry out normal cell functions.
- Molecular biology is an area of biology concerned with the process of gene transcription to yield RNA, the translation of RNA into proteins and the role those proteins play in cellular function. Since around 1960, molecular biologists have developed methods to identify, isolate, and manipulate molecular components in cells including DNA, RNA, and proteins.
- Microscopy is the most common technique used to study the structural details of cells, tissues and organelles present in plants and animals. A microscope is an instrument used to view small things that cannot be seen with the unaided eyes.
- Embryos are small and their study inevitably requires the use of microscopes. Experiments often involve manual interventions using a dissecting microscope. This is a binocular microscope with a magnification in the range of about ×10 to ×50 and a good working distance between the objective lens and the specimen.
- Phase contrast microscopy is a technique that converts small differences of refractive index within the specimen into large differences of light intensity. It is the favored method for examining live cells cultured *in vitro*, or isolated gametes, and is used for small transparent embryos such as the mammalian preimplantation stages.
- Fluorescence microscopy is used for a variety of purposes but they all depend on visualizing the location of a fluorescent substance, or fluorochrome, within a specimen. They include fluorescent antibody staining, fluorescent *in situ* hybridization (FISH), and the visualization of fluorochromes introduced into the specimen to label subsets of cells.
- Confocal microscopy is an advanced microscopy technique which uses a 'pinhole' to eliminate out of focus light. The discrete optical sections can be collected after eliminating out of focus light above and below the plane. High power lasers illuminate the sample. They help in collecting light from the desired focal plane. The beam of laser light gets scanned across the sample and light only from the desired focal plane can enter the detector.
- Microarrays are typically used in the early stage of a research project to identify the genes likely to be involved in a particular system or process. They are made using machines derived from the computer industry and so are sometimes referred to as chips. They come in two basic types: the cDNA array and the Affymetrix oligonucleotide array. cDNA arrays consist of a regular arrangement of closely spaced spots, each of a different cDNA, arranged in a rectangular array on a glass slide.

- Cell-labeling methods are used for a whole variety of different reasons. They are used for fate mapping to show the normal destiny of embryo regions in the course of development. They are used to label single cells in order to do clonal analysis. They are used to label cells into which other substances such as mRNA or antibodies have been introduced. They are used to label whole embryos, which can then be used as donors for labeled grafts.

REVIEW QUESTIONS

1. What is cell biology?
2. What do you mean by molecular biology?
3. What does "microscope" mean?
4. What is confocal microscopy?
5. What is the microinjection method used for?

REFERENCES

1. Avison, M. V. (2007). *Measuring Gene Expression*. Abingdon: Taylor and Francis.
2. Bancroft, J. D., & Gamble, M. (2007). *Theory and Practice of Histological Techniques* (6th ed.). Philadelphia, PA: Churchill Livingstone.
3. Bradbury, H. (1998). *Introduction to Light Microscopy*. Oxford: Bios Scientific Publishers.
4. Buckingham, M. E., & Meilhac, S. M. (2011). Tracing cells for tracking cell lineage and clonal behavior. *Developmental Cell, 21*, 394–409.
5. Carroll, D. J. (2008). *Microinjection: Methods and Applications*. Totowa, NJ: Humana Press.
6. Harland, R. M. (1991). *In-situ* hybridization–An improved wholemount method for *Xenopus* embryos. *Methods in Cell Biology, 36*, 685–695.
7. Kiernan, J. A. (2008). *Histological and Histochemical Methods: Theory and Practice* (4th ed.). Bloxham, UK: Scion Publishing.
8. Kretzschmar, K., & Watt, F. M. (2012). Lineage tracing. *Cell, 148*, 33–45.
9. Murphy, D. B. (2001). *Fundamentals of Light Microscopy and Electronic Imaging*. New York: Wiley-Liss.
10. Nagy, A., Gertsenstein, M., Vintersten, K., & Behringer, R. (2003). *Manipulating the Mouse Embryo: A Laboratory Manual* (3rd ed.). Cold Spring Harbor Laboratory Press.
11. Polak, J. M. (2003). *Introduction to Immunocytochemistry* (3rd ed.). Oxford: BIOS Scientific Publishers.
12. Polak, J. M., & McGee, J. O. D. (1999). *In Situ Hybridization: Principles and Practice*. Oxford: Oxford University Press.
13. Supatto, W., Truong, T. V., Débarre, D., & Beaurepaire, E. (2011). Advances in multiphoton microscopy for imaging embryos. *Current Opinion in Genetics & Development, 21*, 538–548.

CHAPTER

5 Cells into Tissues

LEARNING OBJECTIVES

After studying this chapter, you will be able to:

- Describe embryonic stem cell;
- Define cytoskeleton;
- Understand the basics of small GTP-binding proteins;
- Discuss about extracellular matrix;
- Know about cell movement or motility;
- Understand basic principle of morphogenesis.

KEY TERMS FROM THIS CHAPTER

Cell movement	Cell replacement
Cytoskeleton	Embryonic stem cells (ESCs)
Eukaryotic cells	Extracellular matrix (ECM)
Microtubules	Multicellular lineages
Prenatal genetic diagnosis	Prokaryotes
Protein synthesis	Retinoblastoma proteins
Somatic cells	Tissue engineering

INTRODUCTION

The fundamental structural and functional component of all living forms is the cell. Every cell is made up of cytoplasm that is surrounded by a membrane; many cells also have organelles, each of which has a distinct purpose. The word "small room" is derived from the Latin word *cellula*. The majority of cells can only be seen under a microscope. About 4 billion years ago, on Earth, cells first appeared. Every cell has the ability to replicate, synthesize proteins, and move.

In general, there are two types of cells: prokaryotic cells, which have a nucleoid region but no nucleus, and eukaryotic cells, which have a nucleus. Prokaryotes are single-celled creatures like bacteria, whereas eukaryotes are multicellular creatures like some algae, plants, animals, and fungi, or they can be single-celled like amoebae. Organelles found in eukaryotic cells include ribosomes, which synthesize proteins, chloroplasts, which produce sugars in plants through photosynthesis, and mitochondria, which supply energy for cellular activities.

Tissue is made up of a collection of related cells with extracellular matrix from the same embryonic origin that work together to accomplish a particular task. Tissues are in between cells and an entire organ in the biological hierarchy. As a result, various tissues are functionally grouped together to form organs.

Biological organisms follow this hierarchy:

Cells < Tissue < Organ < Organ System < Organism

Tissue is a collection of cells or fluid that cooperates to carry out a specific function in the body, such as the blood cells that transport waste products and oxygen to the body's cells or the cells in an organ like the kidney or heart.

5.1. EMBRYONIC STEM CELL

Stem cells with pluripotency that originate from the inner mass of a blastocyst—an early stage of pre-implantation embryo—are known as embryonic stem cells, or ESCs (Figure 5.1). Four to five days after fertilization, human embryos reach the blastocyst stage, when they are made up of fifty to one hundred fifty cells. The blastocyst is destroyed when the inner cell mass (embryoblast) is isolated using immunosurgery. This procedure presents ethical questions, such as whether or not embryos in the pre-implantation stage of development are morally equivalent to those in the post-implantation stage.

The therapeutic potential of embryonic stem cells is currently receiving a lot of attention from researchers, with many labs hoping to use them in clinical settings. Potential applications include the management of heart disease and diabetes. The cells are being researched for potential applications in cellular/DNA repair, clinical therapy, and models of genetic disorders. However, there have also been reports of unfavorable outcomes in clinical and research settings, including tumor development and unintended immune responses. (Carla, Marcel, & Harm, 2011).

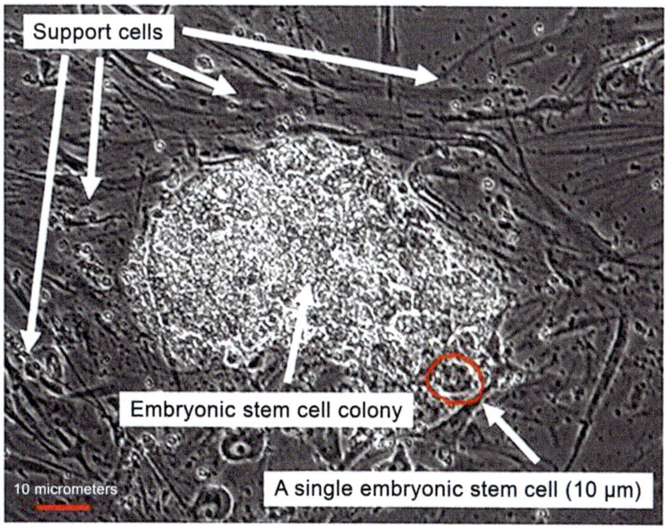

Figure 5.1. *Early embryonic development and pluripotent cells.*

Source: https://kids.frontiersin.org/articles/10.3389/frym.2020.00032

5.1.1. Properties

Originating from the blastocyst stage of early mammalian embryos, embryonic stem cells (ESCs) (Figure 5.2) are unique in that they can self-renew and differentiate into any type of embryonic cell. These characteristics make them useful in the domains of science and medicine. ESCs maintain a high level of telomerase activity, have a normal karyotype, and have exceptional long-term proliferative potential.

5.1.1.1. Pluripotent

The inner cell mass's embryonic stem cells are pluripotent, which means they can differentiate into primitive ectoderm, which can then differentiate during gastrulation into any of the three main germ layers' derivatives: ectoderm, endoderm, and mesoderm. Every one of the adult human body's more than 220 cell types is produced by these germ layers. When given the right signals, ESCs first develop into precursor cells, which then differentiate into the target cell types. Embryonic stem cells are differentiated from adult stem cells by their pluripotency; the latter are multipotent but can only generate a restricted variety of cell types.

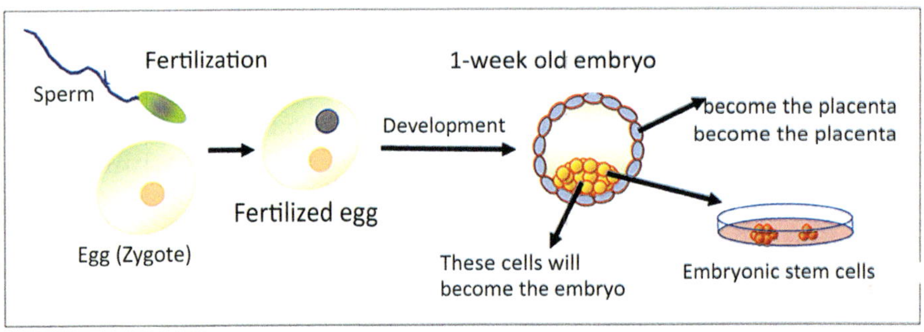

Figure 5.2. *Early embryonic development and pluripotent cells.*

Source: https://kids.frontiersin.org/articles/10.3389/frym.2020.00032

5.1.1.2. Self-Renewal and Repair of Structure

Embryonic stem cells have the capacity to self-renew indefinitely in an undifferentiated state under specific circumstances. Conditions for self-renewal need to keep cells from clumping together and preserve an environment that allows for an unspecialized state. This is usually performed in a laboratory setting using serum-free media supplements containing two inhibitory drugs ("2i"), the GSK-3 inhibitor CHIR99021 and the MEK inhibitor PD03259010, or serum-containing media supplemented with leukemia inhibitory factor.

5.1.1.3. Growth

Because of a shortened G1 phase in their cell cycle, ESCs divide a lot. Rapid cell division is crucial for the early development of embryos because it enables cells to proliferate in number but not in size. The G1/S transition-related proteins cyclin A and cyclin E are always highly expressed in ESCs. Part of the reason for the overactivity of cyclin-dependent kinases, like CDK2, that encourage cell cycle progression is the downregulation of their inhibitors. Retinoblastoma proteins hyperphosphorylate and become inactive in ESCs, causing the expression of proliferation genes to persist. These modifications speed up cell division cycles by inhibiting the transcription factor E2F until the cell is prepared to enter S phase. While maintaining pluripotency has been associated with high levels of pro-proliferative protein expression and a shortened G1 phase, ESCs cultured in serum-free 2i conditions do express hypo-phosphorylated active Retinoblastoma proteins and have an elongated G1 phase. These cells have similar pluripotent properties to ESCs grown in serum-containing media, despite their different cell cycle. Oct4 and Nanog, two pluripotency factors, are involved in transcriptional regulation of the embryonic stem cell cycle. (Lee et al., 2010).

5.1.1.4. Uses

Promising for regenerative medicine and tissue replacement following injury or disease, embryonic stem cell therapies are thought to be highly versatile and have an almost limitless potential for self-renewal. Numerous conditions, including but not limited to spinal cord injuries, age-related macular degeneration, diabetes, neurodegenerative disorders (like Parkinson's disease), AIDS, etc., have demonstrated potential for treatment with pluripotent stem cells. Apart from their potential applications in regenerative medicine, embryonic stem cells offer a potential substitute source of tissue or organs, potentially resolving the issue of organ shortage. However, there are some ethical disagreements about this (see the section below on the ethical debate). In addition to these applications, ESCs can be employed in *in vitro* toxicological testing, early human development, and some genetic diseases.

5.1.2. Utilizations

According to a 2002 article in PNAS, "Human embryonic stem cells have the potential to differentiate into various cell types, and, thus, may be useful as a source of cells for transplantation or tissue engineering."

5.1.2.1. Tissue Engineering

Stem cell use is known to be important in tissue engineering. The cells utilized in tissue engineering must be capable of carrying out particular biological tasks, such as secreting cytokines, communicating with other cells, interacting with them, and correctly organizing extracellular matrix. These particular biological functions are demonstrated by stem cells, which can also self-renew and differentiate into one or more types of specialized cells. One source being explored for tissue engineering applications is embryonic stem cells. Although there are numerous obstacles to overcome before human embryonic stem cells can even be used, their use has opened up a plethora of new possibilities for tissue engineering. Theoretically, tissue engineering would advance dramatically if embryonic stem cells could be modified so that they do not trigger an immune reaction in recipients. There are uses for embryonic stem cells outside of tissue engineering.

5.1.2.2. Cell Replacement Therapies

Differentiating ESCs into diverse cell types for potential application in cell replacement therapies has been the focus of research. Numerous cell types have been developed or are in the process of being developed, such as neurons, hepatocytes, bone marrow cells, islet cells, endothelial cells, and cardiomyocytes. However, there are challenges in deriving these cell types from ESCs, and research has been focused on overcoming these challenges. Research is being conducted, for instance, to eliminate the immature characteristics that set ESCs apart from adult cardiomyocytes and to differentiate them into tissue-specific cardiomyocytes.

5.1.2.3. Drug Discovery

In addition to being a significant organ transplant substitute, ESCs are being utilized in toxicology and as cellular screens to find novel compounds that may be turned into small-molecule medications. Research has demonstrated that *in vitro* models using cardiomyocytes derived from ESCs are reliable for assessing drug responses and forecasting toxicity profiles. Since it has been demonstrated that ESC-derived cardiomyocytes react to pharmacological stimuli, they can be used to evaluate cardiotoxicity, including torsade de pointes.

Hepatocytes derived from stem cells are another valuable model that may be applied during the preclinical phases of drug discovery. Unfortunately, it has been difficult to develop hepatocytes from ESCs, which makes it difficult to test drug metabolism. Thus, the goal of research has been to create hepatocytes derived from ESC that are fully functional and exhibit stable phase I and II enzyme activity.

5.1.2.4. Models of Genetic Disorder

The idea of using embryonic stem cells to model genetic disorders has been the subject of several recent studies. Stem cells have been used to model genetic disorders, either through genetic manipulation of the cells or, more recently, by deriving diseased cell lines identified by prenatal genetic diagnosis (PGD). This method could be very helpful in the study of genetic diseases without a good model system, such as cystic fibrosis, fragile-X syndrome, and other conditions. Prenatal diagnosis testing methods were developed by Russian-American medical researcher Yury Verlinsky, who specialized in embryo and cellular genetics (genetic cytology). These methods allowed for the identification of genetic and chromosomal disorders 1.5 months ahead of time using standard amniocentesis. Nowadays, a lot of expectant mothers and would-be parents use the techniques, particularly those whose partner has a history of genetic abnormalities or who is over 35 (when the risk of genetically related disorders is higher). Furthermore, by giving parents the option to choose an embryo free of genetic disorders, they may be able to use the disease-free offspring's cells to treat siblings who already had comparable illnesses.

5.1.2.5. Repair of DNA Damage

Different approaches are taken by ES cells and differentiated somatic cells to address DNA damage. For example, the primary pathway for repairing double-strand breaks (DSBs) during all cell cycle stages is non-homologous end joining (NHEJ), an error-prone DNA repair process, used by human foreskin fibroblasts, one type of somatic cell. Due to its error-proneness, NHEJ frequently results in mutations in a cell's clonal offspring.

A distinct approach is taken by ES cells in handling DSBs. Mutations resulting from defective DNA repair in ES cells pose a greater threat than in differentiated somatic cells since ES cells give rise to all cell types of an organism, including the germ line cells. Therefore, strong mechanisms are required in ES cells to precisely repair DNA damage and, in the event that repair is unsuccessful, to eliminate the cells that still have unrepaired DNA damage. Therefore, the primary method of DSB repair used by mouse ES cells is high fidelity homologous recombinational repair (HRR). The interaction of the two sister chromosomes that were created during the S phase and are present together during the G2 phase of the cell cycle is what allows for this kind of repair. By utilizing intact data from the other sister chromosome, HRR can precisely repair DSBs in one sister chromosome. Sister chromosomes are absent from cells in the G1 phase of the cell cycle, which occurs after metaphase and cell division but before the subsequent round of replication. When DNA damage occurs, mouse ES cells do not experience cell cycle arrest because they do not have a G1 checkpoint. Instead, they experience apoptosis, or programmed cell death, in reaction to DNA damage. As a backup plan, apoptosis can be employed to eliminate cells that have unrepaired DNA damage to prevent mutation and the development of cancer. In line with this approach, the frequency of mutations in mouse ES stem cells is approximately 100 times lower than in isogenic mouse somatic cells.

5.1.2.6. Clinical Trial

The US Food and Drug Administration (FDA) approved the world's first human ESC clinical trial on January 23, 2009, for the transplantation of oligodendrocytes—a type of brain and spinal cord cell—derived from human ESCs into spinal cord-injured individuals. The study leading to this scientific advancement was conducted by Hans Keirstead and colleagues at the University of California, Irvine and supported by Geron Corporation of Menlo Park, CA, founded by Michael D. West, PhD. A

Rats with spinal cord injuries recovered more quickly in a prior study following a 7-day transplantation of human ESCs that had been propelled into the oligodendrocytic lineage. Since the cells must be injected before scar tissue can form, the phase I clinical study was planned to enroll eight to ten paraplegics who have had their injuries no more than two weeks prior to the trial's start. The doctors made it clear that they did not anticipate that the patients would recover completely and have full mobility following the injections. Based on the outcomes of the trials conducted on rodents, scientists hypothesized that myelin sheath restoration and increased motility might take place. The main goal of this initial trial was to evaluate the safety of these treatments, and if all went according to plan, it was envisaged that it would pave the way for more extensive trials involving participants with more severe disabilities. The FDA expressed concerns about a small number of microscopic cysts found in several treated rat models, which led to the trial's suspension in August 2009. However, the suspension was lifted on July 30, 2010.

At Shepherd Center in Atlanta, researchers enrolled and gave ESCs to their first patient in October 2010. Geron Corporation, the company that created the stem cell therapy, predicted that it would take several months for the stem cells to proliferate and for the efficacy of the GRNOPC1 therapy to be determined.

Geron declared in November 2011 that it was ending the trial and quitting stem cell research due to budgetary constraints, but that it would keep an eye on its current patient base and look for a partner to carry on the research. All of Geron's stem cell assets were purchased by BioTime in 2013, under the CEO Dr. Michael DWest, with the declared goal of resuming Geron's clinical trial utilizing embryonic stem cells for the study of spinal cord injuries.

The California Institute for Regenerative Medicine (CIRM) awarded BioTime company

Asterias Biotherapeutics (NYSE MKT: AST) a $14.3 million Strategic Partnership Award to relaunch the world's first human clinical trial utilizing embryonic stem cells for spinal cord injury. CIRM is the world's largest donor to stem cell-related research and development, with funding provided by the state of California.

Asterias will be able to expand clinical testing of increasing doses in the target population for upcoming pivotal trials, as well as resume clinical development of AST-OPC1 in subjects with spinal cord injury, thanks to the funding provided by the award.

OPCs, or oligodendrocyte progenitor cells, are found in the cell population known as AST-OPC1, which is derived from human embryonic stem cells (hESCs). OPCs and their developed offspring, oligodendrocytes, give nerve cells in the brain and spinal cord vital functional support. The outcomes of a phase 1 clinical trial evaluating a low dose of AST-OPC1 in patients with neurologically complete thoracic spinal cord injury were recently presented by Asterias. The outcomes demonstrated that AST-OPC1 was successfully administered to the spinal cord injury site. Following up for two to three years after AST-OPC1 administration, patients underwent detailed follow-up evaluations, including frequent MRIs and neurological exams, and no evidence of serious adverse events linked to the cells was found. After a year of transplantation, immune monitoring of the subjects failed to reveal any signs of cellular or antibody-based immune responses to AST-OPC1. Serial MRI scans conducted over the course of the two- to three-year follow-up period in four of the five subjects suggest that there may have been less spinal cord cavitation and that AST-OPC1 may have had some beneficial effects in preventing the degeneration of spinal cord tissue. The International Standards for Neurological Classification of Spinal Cord Injury (ISNCSCI) exam was used to evaluate the five trial subjects, and it revealed no unexpected neurological degeneration or improvement.

Asterias will receive funding from the CIRM Strategic Partnership III grant to support the next clinical trial of AST-OPC1 in subjects who have suffered spinal cord injuries. The grant will also support Asterias' efforts to develop new products and scale up manufacturing processes in order to support later-stage trials and eventual commercialization. The trial's FDA approval, the conclusion of a formal agreement between Asterias and CIRM, and Asterias' continuous advancement toward the accomplishment of specific project milestones will all be prerequisites for CIRM funding.

5.1.3. Techniques and Conditions for Derivation and Culture

5.1.3.1. Derivation from Humans

Many embryos are created during *in vitro* fertilization. The excess embryos can be donated by the donor with permission because they are not clinically utilized or appropriate for patient implantation. The inner cell mass, or cells of interest, from the blastocyst stage of the embryo is separated from the trophectoderm, the cells that would differentiate into extra-embryonic tissue. Human embryonic stem cells can be derived from these donated embryos or alternatively they can be extracted from cloned embryos created using a patient's cell and a donated egg through the process of somatic cell nuclear transfer. To achieve separation, two procedures are used:

immunosurgery, which involves binding antibodies to the trophectoderm and removing them with a different solution, and mechanical dissection. The cells that make up the inner cell mass are plated on top of support cells. The inner cell mass cells adhere and proliferate to generate an undifferentiated human embryonic cell line. Every four to seven days, these cells are mechanically or enzymatically separated after being fed on a daily basis. Human embryonic stem cells can be grafted into a three-dimensional scaffold, co-cultured with a serum containing the required signals, or separated from the supporting cells to form embryoid bodies in order for differentiation to take place.

5.1.3.2. Derivation from Other Animals

The inner cell mass of the early embryo, which is removed from the donor mother animal, is where embryonic stem cells originate. A method described by Martin Evans and Matthew Kaufman allows the inner cell mass to grow by delaying the implantation of the embryo. In order to change the donor mother's hormone environment and keep the embryos free in the uterus, the donor mother's ovaries are removed and she is given a progesterone dose. Following 4–6 days of intrauterine culture, the embryos are removed and cultured *in vitro* until the inner cell mass forms structures resembling egg cylinders. These structures are then separated into individual cells and plated on fibroblasts that have been treated with mitomycin-c to stop fibroblast mitosis. One single cell is grown to create clonal cell lines. Evans and Kaufman demonstrated that the cells derived from these cultures exhibited pluripotency as evidenced by their ability to differentiate *in vitro*, form teratomas, and form embryoid bodies.

Gail Martin produced and cultivated her ES cells in a unique way. About 76 hours after copulation, she extracted the embryos from the donor mother and cultured them in a serum-containing medium for the entire night. She performed microsurgery the next day to remove the late blastocyst's inner cell mass. The isolated inner cell mass was cultivated on mitomycin-C-treated fibroblasts in a serum-containing medium, which was further conditioned by endothelial cells. A week or so later, cell colonies began to grow. These cells proliferated in culture and showed signs of pluripotency, as evidenced by their capacity to differentiate *in vitro*, form embryoid bodies, and form teratomas. These cells were called ES cells by Martin.

It is now established that serum and feeder cells supply the bone morphogenetic proteins (BMPs) and leukemia inhibitory factor (LIF) required to stop ES cells from differentiating. These elements have a major impact on how well ES cells are derived. Additionally, it has been shown that the efficiency of different mouse strains in isolating embryonic stem cells varies. The creation of transgenic mice, including knockout mice, is one of the current applications for mouse ES cells. Pluripotent stem cells specific to each patient are required for human treatment. Human ES cell generation is more challenging and fraught with moral dilemmas. Thus, many groups are studying the production of induced pluripotent stem cells (iPS cells) in addition to human ES cell research.

5.1.3.3. Potential Methods for New Cell Line Derivation

Dr. Robert Lanza, the medical director of Advanced Cell Technology in Worcester, Massachusetts, wrote a letter to the online edition of the scientific journal Nature on August 23, 2006, announcing that his team had discovered a means to extract embryonic stem cells without harming the actual embryo. Due to federal funding restrictions at the time, research using embryonic stem cell lines derived before August 2001 was not permitted. However, this technical achievement may allow scientists to work with new lines of embryonic stem cells derived using public funding in the US. 2009 March saw the lifting of the restriction.

Somatic cell nuclear transfer (SCNT) has also been used to generate human embryonic stem cells. Since the method of SCNT is similar to other forms of cloning in that nuclei are transferred from a somatic cell into an enucleated zygote, this technique has also occasionally been referred to as therapeutic cloning. But in this instance, embryonic stem cell lines were created in a lab using SCNT rather than using pregnancy to create living things. The name's "therapeutic" component was added in the hopes that the embryonic stem cells generated by SCNT would prove useful in medicine.

5.1.3.4. Induced Pluripotent Stem Cells

The iPS cell technology was pioneered by Shinya Yamanaka's lab in Kyoto, Japan, who showed in 2006 that the introduction of four specific genes encoding transcription factors could convert adult cells into pluripotent stem cells. He was awarded the 2012 Nobel Prize along with Sir John Gurdon "for the discovery that mature cells can be reprogrammed to become pluripotent."

In 2007, research demonstrated that delivering four factors (Oct3/4, Sox2, c-Myc, and Klf4) to differentiated cells could induce pluripotent stem cells, which are highly similar to embryonic stem cells. The differentiated cells are reprogrammed into pluripotent stem cells using the four previously mentioned genes, enabling the production of pluripotent/embryonic stem cells without the need for an embryo. This observation was originally made in mouse pluripotent stem cells, but it can now be repeated in human adult fibroblasts using the same four genes. The morphology and growth factors of these lab-induced pluripotent cells are equivalent to embryonic stem cells, leading these cells to be known as induced pluripotent stem cells (iPS cells). Reprogramming to these iPS cells is thought to be less contentious because ethical concerns about embryonic stem cells usually relate to their genesis from terminated embryos.

This could make it possible to create ES cell lines tailored to individual patients, which might then be utilized for cell replacement treatments. Furthermore, this will make it possible to create ES cell lines from patients who have a range of genetic diseases, which will provide researchers with invaluable models to research those diseases.

On December 6, 2007, the online edition of Science journal reported that a research team led by Rudolf Jaenisch of the Whitehead Institute for Biomedical Research in Cambridge, Massachusetts, had successfully cured mice of sickle cell anemia using iPS cell technology. This was a first indication that the technology can quickly lead to new cures. A California-based business called Stemagen declared on January 16, 2008, that they had successfully produced the first fully developed cloned human embryos using adult skin cells. These embryos can be taken out to obtain embryonic stem cells that match the patient.

5.2. CYTOSKELETON

All cells, including bacteria and archaea, have cytoplasm that contains a complex, dynamic network of interconnecting protein filaments known as the cytoskeleton. It is made up of similar proteins in all eukaryotic organisms and extends from the cell nucleus to the cell membrane. It is made up of three primary parts: microtubules, intermediate filaments, and microfilaments. Depending on what the cell needs, these parts can all rapidly assemble or disassemble.

The cytoskeleton is capable of performing a wide range of tasks. Its main job is to give the cell its shape and mechanical resistance against deformation. It also stabilizes entire tissues by joining forces with other cells and extracellular connective tissue. Additionally, the cytoskeleton can contract, causing the cell and its surroundings to change and permitting cell migration. Additionally, it participates in a variety of cell signaling pathways, the uptake of extracellular material (endocytosis), chromosome segregation during cellular division, the cytokinesis stage of cell division, intracellular transport (such as the movement of vesicles and organelles within the cell), and it can serve as a template for the formation of a cell wall. Moreover, it has the ability to form specialized structures like podosomes, lamellipodia, cilia, and flagella. Depending on the organism and type of cell, the cytoskeleton's composition, capabilities, and dynamic behavior can vary greatly. The cytoskeleton can alter even within a single cell due to interactions with other proteins and the network's past. (Fletcher & Mullins, 2010).

Muscle contraction is a major example of an activity carried out by the cytoskeleton. Groups of highly specialized cells cooperating to carry out this task. The microfilament is a crucial part of the cytoskeleton that aids in revealing the actual purpose of this muscle contraction. Actin, the most prevalent cellular protein, makes up microfilaments. Parallel actin filaments are subjected to collective forces by myosin molecular motors within each muscle cell during muscle contraction. Increased calcium release from the sarcoplasmic reticulum follows nerve impulses that initiate muscle contraction. With the aid of two proteins, tropomyosin and troponin, increases in calcium in the cytosol trigger the start of muscle contraction. Actin and myosin interaction is inhibited by tropomyosin; troponin detects an increase in calcium and releases the inhibition. This action contracts the muscle cell and, because many muscle cells contract simultaneously, the entire muscle.

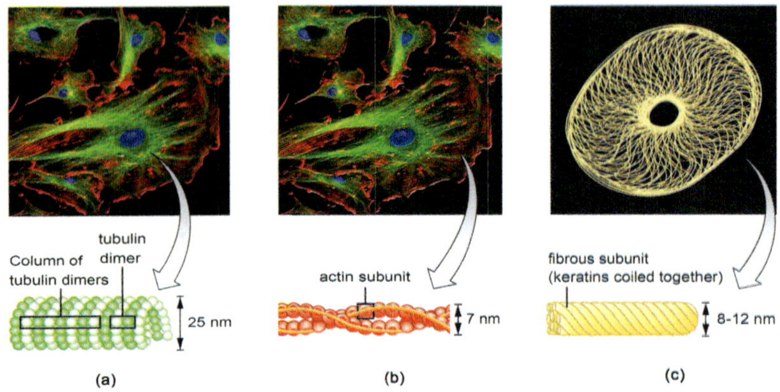

Figure 5.3. *The cytoskeleton consists of (a) microtubules, (b) microfilaments, and (c) intermediate filaments.*

Source: https://upload.wikimedia.org/wikipedia/commons/c/cb/0317_Cytoskeletal_Components.jpg.

5.2.1. Eukaryotic Cytoskeleton

The three primary cytoskeletal filament types found in eukaryotic cells are microfilaments, microtubules, and intermediate filaments. Neurofilaments are the term for the intermediate filaments found in neurons. Each type has a unique intracellular distribution and shape that result from the polymerization of a particular type of protein subunit. Microfilaments, which have a diameter of 7 nm, are actin polymer polymers. Tubulin makes up microtubules, which have a diameter of 25 nm. Depending on the kind of cell they are found in, different proteins make up intermediate filaments, which typically have a diameter of 8–12 nm. The cytoskeleton gives the cell its shape and structure and increases the amount of macromolecular crowding in this compartment by keeping some macromolecules out of the cytosol. Cellular membranes are in close and extensive interaction with cytoskeletal components.

Studies on neurodegenerative conditions like ALS, Parkinson's disease, Alzheimer's disease, and Huntington's disease suggest that the cytoskeleton plays a role in these illnesses. Parkinson's disease is characterized by neuronal degeneration, which causes stiffness, tremors, and other non-motor symptoms. Studies have indicated that the cytoskeleton's microtubule assembly and stability are impaired, leading to the gradual deterioration of neurons. Tau proteins, which stabilize microtubules, malfunction in Alzheimer's disease and cause cytoskeleton pathology as the disease progresses. Another theory linking excess glutamine in the Huntington protein—which links vesicles onto the cytoskeleton—to the onset of Huntington's disease. Defects in the cytoskeleton as well as motor neuron degradation lead to amyotrophic lateral sclerosis, a condition that causes paralysis.

Stuart Hameroff and Roger Penrose suggest a role of microtubule vibrations in neurons in the origin of consciousness. Accessory proteins, such as motor proteins, are crucial for the controlled assembly of cytoskeletal filaments in specific locations because they control and connect the filaments to other cell components and to one another.

Numerous small-molecule cytoskeletal medications that interact with microtubules and actin have been found. Numerous of these substances have therapeutic uses in addition to being helpful in the study of the cytoskeleton.

5.2.1.1. Microfilaments

Actin filaments, also referred to as microfilaments, are made of linear polymers of G-actin proteins. When the growing (plus) end of a filament presses against a barrier, like the cell membrane, it produces force. Additionally, they serve as pathways for myosin molecules, which adhere to the microfilament and "walk" along it. Actin is typically the main protein that makes up microfilaments. The microfilament (actin filament) is created by the combination of the G-actin monomer and a polymer. Following their assembly, these subunits form two chains that combine to form what are known as F-actin chains. Contractile forces are produced in so-called actomyosin fibers by myosin motoring along F-actin filaments in both muscle and the majority of non-muscle cell types. The Rho family of small GTP-binding proteins, which includes Rac for lamellipodia, Cdc42 for filopodia, and Rho itself for contractile acto-myosin filaments (stress fibers), controls actin structures.

5.2.1.2. Intermediate filaments

Many eukaryotic cells have intermediate filaments as a component of their cytoskeleton. These filaments, which have a diameter of 10 nanometers on average, are more strongly bound and stable than microfilaments, which are more diverse cytoskeleton constituents. They serve the same purpose as actin filaments in maintaining the shape of cells by bearing tension (microtubules, on the other hand, resist compression but are also capable of bearing tension during mitosis and centrosome positioning). Intermediate filaments anchor organelles and act as structural elements of the nuclear lamina, arranging the internal three-dimensional structure of the cell. Additionally, they take part in a few cell-matrix and cell-cell junctions. All animals and tissues have nuclear lamina. Certain creatures, such as fruit flies, lack cytoplasmic intermediate filaments. These are tissue-specific in animals that express cytoplasmic intermediate filaments. The epithelial cells' keratin intermediate filaments shield the skin from various mechanical stresses that it may experience. Additionally, they shield organs from chemical, oxidative, and metabolic stresses. By lessening the likelihood of stress, strengthening epithelial cells with these intermediate filaments may delay the onset of apoptosis, or cell death.

While they are involved in certain cell functions, intermediate filaments are most commonly recognized for serving as the scaffolding or support structure for the cell and nucleus. The intermediate filaments, in conjunction with desmosomes and proteins, establish connections between cells and serve as anchors for cell-matrix junctions, which are essential for intracellular communication and other essential cellular processes. With the help of these connections, a cell can interact with the desmosomes of other cells to modify tissue structures in response to cues from its surroundings. Serious medical conditions like early aging, desmin mutations that compromise organs, Alexander Disease, and muscular dystrophy have all been linked to mutations in the IF proteins.

5.2.1.3. Microtubules

Microtubules are hollow cylinders with a diameter of roughly 23 nm (with a lumen diameter of about 15 nm). They are typically made up of 13 protofilaments, which are polymers of alpha and beta tubulin. They bind GTP to facilitate polymerization, exhibiting a very dynamic behavior. They are frequently arranged according to the centrosome.

They form the centrioles in nine triplet sets (star-shaped) and the cilia and flagella in nine doublets oriented about two extra microtubules (wheel-shaped). The latter configuration, where each doublet is joined to the next by the protein dynein, is known as a 9+2 arrangement. Since cilia and flagella are both structural elements of the cell and are kept in place by microtubules, they can be categorized as cytoskeleton components. Motile and non-motile cilia are the two different forms of cilia. Compared to flagella, cilia are shorter and more abundant. In contrast to the non-motile cilia, which gather sensory data for the cell by processing signals from neighboring cells or fluids, the motile cilia beat or wave rhythmically. Furthermore, the cilia and flagella's beating (movement) is regulated by the microtubules. Moreover, the molecular motors are the dynein arms that are joined to the microtubules. The microtubules slide past one another to generate motion in the cilia and flagella, a process that needs ATP. They play key roles in:

- intracellular transport (associated with dyneins and kinesins, they transport organelles like mitochondria or vesicles).
- the axoneme of cilia and flagella.
- the mitotic spindle.
- synthesis of the cell wall in plants.

5.2.1.4. Septins

Among the highly conserved GTP-binding proteins in eukaryotes are a class of proteins called septins. Various septins combine to form protein complexes. These can come together to form rings and filaments. Septins are therefore regarded as a component of the cytoskeleton. In cells, septins have two roles: they act as a specific site of attachment for other proteins and stop specific molecules from diffusing from one area of the cell to another. Yeast cells use scaffolding to partition the cell into different sections and offer structural support during cell division. According to recent studies conducted on human cells, septins enclose bacterial pathogens in cages that render them immobile and stop them from spreading to other cells.

5.2.1.5. Spectrin

In eukaryotic cells, the intracellular side of the plasma membrane is lined by the cytoskeletal protein strin. Spectrin organizes into pentagonal or hexagonal structures, acting as a scaffold and contributing significantly to the preservation of the cytoskeletal system and integrity of the plasma membrane.

5.2.1.6. Yeast Cytoskeleton

Actin forms a cytokinetic ring, cortical patches, actin cables, and the cap in budding yeast, an important model organism. Cortical patches, which are distinct actin bodies on the membrane, are essential for endocytosis and, in particular, for recycling glucan synthase, which is necessary for the synthesis of cell walls. Actin cables, which are bundles of actin filaments, play a role in the positioning of mitochondria as well as the movement of vesicles towards the cap, which is home to several proteins that polarize cell growth. The site of cell division is surrounded by the cytokinetic ring, which forms and contracts.

5.2.2. Prokaryotic Cytoskeleton

Before the work of Jones and colleagues, 2001, it was thought that the cell wall determined the shape of many bacterial cells, such as spirals and rods. Upon examination, numerous malformed bacteria were discovered to possess mutations connected to the formation of a cell envelope. It was previously believed that the cytoskeleton was exclusive to eukaryotic cells, but prokaryotes have homologues for every major protein in the eukaryotic cytoskeleton. Harold Erickson points out that, prior to 1992, it was thought that only eukaryotes possessed cytoskeleton elements. On the other hand, studies conducted in the early 1990s revealed that eukaryotic microtubules and microfilaments originated from homologues of actin and tubulin found in bacteria and archaea. Strong evidence that the eukaryotic and prokaryotic cytoskeletons are truly homologous can be found in their similar three-dimensional structures and similar roles in maintaining cell shape and polarity, even though the evolutionary relationships between them are too distant to be discernible from protein sequence comparisons alone. It was independently discovered by three laboratories that the tubulin signature sequence was present in all α-, β-, and γ-tubulins, and that this protein, FtsZ, was a crucial component of bacterial cytokinesis. It's possible, though, that some bacterial cytoskeleton structures haven't been identified yet.

5.2.2.1. FtsZ

The prokaryotic cytoskeleton's first protein to be identified was FtsZ. In the presence of guanosine triphosphate (GTP), FtsZ forms filaments similar to tubulin, but these filaments do not group into tubules. FtsZ is the first protein to arrive at the division site during cell division and is crucial for attracting additional proteins that work together to create the new cell wall in the space between the dividing cells.

5.2.2.2. MreB and ParM

MreB and other prokaryotic actin-like proteins are involved in the preservation of cell shape. Actin-like proteins are encoded by genes found in all non-spherical bacteria. These proteins orient the other proteins involved in cell wall biosynthesis in a helical network beneath the cell membrane.

Actin-like protein ParM is involved in a distinct system that is encoded by certain plasmids. Dynamic instability is seen in ParM filaments, which may partition plasmid DNA into the dividing daughter cells via a process resembling that of microtubules in eukaryotic mitosis.

5.2.2.3. Crescentin

A third protein called crescentin, which is connected to the intermediate filaments of eukaryotic cells, is found in the bacteria Caulobacter crescentus. Although the precise mechanism by which crescin preserves the shape of bacteria's helical and vibrioid forms is still unknown, it plays a role in these cell types as well. Additionally, following the disruption of peptidoglycan synthesis, curvature could be explained by the displacement of crescentic filaments.

5.2.3. The Cytoskeleton and Cell Mechanics

The cytoskeleton is a dynamic, highly anisotropic network that is always changing to adapt to the shifting cellular microenvironment. By differentially polymerizing and depolymerizing its constituent filaments—primarily actin and myosin, but also microtubules and intermediate filaments—the network affects cell mechanics and dynamics. This produces forces that are crucial in providing the cell with information about its microenvironment. Particularly, it has been demonstrated that forces like tension, stiffness, and shear forces affect cell motility, migration, and fate. To detect and react to these forces, the cell modifies its cytoskeleton through a process known as mechanotransduction.

Focal adhesions, which essentially link the extracellular matrix (ECM) and the intracellular cytoskeleton, are critical to mechanotransduction. The cell can integrate extracellular forces into intracellular ones through focal adhesions because the proteins there undergo conformational changes that trigger signaling cascades. Proteins that are thought to be important sensors in the mechanotransduction pathway include focal adhesion kinase (FAK) and Src. These proteins have been demonstrated to transduce force signals in response to cellular activities like proliferation and differentiation. Mechanotransduction causes the cytoskeleton to reorganize in order to adapt to the force stimulus and guarantee that the cell reacts appropriately.

The cytoskeleton adapts the cell's mechanics in reaction to forces that are sensed. For instance, as plasma membrane tension rises, ion channels are more likely to open, increasing ion conductance and the likelihood of ion influx or efflux during cellular changes. Furthermore, a force's direction, propagation, and effects on cell dynamics are all determined by the mechanical characteristics of the cell. If a membrane protein is subjected to a force that is directed specifically at it, for example, it will not significantly affect the cortical actin network because it is not tightly coupled to the cytoskeleton. Membrane proteins that are more strongly linked to the cytoskeleton, however, will cause a more notable reaction. Thus, the cytoskeleton's anisotropy helps to more precisely guide how cells react to intra- or extracellular cues.

5.2.4. Long-Range Order

Investigations are still ongoing into the precise pathways and mechanisms that the cytoskeleton uses to detect and react to forces. Nonetheless, it is well known that the cytoskeleton contributes to mechanotransduction by creating long-range order. With a diameter of approximately 10–50 μm, cells are thousands of times bigger than the cytoplasmic molecules needed to synchronize cellular functions. In the absence of an organizing network, communication between various cytoplasmic components is challenging due to the size of cells relative to vital biomolecules. Furthermore, biomolecules need to polymerize to lengths that are equivalent to a cell's length; however, the resulting polymers may be incredibly disorganized and ineffective at signal transmission across the cytoplasm. Therefore, the cytoskeleton is required to arrange the polymers and guarantee that they can communicate efficiently throughout the cell.

5.2.5. Common Features and Differences Between Prokaryotes and Eukaryotes

The proteins that make up the cytoskeleton can naturally form longitudinal arrays, or fibers, in all living things. There are four classes that these filament-forming proteins fall under. Walker A cytoskeletal ATPases (WACA-proteins), actin-like, tubulin-like, and intermediate filaments.

In prokaryotes, tubulin-like proteins are FtsZ, TubZ, and RepX, while in eukaryotes, tubulin is tubulin. Actin in eukaryotes and MreB and FtsA in prokaryotes are examples of actin-like proteins. MinD is an illustration of a WACA-protein, which is primarily present in prokaryotes. Examples of intermediate filaments include desmin, lamins, keratins, vimentin, and neurofilaments. Eukaryotes are the only organisms that have been found to contain intermediate filaments.

While there is some amino acid sequence similarity among tubulin-like proteins, there is a more noticeable similarity in their protein fold and GTP binding site. This also applies to the structure and ATP binding domain of actin-like proteins.

In prokaryotes and eukaryotes, cytoskeletal proteins are typically associated with cell shape, DNA segregation, and cell division. The functions that different proteins perform vary greatly. For instance, tubulin is used in all eukaryotes for DNA segregation, but in prokaryotes, WACA proteins, actin-like proteins, or tubulin-like proteins can be employed instead. Actin mediates cell division in eukaryotes, whereas tubulin-like (often FtsZ-ring) proteins and occasionally (Thermoproteota) ESCRT-III mediate cell division in prokaryotes. However, ESCRT-III plays a role in the final stage of division in eukaryotes.

5.3. SMALL GTP-BINDING PROTEINS

Monomeric G proteins, or small GTP-binding proteins, have molecular masses between 20–40 kDa. Around 1980, the v-Ha-Ras and v-Ki-Ras oncogenes of sarcoma viruses were identified as the Ha-Ras and Ki-Ras genes. Afterwards, their cellular oncogenes in humans were discovered, and some human carcinomas also contained their mutations. It was later demonstrated that the mutant forms promoted the growth and transformation of cultured cells. Furthermore, it has been demonstrated that the mutant forms cause neuronal cells to differentiate. Numerous scientists in a wide range of fields, including cancer research, were drawn to these findings. Ultimately, it was demonstrated that these Ras proteins were connected to the heterotrimeric G proteins, including Gi and Gs, as well as the G proteins involved in protein synthesis, including EF-Tu, or elongation factor Tu.

In 1985, the YPT1 gene was found to be an open reading frame between the actin and tubulin genes in the yeast Saccharomyces cerevisiae (*S. cerevisiae*); this gene was later identified as the homolog of the Ras gene in Aplysia. small G protein (Arf protein), which was purified as a cofactor for the cholera toxin-catalyzed ADP-ribosylation of Gs in 1984, was found to encode a small G protein in 1986. cerevisiae) in 1983. In 1987, it was discovered that the SEC4 gene, which had been identified in 1980 as a yeast gene involved in secretion, encodes a small G protein. These findings implied the existence of a large family of small G proteins that resemble Ras. In actuality, a large number of small G proteins were methodically isolated using biochemical and molecular biological techniques.

Currently, a superfamily made up of more than 100 small G proteins has been discovered in eukaryotes, ranging from yeast to humans. The Ras, Rho, Rab, Sar1/Arf, and Ran families are the minimum number of structurally distinct groups into which the members of this superfamily can be divided (Table 5.1.). Sequence analysis against the entire genome sequence of the yeast *S. cerevisiae* has identified four members of the Ras family, six members of the Rho family, eleven members of the Rab family, seven members of the Sar1/Arf family, and two members of the Ran family. Many small G proteins have recently had their functions clarified. For example, the Ras subfamily members (Ras proteins) of the Ras family primarily regulate gene expression; the Rho/Rac/Cdc42 subfamily members (Rho/Rac/Cdc42 proteins) of the Rho family regulate both gene expression and cytoskeletal reorganization; the Rab and Sar1/Arf family members (Rab and Sar1/Arf proteins) regulate intracellular vesicle trafficking; and the Ran family members (Ran) for microtubule organization during the M phase and nucleocytoplasmic transport during the G1, S, and G2 phases of the cell cycle. Small G protein activation and action mechanisms have been progressively clarified through the identification of numerous upstream regulators and downstream effectors.

Figure 5.4 illustrates dendrogram of the small G protein superfamily.

Table 5.1. *The Small G Protein Superfamily*

	Ras Family		**Rho Family**		**Rab Family**			**Sar1/Arf Family**	**Ran Family**
Mammal	Ha-Ras	Rheb	RhoA	Rnd2/	Rab1A	Rab11A	Rab26	Arf1	Ran
	Ki-Ras	κ B - Ras1	RhoB	Rho7	Rab1B	Rab11B	Rab27A	Arf2	
	N-Ras	κ B - Ras2	RhoC	Tc10	Rab2	Rab12	Rab27B	Arf3	
	R-Ras		RhoD		Rab3A	Rab13	Rab28	Arf4	
	M-Ras		RhoE/		Rab3B	Rab14	Rab29	Arf5	
	RalA		Rnd3/		Rab3C	Rab15	Rab30	Arf6	
	RalB		Rho8		Rab3D	Rab16	Rab31	Sar1a	
	Rap1A		RhoG		Rab4	Rab17	Rab32	Sar1b	
	Rap1B		RhoH/		Rab5A	Rab18	Rab33A	Arl1	
	Rap2A		TTF		Rab5B	Rab19	Rab33B	Arl2	
	Rap2B		Rac1		Rab5C	Rab20		Arl3	
	Tc21		Rac2		Rab6	Rab21		Arl4	
	Rit		Rac3		Rab7	Rab22		Arl5	
	Rin		Cdc42		Rab8	Rab23		Arl6	
	Rad		Rnd1/		Rab9	Rab24		Arl7	
	Kir/Gem		Rho6		Rab10	Rab25		Ard1	
Yeast	Ras1		Rho1		Ypt1	Ypt52		Arf1	Gsp1
	Ras2		Rho2		Sec4	Ypt53		Arf2	Gsp2
	Rsr1		Rho3		Ypt31/	Ypt6		Arf3	
	Yct7		Rho4		Ypt8	Ypt7		Sar1	
			Cdc42		Ypt32/	Ypt10		Arl1	
			Yns0		Ypt9	Ypt11		Arl2	
					Ypt51/ Vps21			Cin4	

Source: https://journals.physiology.org/doi/full/10.1152/physrev.2001.81.1.153.

In terms of nomenclature, small G proteins are used here instead of small GTPases, as small G proteins are known to have both GTPase and GDP/GTP-binding activities. GTPase activity is frequently required for small G proteins to cease functioning, but it is not always necessary for them to do so. In this sense, the term GTPase is not accurate. Small GTP-binding proteins should be used instead of G proteins, which are heterotrimeric G proteins; however, small G proteins are used in this context only for simplicity's sake. Here, G proteins that are involved in protein synthesis, small G proteins, and heterotrimeric G proteins are used. Though guanine nucleotide exchange factor (GEF) is more commonly used, this article refers to guanine nucleotide exchange protein (GEP) because all GEFs discovered so far are proteins, making GEPs a more accurate term. (Aktories, Rosener, Blaschke, & Chhatwal, 1988).

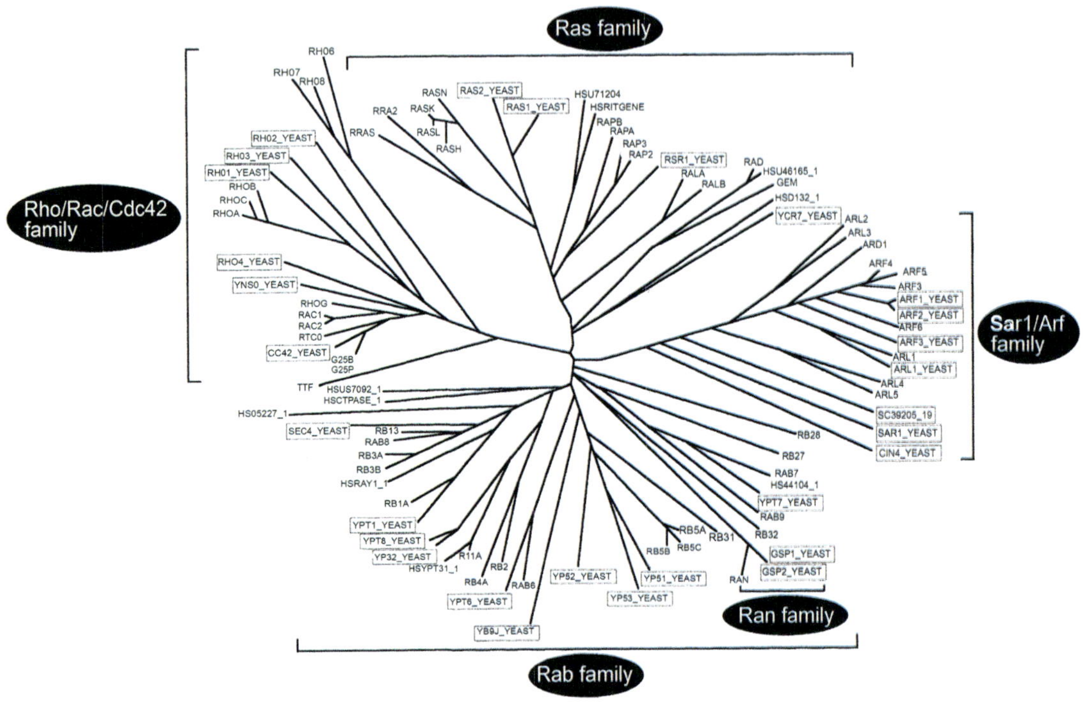

Figure 5.4. *Dendrogram of the small G protein superfamily.*

Source: https://journals.physiology.org/doi/full/10.1152/physrev.2001.81.1.153.

5.4. EXTRACELLULAR MATRIX

Extracellular macromolecules and minerals, including collagen, enzymes, glycoproteins, and hydroxyapatite, form the extracellular matrix (ECM), also known as the intercellular matrix (ICM), a network in biology that supports neighboring cells structurally and biochemically. The composition of ECM differs amongst multicellular structures due to the independent evolution of multicellularity in distinct multicellular lineages; nonetheless, the ECM shares common functions such as cell adhesion, cell-to-cell communication, and differentiation.

The basement membrane and the interstitial matrix are components of the animal extracellular matrix. Animal cells have interstitial matrix in the spaces between them, or intercellular spaces. Filling the interstitial space, gels of fibrous proteins and polysaccharides serve as a compression buffer against the strain applied to the ECM. Different types of epithelial cells rest on basement membranes, which are ECM deposits that resemble sheets. Animal connective tissue has different types of ECM: bone tissue's ECM is composed of collagen fibers and bone mineral; loose connective tissues ECM is made up of reticular fibers and ground substance; and blood plasma is the ECM of blood.

Plant ECM comprises more complex signaling molecules as well as components of the cell wall, such as cellulose. Certain unicellular organisms develop multicellular biofilms, wherein the cells are encased in an ECM mainly made up of extracellular polymeric materials (EPS).

5.4.1. Structure

Resident cells produce components of the ECM and secrete them into the extracellular matrix through exocytosis. After being secreted, they combine with the current matrix. GAGs and fibrous proteins form an interlocking mesh that makes up the ECM.

5.4.1.1. Proteoglycans

GAGs are polymers of carbohydrates that are primarily bound to extracellular matrix proteins to form proteoglycans. The net negative charge of proteoglycans draws positively charged sodium ions (Na^+), which in turn draws water molecules through osmosis to keep the ECM and residing cells hydrated. Proteoglycans might also aid in the ECM's ability to hold and retain growth factors. The various forms of proteoglycan that are present in the extracellular matrix are discussed below.

1. **Heparan Sulfate:** All animal tissues contain the linear polysaccharide heparan sulfate (HS). It takes the form of a proteoglycan (PG) where two or three HS chains are joined to ECM or cell surface proteins in close proximity. In this state, HS binds to a wide range of protein ligands and controls a multitude of biological functions, such as angiogenesis, blood coagulation, tumor metastasis, and developmental processes. Heparan sulfate is primarily bound to the multi-domain proteins perlecan, agrin, and collagen XVIII in the extracellular matrix, particularly in basement membranes.
2. **Chondroitin Sulfate:** The tensile strength of tendons, ligaments, cartilage, and the aortic walls is enhanced by chondroitin sulfates. They may also have an impact on neuroplasticity.
3. **Keratan Sulfate:** Unlike many other GAGs, keratan sulfates do not contain uronic acid and have a variable sulfate content. They can be found in animal horns, cartilage, bones, and corneas.

5.4.1.2. Non-Proteoglycan Polysaccharide

- **Hyaluronic Acid:** Unlike other GAGs, hyaluronic acid (also known as hyaluronan) is a polysaccharide that is not found as a proteoglycan. It is made up of alternating residues of D-glucuronic acid and N-acetylglucosamine. By absorbing large amounts of water, hyaluronic acid in the extracellular space gives tissues the ability to withstand compression by acting as a counteracting turgor (swelling) force. Thus, hyaluronic acid is abundant in the ECM of load-bearing joints. It also makes up a significant portion of the interstitial gel. During biosynthesis, hyaluronic acid is translocated out of the cell and is present on the inner surface of the cell membrane.

When it comes to inflammation, tumor development, healing processes, and embryonic development, hyaluronic acid functions as an environmental cue to control cell behavior. It communicates with CD44, a particular transmembrane receptor.

5.4.1.3. Proteins

1. **Collagen:** The most prevalent type of protein in the ECM is collagen. As collagen makes up 90% of the protein content of the bone matrix, it is actually the most prevalent protein in the human body. Collagens are fibrillar proteins found in the ECM that provide resident cells with structural support. Procollagen, the precursor form of collagen that is exocytosed, is subsequently broken down by procollagen proteases to facilitate extracellular assembly. Genes that encode collagen are defective in disorders like osteogenesis imperfecta, epidermolysis bullosa, and Ehlers Danlos syndrome.
2. **Elastin:** Unlike collagens, elastins give tissues their elasticity, enabling them to stretch when necessary and then revert to their initial shape. This is beneficial for the skin, ligamentum nuchae, lungs, blood vessels, and tissues with high elastin content. Fibroblasts and smooth muscle cells produce elastins. Tropoelastins are secreted inside chaperone molecules, which

release the precursor molecule when they come into contact with a mature elastin fiber. Elastins are highly insoluble. After that, tropoelastins are deaminated and integrated into the elastin strand. Williams syndrome and other disorders like cutis laxa are linked to inadequate or nonexistent elastin fibers in the extracellular matrix (ECM).

5.4.1.4. Extracellular Vesicles

Huleihel and associates in 2016. revealed that ECM bioscaffolds contained DNA, RNA, and Matrix-bound nanovesicles (MBVs). The size and shape of MBVs were discovered to be in line with exosomes that have already been reported. MBVs carry a variety of protein molecules, lipids, fragments of DNA, and miRNAs. MBVs can change the activation state of macrophages and affect various cellular characteristics like proliferation, migration, and cell cycle, just like ECM bioscaffolds can. It is currently thought that MBVs are a crucial and essential part of ECM bioscaffolds.

5.4.2. Cell Adhesion Proteins

1. **Fibronectin**: Glycoproteins called fibronectins help cells move through the extracellular matrix (ECM) by connecting them to collagen fibers. Collagen and cell-surface integrins are bound by fibronectins, which causes the cytoskeleton of the cell to reorganize and promote cell migration. Cells secrete fibronectins in an inactive, unfolded state. Fibronectin molecules unfold when they bind to integrins, enabling them to form dimers and perform as intended. By attaching to platelets during blood clotting and promoting cell migration to the injured area during wound healing, fibronectins also aid at the site of tissue damage.
2. **Laminin:** Proteins called laminins are present in almost every animal's basal laminae. Laminins in the basal lamina form networks of web-like structures that withstand tensile forces, as opposed to collagen-like fibers. They support cell adhesion as well. Collagen and nidogens are among the other ECM constituents that laminins bind.

5.4.3. Development

The development of the diverse extracellular matrix types present in the wide variety of tissue types is facilitated by a multitude of cell types. The characteristics of the connective tissue are determined by the local components of ECM.

The most prevalent cell type in the ECM of connective tissue, fibroblasts produce, preserve, and establish a structural framework. They also secrete the ground substance and other precursor components of the ECM. In cartilage, chondrocytes are responsible for the production of the cartilaginous matrix. Osteoblasts are the cells that form new bones.

5.4.4. Function

The ECM is a multifunctional matrix that can be used for a variety of tasks, including support, tissue segregation, and intercellular communication regulation. The dynamic behavior of a cell is regulated by the extracellular matrix. Furthermore, it functions as a local store for a variety of cellular growth factors by sequestering them. Protease activities can be activated by changes in physiological conditions, leading to the local release of these stores. This eliminates the need for de novo synthesis and enables the quick and localized activation of cellular processes *via* growth factors.

Growth, wound healing, and fibrosis are among the processes that depend on the extracellular matrix's formation. Comprehending the structure and composition of ECM is crucial for understanding the intricate dynamics of tumor invasion and metastasis in cancer biology. This is because enzymes like serine proteases, threonine proteases, and matrix metalloproteinases frequently degrade extracellular matrix during metastasis.

Gene expression, cell migration, and differentiation are all significantly impacted by the ECM's stiffness and elasticity. In a process known as durotaxis, cells actively detect the stiffness of the extracellular matrix (ECM) and migrate preferentially towards stiffer surfaces. Additionally, they recognize elasticity and modify their gene expression accordingly; this has drawn more attention in recent years due to its implications for cancer progression and differentiation.

Hyaluronan, the primary ECM component in the brain, is exhibited by the matrix in terms of both structural and signaling properties. Hyaluronan with a high molecular weight functions as a diffusional barrier that can affect diffusion locally in the extracellular space. Hyaluronan fragments are liberated from the matrix and enter the extracellular space. There, they act as pro-inflammatory molecules and direct the response of immune cells, including microglia.

5.4.4.1. Cell adhesion

Components of the extracellular matrix are bound by numerous cells. Two mechanisms exist for cell adhesion: focal adhesions, which attach the ECM to the cell's actin filaments, and hemidesmosomes, which attach the ECM to intermediate filaments like keratin. Certain cell-surface cellular adhesion molecules (CAM) called integrins control this cell-to-ECM adhesion. Integrins are proteins on the surface of cells that bind other cells' integrin proteins as well as ECM structures like laminin and fibronectin.

ECM macromolecules can more easily bind to transmembrane integrins when fibronectins are present. Fibronectin's attachment to the extracellular domain triggers the initiation of both intracellular signaling pathways and association with the cytoskeleton of the cell through a series of adaptor molecules, including actin.

5.4.5. Clinical Significance

It has been discovered that extracellular matrix promotes tissue healing and regrowth. Researchers now think that Matrix-bound nanovesicles (MBVs) are a major factor in the healing process, even though the exact mechanism of action by which extracellular matrix promotes constructive remodeling of tissue is still unknown. For instance, the extracellular matrix enables stem cells to grow and regenerate every part of the human body in fetuses, and they are also able to regenerate any damage sustained while still inside the womb. It has long been accepted by scientists that the matrix ceases to exist once it has fully developed. Although it has been used to aid in the healing of torn ligaments in horses, more research is being done on it as a tool for human tissue regeneration.

Regarding tissue engineering and injury repair, the extracellular matrix has two primary functions. Initially, it stops the immune system from being activated by the damage and producing scar tissue and inflammation in response. Secondly, it helps the surrounding cells to mend the tissue rather than causing scarring.

The necessary ECM for medical applications is typically taken from pig bladders, which are a readily available and mostly underutilized source. By sealing the hole in the stomach's lining, it is currently used frequently to treat ulcers. However, numerous universities and the US government are currently conducting additional research on its potential applications for wounded soldiers. Testing was underway on a Texas military base as of early 2007. Veterans of the Iraq War whose hands were injured in the conflict are receiving a powdered version from scientists.

Not every ECM gadget originates from the bladder. Atrial septal defects (ASD), patent foramen ovale (PFO), and inguinal hernias are being repaired with extracellular matrix derived from the submucosa of the pig small intestine. After a year, the body replaced 95% of the collagen ECM in these patches with the typical heart soft tissue.

Extracellular matrix proteins have the ability to stimulate the differentiation of epithelial, endothelial, and smooth muscle cells *in vitro* and are frequently employed in cell culture systems to preserve stem and precursor cells in an undifferentiated state during cell culture. Proteins found in extracellular matrix can also be utilized to facilitate 3D cell culture *in vitro* in order to model the development of tumors.

ECM biomaterials are a class of biomaterials made by processing human or animal tissues so that some of the matrix is retained.

5.5. CELL MOVEMENT

Many biological processes, including wound healing, cancer metastasis, immune response, and organism development (morphogenesis), depend on the highly dynamic phenomenon of cell motility. For instance, during morphogenesis, dividing cells are directed to particular locations to form tissues and organs. In order for wound healing to take place, cells called fibroblasts—which are connective tissue cells—and neutrophils—which are white blood cells and bacteria-consuming cells—must migrate to the wound site in order to destroy the microorganisms causing infection. (Dogterom & Yurke, 1997).

Cells in each of these instances crawl to their destination. Other forms of motility include the majority of sperm cells' ability to swim and the ability of certain bacteria to move by rotating their flagellar motors. However, the majority of motile eukaryotic animal cells use cell crawling as a common mechanism.

Even though Van Leeuwenhoek first noticed movement in cells in 1675 when he observed them crawling across his microscope slide, the molecular mechanisms underlying cell movement have only recently come under scientific scrutiny. Developments in fluorescence microscopy, molecular biology, and biochemistry have made it possible to identify the key proteins responsible for these processes and to uncover the mechanisms underlying motility. However, neither the physical mechanisms involved nor the question of whether these proteins are able to produce the forces necessary for motility can be sufficiently explained by these experimental techniques alone. Significant progress was made in this direction when biophysical studies were able to determine the locations of various force-generating proteins, measure the precise forces produced by some of these proteins *in vitro*, and measure the forces related to movement *in vivo*. Complementing the experimental work, theoretical investigations and computational modeling have helped quantify the ways in which forces generated at the molecular level and the suggested mechanisms combine to produce whole cell movement. The physical underpinnings of the forces that control a single cell's ability to crawl are highlighted in this review, which outlines the current state of knowledge on cell movement as a result of these experimental and theoretical investigations. In order to make the processes outlined more clear, it also includes footnotes referencing a number of websites with movies and animations.

A cell experiences both internal forces produced by the cytoskeleton and external forces such as the viscous force or resistance from the surrounding medium and cell-substrate interaction forces as it moves on a substrate (the extracellular matrix if the cell moves inside an organism or a cover slide if it moves outside an organism). The cytoskeleton plays a crucial role in producing these forces that propel motility and in coordinating the entire movement process in the majority of animal cells. Actin, microtubules, and intermediate filaments are the three different biopolymer types that make up the cytoskeleton, which is a polymer network. The persistence length L_p, which characterizes the rigidity of these biopolymers, serves as a key differentiator. The persistence length, which rises with increasing stiffness, is the distance that thermal forces cause the filament to bend.

L_p ~17 μm, actin filaments (AFs) are semiflexible polymers. They have a diameter of about 7 nm, are functionally polar, and are composed of dimer pairs of globular actin monomers. This indicates that they have two separate ends, known as the plus end and the minus end, respectively, which grow quickly and slowly. Actin monomer concentrations at the minus and plus ends are approximately six times higher, with a critical value of ~0.6 μM and ~0.1 μM, respectively. The filament end of an AF binds monomers and grows by polymerization when it is exposed to a concentration of monomeric actin above its critical concentration. On the other hand, monomers separate from the filament end and the filament shrinks through depolymerization when the concentration is below the critical concentration. All AFs need to do to grow asymmetrically is to have these two distinct critical actin concentrations at the opposite ends of the filament. When the actin monomer concentration is in the middle, only the plus end grows and the minus end shrinks. This process, called treadmilling, is a crucial part of how polymerizing AFs can produce force. It occurs when the length of the filament remains relatively constant and the polymerized monomers within the AF transfer momentum forward due to asymmetric plus end polymerization. Depending on the filament length, microtubules (MTs) can have a length of 100–5000 μm. MTs are the stiffest biopolymers. MTs are rod-shaped polymers, about 25 nm in outer diameter. Subunits of the tubulin protein assemble into protofilaments. Usually, 13 of these protofilaments line up to form a hollow tube, which gives MTs their remarkable stiffness. Actin-like MT dynamics include functional polarity, treadmilling, and the ability to polymerize and exert force. Intermediate filaments (IFs) have a much higher degree of flexibility (L_p~0.3–1.0 μm) than AFs and MTs. Their diameter falls between that of AFs and MTs, ranging from 8 to 12 nm. Different cell types have different IFs, and there are several classes of IFs, including vimetin, desmin, keratin, lamin, and neurofilaments. IFs are thought to be more static in nature than AFs and MTs because they are not polarized, do not treadmill, and, once polymerized, do not typically depolymerize under physiological conditions. (Block et al., 2003).

By joining these filaments via entanglements and also crosslinking, bundling, binding, motor, and other proteins, these three types of biopolymers create the cytoskeleton, an internal scaffolding of the cell that is ordered and coherent. Subsequently, these cytoskeletal assemblies cooperate as a composite, dynamic material in the processes of cell division, transport of organelles, and motility. When it comes to motility, the actin cytoskeleton is thought to be the primary engine that propels cell protrusion, the initial stage of movement, even though the other polymer assemblies within the cell also help to coordinate movement and power translocation. It is also necessary to accomplish the two other stages of movement, which are the translocation of the majority of the cell and the adhesion of the leading edge and deadhesion at the cell body and rear. The actin cytoskeleton is incredibly dynamic, and cells can easily rearrange their actin structures to change how they move in response to their surroundings. Rapid changes in the cell's elastic properties are made possible by the actin cytoskeleton's constant restructuring and the movement of actin structures from one to another. This dynamic response is essential for movement.

In vivo, AFs can organize into a variety of structures, including bundles and networks. At the leading edge of cells, mesh-like actin networks made up of short, crosslinked AFs are predominantly observed. It is thought that pushing the cell forward requires the expansion of these meshworks, or the ongoing synthesis of new actin networks at the leading edge. Several accessory proteins assist in the formation of this network. Nucleator proteins, such as the arp2/3 complex, are able to start the polymerization and assembly of new actin filaments when activated proteins, such as WASp, are present. Proteins that promote actin depolymerization, such as cofilin, can also help networks expand. Actin filaments are severed by cofilin, also referred to as Actin Depolymerizing Factor ADF, which generates new plus ends necessary for the formation of new actin filaments. While crosslinking and bundling proteins, such as filamin, α-actinin, and fascin, aid in the formation of connected actin networks, actin binding proteins, such as profilin and thymosineβ-4, maintain a constant actin monomer pool for polymerization. While severing and fragmenting proteins, such as gelsolin and severin, cut actin filaments and networks, capping proteins, such as CapZ, control filament length by attaching to actin filament ends and preventing further polymerization. Together, these proteins facilitate the formation of actin networks and lead to leading edge motility in multiple stages.

Actin bundles play crucial structural and sensory roles in cell movement. They are made up of parallel arrays of individual AFs that are tightly packed and crosslinked by proteins like fascin, fimbrin, and scruin. These actin bundles frequently link distal points of adhesion, allowing tension to spread throughout the cell and facilitating force application and movement of the cell on the substrate. With their ability to distribute forces and positively reinforce adhesion sites, they are aptly referred to as stress fibers. Actin bundles called filopodia may protrude past the lamellar edge in certain cells, serving as mechanical and chemical sensors as well as helping the cell move through tissue.

AFs interact with myosin motors to produce motility forces as well. The head, neck, and tail regions make up myosin motors; some have just one head and neck, while others have two. While the tail region is primarily thought to be for connecting to cargo, such as other myosins, vesicles, or filaments, the head/neck region is in charge of attachment and force production. The three basic steps that myosin motors use to work on actin filaments are binding, power stroke, and unbinding. The acto-myosin contractile force, which is produced as a result of this process' constant repetition, is believed to be crucial for propelling the majority of the cell forward during movement.

The microtubules in the majority of cell types exhibit a hub and spoke arrangement as they primarily extend radially from the centrosome to the actin network at the cell periphery, with their plus ends pointing towards the cell edge. These microtubules help to determine which way cells move.

Cells are given structural integrity by intermediate filaments, which form a fibrous network that spans the interior of the cell and joins the nucleus to the cell membrane. Long believed to be less important for cell movement because of their more static characteristics, IFs are actually more dynamic than previously believed. This is because rapid cytoskeleton reorganization is necessary for cell movement, even though IFs may not be essential for it.

5.5.1. The Process of Cell Movement

When a cell receives an external signal from its surroundings, it starts to move. This signal can be chemical, physical, diffusible, or non-diffusible. It is picked up by receptor proteins on the cell membrane and sent into the interior of the cell by signaling cascades. It is thought that a cell can detect the direction of a signal by spatially identifying external gradients, such as those present in slime mold, yeast, or white blood cells. This is achieved by receptor proteins becoming more concentrated on the side of the cell where the signal is present.

In most cells, there are three stages to the process of cell movement (Figure 5.5), which entails the actin cytoskeleton's constant restructuring. The actin network at the membrane's leading edge is first oriented and reorganized (grown) by a cell in order to advance the membrane. Second, it deadheres (releases) at the cell body and rear of the cell after adhering to the substrate at the leading edge. Ultimately, the cell is propelled forward by contractile forces, which are primarily produced by the acto-myosin network.

Once the signal is detected, the cell polymerizes actin to initiate movement in response to the signal. Actin, for instance, polymerizes in the cell's vicinity of the signal if it is a chemoattractant; conversely, if it is a chemorepellant, the cell moves away by polymerizing actin on the other side. The cell continuously senses the direction of the signal as the extending edge advances and modifies its motion in response. A cell that follows an object it wants to swallow or a cell that moves in reaction to a chemoattractant are two excellent examples of this signal tracking in action. Adhesion molecules gathered in the extending region assist in attaching the leading edge to the substrate shortly after it starts to protrude. At the leading edge, cell-substrate attachments are formed when actin bundles use adhesion molecules to connect the cytoskeleton to specific substrate sites. The leading edge that protrudes cannot retract thanks to these attachments. The cell deadheres at the cell body and rear as it adheres at the leading edge, maybe as a result of the attachments' disintegration or contraction (actin bundles) (Gittes et al., 1993).

Ultimately, contractile forces generated by myosin motors sliding on actin filaments, which are located in the cell body and at the back, pull the remaining portion of the cell forward. As the cell moves on the substrate, all of the previously mentioned stages or processes are ongoing, with the actin cytoskeleton changing from a solution-like viscous material (sol) to an elastic material that resembles a solid (gel). Cell movement depends on these transitions, which are known as gel-sol transitions. They are most likely brought on by the continuous depolymerization and disassembly at the cell's back and net actin polymerization and network assembly at the leading edge. Motor drive may also be involved in the actin network's transition to a more fluid-like state. Localized alterations in the cell's elasticity during motion are caused by these processes. In summary, numerous proteins mechanically alter the cytoskeletal structure and the force production centers to coordinate the entire process of cell movement both spatially and temporally.

5.5.2. The Forces of Cell Movement

1. **Polymerization Mechanics**: The leading edge protrudes first in the movement of the cell. The primary active mechanism thought to propel

the leading edge is undoubtedly the polymerization of actin filaments in the direction of the cell membrane. However, leading edge propulsion is undoubtedly a multi-step, intricate process. Actin filament polymerization by itself, in the absence of any auxiliary motors, can produce sufficient force to move the leading edge of a cell. Furthermore, filament-substrate adhesions are necessary to stop actin filaments from polymerizing backward.

Ratchet models and autocatalytic models are the two primary model types that have been put forth to explain the force generation caused by actin polymerization (polymerization force).

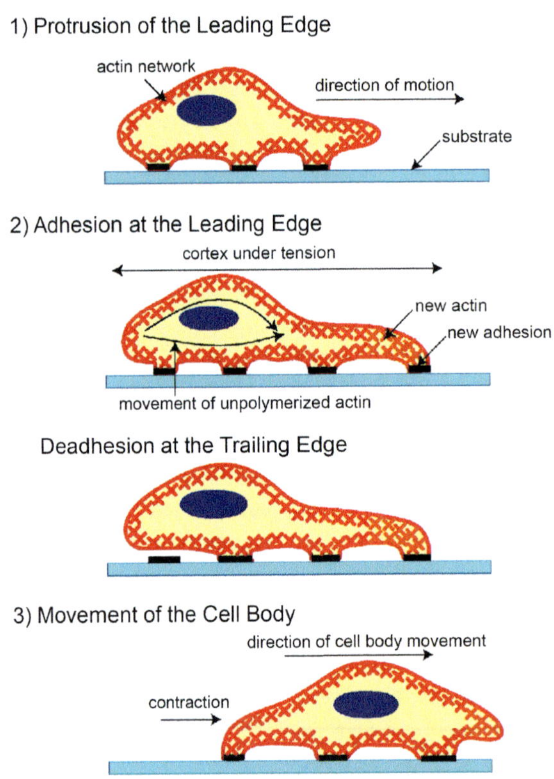

Figure 5.5. *A schematic of the three stages of cell movement: after determining its direction of motion, the cell extends a protusion in this direction by actin polymerization at the leading edge. It then adheres its leading edge to the surface on which it is moving and de-adheres at the cell body and rear. Finally, it pulls the whole cell body forward by contracile forces generated at the cell body and rear of the cell.*

Source: Revathi Ananthakrishnan and Allen Ehrlicher, The Forces Behind Cell Movement, ISSN 1449–2288 www.biolsci.org 2007, 3(5), 303–317.

The following sums up the fundamental operation of ratchet models. An actin filament bumping into the membrane would stop polymerizing and be unable to push against it or exert force if the membrane were fixed and immovable. However, because of the membrane's small size scale and relative flexibility, it experiences random thermal fluctuation and perpetual Brownian motion instead of remaining stationary. Furthermore, an actin filament is an elastic filament that can bend in response to a load rather than a rigid, immobile rod that ceases growing once it reaches the membrane. A 2 point 7 nm-sized actin monomer can easily insert itself between the filament and membrane if the filament bends away from the membrane at an angle greater than about 30° and is long enough (more than 70 nm) without buckling. The membrane can then be pushed forward by the extended filament by applying an elastic force to it. Every monomer addition causes the membrane to ratchet, suggesting that this mechanism prevents the membrane from moving backward and guarantees that the cell edge is only moving forward overall. The Elastic Brownian Ratchet Model is the name of this model, which was proposed to explain force generation by a single actin filament that is polymerizing. The Tethered Elastic Brownian Ratchet Model is an extension of the Elastic Brownian Ratchet Model. It takes into account the transient attachment of actin filaments to the membrane or surface and distinguishes between two types of filaments: attached filaments, which are unable to exert force on the surface, and working filaments, which are not attached to the surface and can exert force on it. When attached filaments separate from the surface, they become working filaments, and when working filaments are capped, they become non-working (i.e., attached) filaments. It is important to note that the Tethered Elastic Brownian Ratchet Model, which is based on the Dendritic-Nucleation Model and its branching mechanism, makes the assumption that new actin filaments/branches are generated independently of existing branches. Based on this supposition, the model aims to comprehend the production of force from a population of filaments within an actin network. Figure 5.6 illustrates the schematic showing how the cell adheres to the substrate.

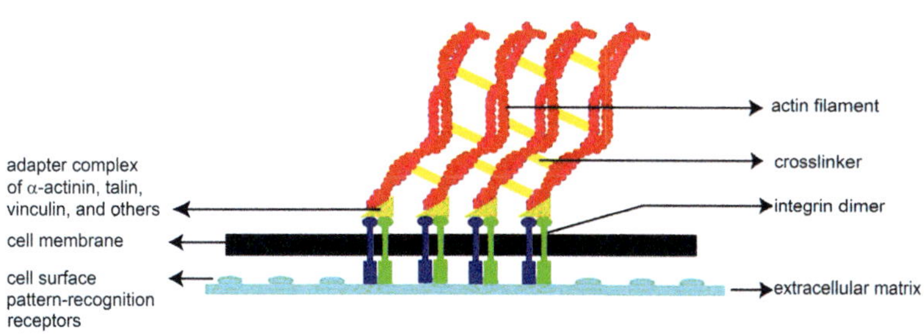

Figure 5.6. *A schematic showing how the cell adheres to the substrate. Cell-substrate attachments are formed when actin bundles connect to the substrate at certain sites via adhesion molecules such as vinculin, talin and integrin.*

Source: Revathi Ananthakrishnan and Allen Ehrlicher, The Forces Behind Cell Movement, ISSN 1449–2288 www.biolsci.org 2007, 3(5), 303–317.

The greatest force produced by a single polymerizing actin filament can be estimated theoretically using the above-mentioned ratchet mechanism and thermodynamic principles to determine the energy of monomer addition to a point of polymerization. By calculating the energy needed to add a monomer to the filament tip, one can estimate the maximum force (stall force) that a filament can develop by equating the work required to push a load through a distance δ. This is given by the thermodynamic expression $F_{max} = (k_B T/\delta)\ln(A_* k_{on}/k_{off})$, where $k_B T$ is thermal energy, δ is the size of an actin monomer, kon and koff are the association and dissociation constants for an actin monomer adding to or falling off from a polymer tip, and A is the monomeric actin concentration. If typical values are substituted in the above equation for an actin filament and for the monomeric actin concentration in the cell, the maximum force generated by a single actin filament is 5–7 pN. Although this calculation is simplistic, it can be used to estimate the force generated at the leading edge. Aggregating the predicted force of ~5–7 pN across the hundreds of actin filaments per micron thought to exist at the leading edge of a cell, an estimate for the polymerization force at the leading edge is several piconewtons per micron (or is on the order of nanonewtons per micron) – a force large enough to tackle the membrane load or resistance.

As mentioned, however, the contribution of motor proteins to force generation is not taken into account in the force calculations using the ratchet models. Recent work by Dickinson and others concludes that the forces predicted by the ratchet models alone are insufficient to account for the observed polymerization forces in these systems, based on studies in the bacterial pathogen Listeria and other experimental systems. It demonstrates how much greater forces can be produced by taking into account the function of end tracking motors, which are motors located at the tips of actin filaments and are capable of effectively converting free energy into work even in the event that filaments are permanently tethered.

Actin filament and actin network force generation is also investigated by the Autocatalytic Model. The model, which is the same as the foundation of the Elastic Brownian Ratchet Model, assumes that the load diffuses and that the probability of monomer addition to a polymer tip includes a Boltzmann factor in order to explain force generation by a single filament. Two methods have been used to study the problem of force generation by a network: a deterministic method that uses an explicit rate equation for filament orientation distribution, and a numerical method that uses stochastic simulations of an actin network growing against a load while tracking the positions and orientation of all actin filaments over time. The primary premise shared by both methods is that actin branches regenerate from preexisting ones, in contrast to the Tethered Elastic Brownian Ratchet Model's premise. Both Autocatalytic Model approaches ignore the actin filaments' temporary attachment to the surface. The dependence of growth velocity on capping rate, load force, branching, and other parameters is studied using stochastic simulations. The simulations demonstrate that the growth velocity has an inverse linear relationship with the capping rate for a fixed actin concentration. It also equals zero when the number of branches is less than 1.5. However, these simulations could only be run for a limited range of parameters that are computationally viable. In order to investigate a wider parameter range and rigorously test the generality of the earlier numerical results, the deterministic model was created. Furthermore, the Tethered Elastic Brownian Ratchet Model's underlying assumption—that new branches are generated independently of existing branches—can be readily accommodated by modifying the deterministic equations. Thus, Carlsson (2003) compares the predictions of the autocatalytic and ratchet models using the same fundamental set of equations.

2. **Adhesion Mechanics:** The actin network polymerizes and the extending network adheres to the substrate through integrins and other adhesion molecules, causing the leading edge of the cell to protrude. Research is still ongoing to determine the precise distribution of cell-substrate attachments, the force that protruding cells in different cell types adhere to the underlying substrate, and changes or gradients in this adhesion force within the leading edge as one moves perpendicularly away from the tip of the leading edge. However, the adhesive force and the spatial distribution pattern of attachments differ between cell types and are significant determinants of each cell type's protrusion and translocation rates, or the speed at which the majority of the cell or cell body moves. Part of the polymerization force is converted into protrusion by the adhesions close to the leading edge of the cell; the remaining portion of the force is used to propel the actin network. Take into consideration the example of a neuronal growth cone to comprehend what actin flow means. In this instance, the actin network there is propelled forward as a result of polymerization at the leading edge. Centripetal actin flux, also referred to as actin flow, is the relatively constant rate at which polymerizing actin filaments move away from the leading edge during this process.

Retrograde flow is a phenomenon that regulates the actin flow in many migrating cells, including neuronal growth cones, fibroblasts, and fish keratocytes. This phenomenon is crucial in determining the cell's translocation rate through adhesion mechanics, as will be discussed in the section that follows. The variable movement of actin filaments rearward with respect to the substrate—typically in a direction opposite to that of the cell—is referred to as retrograde flow. Retrograde flow is primarily driven by myosin motors and actin polymerization. Myosin motors' role in retrograde flow is still being investigated because it is unknown which myosin subtypes are essential in many systems. Experimental studies on fibroblasts and dictyostellium have shown that actin polymerization contributes to the generation of retrograde flow by providing a constant actin source and by pushing the lamellipodial actin network backward. Retrograde flow is thought to be produced by myosin motors through contractile cytoskeletal forces, or inward pulling forces. The following are some observations about myosin motors and retrograde flow: it has been found that myosin II motors are crucial for producing retrograde flow in fish keratocytes. The different roles that myosin plays in growth cone movement are highlighted by the findings that myosin IIA, IIB, and Ic contribute to retrograde flow in neuronal growth cones, while myosin IIB and V contribute to growth cone spreading. Similarly, it has been determined that myosin IIA plays a critical role in retrograde flow and cell spreading in fibroblasts; however, the deletion of myosin IIB has no effect on either of these processes. In conclusion, all of these studies show that movement and retrograde flow are critically influenced by both molecular motors and polymerization; however, the relative contributions of polymerization and myosin motors to retrograde flow may vary depending on the type of cell.

3. **Translocation of Cell Body and Rear of the Cell or Retraction Mechanics**: The rear of the cell and the cell body are moved as the last stage of movement. It is generally accepted that the retraction force needed for this process is produced by myosin motors, such as myosin II or Ic, sliding on actin filaments or actin bundles

in the cell body and rear. The force produced can be transformed into traction forces that allow the cell to move forward because the actin filaments and bundles are attached to both the substrate and the cell membrane. Based on multiple studies investigating the function of myosin motors in retraction, the acto-myosin network contraction is hypothesized to be the source of the retraction force. It should be noted, though, that contractility can occur elsewhere due to cortical tension surrounding the cell, not just at the rear. An imbalance between adhesion and contraction at the back lifts the back. For instance, by knocking out myosin II in dictyostellium cells, one experimental study examined the function of myosin II motors in retraction. The study shows that myosin minus cells move slower and do not re tract as well as their normal counterparts do on adhesive surfaces. Another study using keratocytes from fish demonstrates that when myosin contraction is inhibited, the cell body stops and a lamellar fragment is produced that moves forward. These investigations show that the movement of the cell body and the retraction of the trailing edge are significantly facilitated by conventional myosins.

Estimates for the acto-myosin contractile force are also provided by studies on the acto-myosin network; given the thousands of myosin motors at the back, each of which generates approximately 1 pN of force, it is comparable in magnitude to the polymerization force at the leading edge. The contractile force generated by myosin is probably the main source of the retraction force, though it might not be the only one. Additional explanations for the retraction force include transport mechanisms, which entail the cell being pulled by myosin motors along actin filaments that are oriented in a track-like fashion. Another way that the actin gel in some cell types can produce contractile forces is through a process called solation, which is a decrease in gel structure that may indicate that the network is disassembling due to filament severing, crosslink breaking, or filament unbundling. A few piconewtons of contractile force can be produced by a solating actin gel per actin filament.

Recent studies have been conducted to better understand the retraction force in general and the generation of a contractile force by solation. Studies have used crawling nematode sperm cells to simulate retraction *in vitro*. These cells provide an excellent model system in which to study crawling because they are simple prokaryotic cells that use major sperm protein (MSP) to move like eukaryotic cells do. Unlike eukaryotic cells, which use actin for a variety of functions, these cells use MSP exclusively for cell movement. The study specifically used vesicles that were encircled by cytoplasm and included MSP. These vesicles were derived from the cutting-edge membrane of nematode sperm cells. When Yersinia thyrosine phosphatase (YOP) was present, the researchers found that the disassembly or unbundling of MSP bundles attached to the vesicle resulted in the contraction of individual filaments and a contractile force.

They discovered that retraction forces can be explained solely by taking into account cytoskeletal dynamics (without taking into account motors), which suggests a different mechanism for producing a contractile force at the back of the cell. Motivated by these data, a recent model further elucidates how the force required to pull the cell body forward is generated exclusively by cytoskeletal disassembly. The model comprises two steps, which can be qualitatively described as follows: The first step is the detachment of filaments from the bundle. With an effective persistence length of N^2L_p, a bundle of N filaments, each with a persistence length of L_p, is significantly stiffer than a single filament. Its effective persistence length and rigidity (which is directly proportional to L_p) decrease when the filaments separate from the bundle (caused by YOP), which causes the bundle to contract as a result of an increase in entropy. Depolymerization, or the separation of monomers from filaments, is the subject of the second step. Therefore, the retraction forces resulting from disassembly can be quantified using this kinetic model for disassembly that incorporates polymer entropy. Future research is necessary to determine the significance and relative role of this mechanism in producing retraction forces *in vivo*, for example, in relation to forces generated by myosin.

5.6. MORPHOGENESIS

The biological process called morphogenesis is what gives a cell, tissue, or organism its shape. It is one of the three core concepts of developmental biology, along with the regulation of cell differentiation and tissue growth.

The method regulates how cells are arranged in space while an organism is developing embryonically. Morphogenesis can occur in an adult organism as well, for example, when stem cells are responsible for the regular upkeep of tissue or the regeneration of damaged tissue. One example of severely aberrant and pathological tissue morphogenesis is cancer. The development of unicellular organisms without an embryonic stage in their life cycle is also referred to as morphogenesis. For new forms to evolve, morphogenesis is necessary.

Morphogenesis is a mechanical process that involves forces that cause mechanical stress, strain, and cell movement. Genetic programs based on the spatial arrangement of cells within tissues can induce morphogenesis. Dysmorphogenesis is the term for abnormal morphogenesis.

5.6.1. Genetic and Molecular Basis

The process of morphogenesis involves several kinds of molecules. Soluble molecules known as morphogens have the ability to diffuse and convey signals via concentration gradients that regulate cell differentiation. Usually, morphogens work by attaching themselves to particular protein receptors. Transcription factor proteins are a significant class of molecules involved in morphogenesis that interact with DNA to determine a cell's fate. These secondary gene products can control the expression of still other genes in a regulatory cascade of gene regulatory networks. They can be coded for by master regulatory genes and can either activate or deactivate the transcription of other genes. Classes of molecules that regulate cellular behaviors like migration or, more broadly, their characteristics like adhesion or contractility, are at the end of this cascade. For instance, during the process of gastrulation, clusters of stem cells cease to adhere to one another, migrate, and occupy new locations within the embryo where they reactivate particular cell adhesion proteins to form new tissues and organs. The developmental signaling pathways Wnt, Hedgehog, and ephrins are associated with morphogenesis (Kouros-Mehr & Werb, 2006).

5.6.2. Cellular Basis

Ignoring the mechanisms of control, morphogenesis results from cellular motility and proliferation at the tissue level. Moreover, morphogenesis entails modifications to the cellular architecture or the interplay of cells within tissues. Tissue elongation, thinning, folding, invasion, or division into discrete layers can all be caused by these alterations. This latter scenario is commonly known as "cell sorting.". Cells move to sort into clusters that maximize contact between cells of the same type, a process known as cell sorting out. Malcolm Steinberg's differential adhesion hypothesis suggests that the reason cells are able to do this is because of differential cell adhesion. More dramatic cellular differentiation events, in which epithelial cells transform into mesenchymal cells, can also cause tissue separation. Changes in the cell's adhesive and contractile qualities usually cause mesenchymal cells to migrate out of the epithelial tissue. Cells can migrate away from an epithelium after the epithelial-mesenchymal transition and then associate with other similar cells in a new location. Cellular morphogenesis in plants is closely related to the molecular makeup and mechanical characteristics of the cell wall.

5.6.2.1. Cell-to-Cell Adhesion

Differential affinities confine cells to distinct layers during embryonic development. When cells have the same cell-to-cell adhesion molecules, this can happen. Homotypic cell adhesion, for example, can preserve the boundaries between cell types with distinct adhesion molecules. Additionally, cells can separate based on variations in adhesion, meaning that two populations of cells with varying concentrations of the same adhesion molecule can still separate. In a mixed cell aggregate, the most adherent cells in the culture move to the center. Furthermore, cell contractility frequently regulates cell-cell adhesion, exerting forces on the cell-cell contacts to facilitate the sorting of two cell populations that have equal amounts of the same adhesion molecule. Cell adhesion molecules (CAMs) are the molecules that cause adhesion. There are many different kinds of cell adhesion molecules, and cadherins are a well-known class of these molecules. Numerous cadherins exist, and they are expressed by various cell types. Similar to other cadherins, E-cadherin, which is present on many epithelial cells, binds to other E-cadherin molecules preferentially. Typically, mesenchymal cells express N-cadherin and other cadherin types.

5.6.2.2. Cell Contractility

Through the use of cell contractility, tissues are able to alter their shape and divide into separate layers. Myosin has the same ability to contract various cytoplasmic regions to alter their structure and form as it does in muscle cells. In the model organisms Caenorhabditis elegans, Drosophila, and Zebrafish, myosin-driven contractility is observed during the separation of germ layers in embryonic tissue morphogenesis. In the process of embryonic morphogenesis, periodic pulses of contraction are common. A bistable organelle located at the apical end of each cell initiates the alternating contraction and expansion of cells in a model known as the cell state splitter. Microtubules and microfilaments are in mechanical opposition to one another within the organelle. It reacts to morphogenetic movements' local mechanical perturbations. Following cell differentiation, these induce traveling waves of contraction or expansion during embryonic differentiation over putative tissues that determine cell type. First put forth to explain neural plate morphogenesis during axolotl gastrulation, the cell state splitter model was later extended to encompass all of morphogenesis.

5.6.2.3. Branching Morphogenesis

The respiratory tree is formed when a bronchus splits off into bronchioles during lung development. Each bronchiolar tube's tip bifurcates to cause branching, which in turn forms the bronchi, bronchioles, and eventually the alveoli through the process of branching morphogenesis.

There is also evidence of branching morphogenesis in the mammary gland's ductal formation. While primitive duct formation starts during development, the duct system's branching formation starts later during puberty in response to estrogen and continues to develop in step with the development of the mammary glands.

CASE STUDY

THE SMALL GTP-BINDING PROTEIN RHO REGULATES CORTICAL ACTIVITIES IN CULTURED CELLS DURING DIVISION

Cytokinesis is believed to involve localized cortical contraction through interactions of actin and myosin II filaments. After chromosomal separation at anaphase onset, cortical actin and myosin II begin to move toward the equatorial region, where they form a contractile cleavage furrow. However, the cleavage process may involve more than actin–myosin II interactions along the equator. For example, detailed three-dimensional studies of the organization of actin and myosin do not support a simple "purse string" model in cultured mammalian cells and Dictyostelium. In addition, myosin II–null Dictyostelium are capable of initiating and completing a furrow when attached to a substrate, and the budding yeast Saccharomyces cerevisiae can undergo cytokinesis in the absence of myosin II or an intact F-actin cytoskeleton.

Equally important are the signaling events that coordinate cortical dynamics with chromosomal separation. Previous studies have indicated that Rho, a ubiquitous protein of the Ras family, plays an important role in regulating the actin cytoskeleton. In interphase Swiss 3T3 fibroblasts, Rho regulates the actin cytoskeleton by promoting the assembly of stress fibers and focal adhesions, possibly through elevating the level of phosphorylation of the myosin II regulatory light chai. However, there are additional, poorly characterized effectors of Rho, which might account for the variability of observations among different cell types.

Several studies with embryos indicate that Rho is required for the reorganization of actin cytoskeleton and the maintenance of cleavage activities during cell division. Inhibition of Rho prevented cytoplasmic division from taking place and caused existing furrows to regress. Inhibition of one of the downstream effectors of Rho, a novel protein kinase that localizes to the cleavage furrow, also caused inhibition of cytokinesis. These observations, together with the disassembly of stress fibers and focal contacts during cell division, lead to the hypothesis that Rho may be deactivated as cells enter mitosis. Its reactivation during cytokinesis may then be responsible for the generation and maintenance of cell ingression. We have investigated the role of Rho in cultured cells during division by microinjecting C3 ribosyltransferase, a Clostridium botulinum toxin that specifically inactivates Rho by ADP ribosylating its Asp41 residue, into mitotic normal rat kidney (NRK) epithelial cells, Swiss 3T3 fibroblasts, and HeLa cells. We found that C3 inhibited the cleavage of only poorly adherent HeLa cells. The majority of firmly attached NRK cells cleaved successfully, but showed irregular ingressions without an apparent concentration of F-actin or myosin II along the equator. Our results suggest that Rho may be involved in defining the spatial and temporal pattern of cleavage. In addition, the intriguing adhesion-dependent responses of dividing cells to C3 provide new insights into the mechanism of cytokinesis.

Materials and Methods

Cell Culture

A well spread subclone (NRK-52E) of NRK epithelial cells, Swiss 3T3 fibroblasts, and HeLa cells (American Type Culture Collection) were cultured on glass coverslips as described previously. NRK cells were grown in Kaighn's modified F12 medium (Sigma Chemical Co.) containing 10% FCS (JRH Biosciences), 1 mM l-glutamine, 50 µg/ml streptomyosin, and 50 U/ml penicillin. 3T3 fibroblasts and HeLa cells were cultured in Dulbecco's modified Eagle's medium (Sigma Chemical Co.) supplemented with 10% donor calf serum (JRH Biosciences), l-glutamine, and antibiotics as above.

Protein Preparation and Microinjection

C3 transferase was either purchased from Calbiochem or received as a gift from Dr. S. Narumiya (Kyoto University, Kyoto, Japan). The cells responded the same to C3 from both sources. It was stored as 1.0 mg/ml aliquots in 5 or 10 mM Hepes at –80°C and was microinjected at concentrations ranging from 0.2 to 1.0 mg/ml with fluorescein dextran as a marker (Molecular Probes). Injection of fluorescein dextran alone had no effect on mitosis or cytokinesis. Rhodamine tubulin was prepared as described by Wheatley and Wang, and cells were microinjected as described previously.

pGEX RhoA, RhoA^{V14}, and RhoA^{L63} constructs were expressed in the Escherichia coli strain XL-1 blue and proteins were purified essentially as described in Self and Hall. Proteins were stored at –70°C in aliquots of 50 µl and activities were verified based on their ability to induce cell rounding as described by Allen et al., (1997). Wild-type RhoA was microinjected at a concentration of 2.0 mg/ml, and the dominant positive mutant proteins were microinjected at 0.1–0.5 mg/ml.

Labeling of Surface Receptors and Drug Treatments

Carboxylated polystyrene beads were prepared as described previously. Briefly, 0.1-µm yellow-green fluorescent beads (Molecular Probes) were washed three times by centrifugation in an Airfuge (Beckman Instruments) and resuspended by sonication in PBS with 1% BSA (Sigma Chemical Co.). Surface receptors were labeled by incubating cells with beads at 10–100× dilution for 2 min, followed by at least five rinses with medium.

Cells were treated with cytochalasin D by replacing the medium with 2 or 5 µM cytochalasin D in culture medium. Depolymerization of microtubules was induced by treating cells with 2.5 µM nocodazole. Both cytochalasin D and nocodazole were diluted into the medium immediately before use, from frozen stock solutions of 2 and 10 mM, respectively, in DMSO. In most experiments cells were microinjected with C3 at prometaphase or metaphase, then treated with the drugs at specified stages of division

Fixation and Staining

Fixation solutions for actin, tubulin, and myosin were prepared with cytoskeleton buffer. Actin filaments were preserved by fixing cells in warm 0.5% glutaraldehyde and 0.2% Triton X-100, followed by postfixation in 1% glutaraldehyde. Cells were treated with 0.5 mg/ml $NaBH_4$ to reduce autofluorescence and stained with 200 nM FITC-phalloidin or TRITC-phalloidin (Molecular Probes) for 30 min. Microtubules were preserved similarly, except Triton X-100 was at a concentration of 0.1%. Tubulin was immunostained using anti–β-tubulin mAb (Amersham Corp.) at a dilution of 1:10 and FITC or TRITC-conjugated secondary antibodies (Sigma Chemical Co.) at a dilution of 1:100. Myosin II was preserved by fixing cells for 1 min in warm 1% formaldehyde, 0.1% glutaraldehyde, and 0.3% Triton X-100, followed by postfixation for 15–20 min in 0.5% glutaraldehyde. Myosin II was immunostained with an anti–platelet myosin II antiserum (a gift from Dr. K. Fujiwara, National Cardiovascular Center, Osaka, Japan) at a dilution of 1:100.

Spindle poles were located by methanol fixation and subsequent staining with anti–γ tubulin (Sigma Chemical Co.) at a dilution of 1:1,000. Fixation and staining of Telophase Disc 60 protein (TD60) were performed as described by Wheatley and Wang.

Microscopy and Data Collection

Imaging of cells was done using either an Axiovert 10 or an IM35 inverted microscope (Carl Zeiss), with a ×40, N.A. 0.75 plan-achroplan phase-contrast lens or a ×100, N.A. 1.3 neofluar lens. Images were acquired and processed using a cooled CCD camera (model TE/CCD-576EM; Princeton Instruments) and custom designed hardware and software. Optical sectioning and three-dimensional reconstruction were performed with custom software and a computer-controlled stepping motor.

Results

C3 Transferase Induces Poorly Regulated Cleavage Activities

The effects of Rho on cell cortical activities were investigated by microinjecting 0.2 mg/ml C3 transferase, a specific inhibitor of Rho proteins. Identical results were observed using 1.0 mg/ml C3, indicating that the injection exerted saturating effects on cells. Upon exposure to the toxin, interphase 3T3 fibroblasts and isolated NRK cells retracted and produced long, dendritic processes, similar to what was documented in previous studies.

Both interphase and mitotic NRK cells within a monolayer showed a transient increase in surface area upon microinjection with C3 (Figure 1). Mitotic cells then developed an elongated cleavage furrow between separating chromosomes as if the cleavage activity were spread over a wide area. In addition, while most injected NRK cells completed cytokinesis, cleavage became poorly coordinated with mitosis. Ingressions in many cells started at random sites well before the onset of anaphase. Moreover, 57% of C3-injected cells developed randomly placed ectopic furrows, which lead to the formation of anuclear fragments. Cytokinesis failed in 45% of injected cells, not due to the lack of ingression but apparently to the disorganization of cortical activities. As for cleavage in normal cells, ingression in the presence of C3 was inhibited by cytochalasin D.

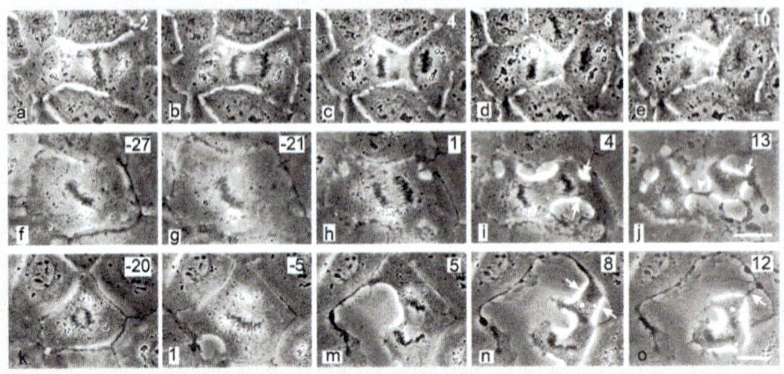

Figure 1. *Time lapse images of control NRK cells and cells microinjected with C3 transferase during prometaphase or metaphase.*

Source: https://rupress.org/jcb/article/144/2/305/16023/The-Small-GTP-binding-Protein-Rho-Regulates.

The effects of C3 were dependent on the timing of microinjection. When C3 was microinjected into NRK cells already undergoing cytokinesis, the cleavage continued to completion 100% of the time. In 53% of these cells cytokinesis showed no detectable abnormalities (Figure 2). The remaining cells showed some abnormal cortical activities outside the equator, without impairing the morphology or progression of the cleavage furrow.

Both transient spreading and irregular ingression induced by C3 appeared to depend on cell–cell adhesions and associated mechanical interactions. Isolated NRK cells and cells located at the edge of a colony showed neither transient increase in surface area nor extensive ectopic cortical activities along the free boundary, although the cleavage furrows were still elongated.

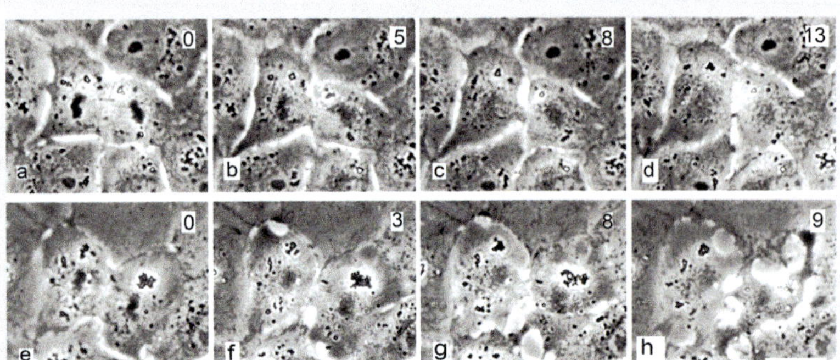

Figure 2. *Time lapse images of NRK cells microinjected with C3 transferase after the initiation of cytokinesis.*

Source: https://rupress.org/jcb/article/144/2/305/16023/The-Small-GTP-binding-Protein-Rho-Regulates.

To further investigate the role of cell adhesion, C3 was injected into 3T3 fibroblasts that round up during mitosis (Figure 3). Unlike NRK cells, no spreading was observed immediately after the microinjection of C3. The initiation of cytokinesis in C3-injected cells appeared indistinguishable from that of control 3T3 cells. However, the effects of C3 became clear as the cell started to spread out. The furrow elongated as in NRK cells, while other regions of the cell developed additional sites of ingression or ectopic furrows. Control cells never developed such broad or ectopic furrows. All C3-injected 3T3 cells completed cytokinesis.

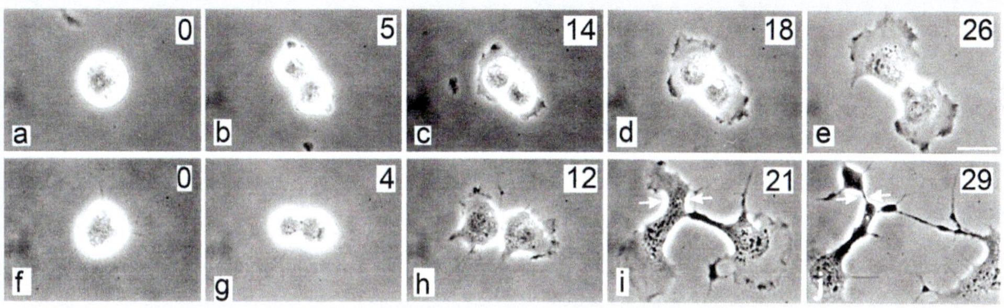

Figure 3. *Time lapse images of control 3T3 fibroblasts and cells microinjected with C3 at metaphase.*

Source: https://rupress.org/jcb/article/144/2/305/16023/The-Small-GTP-binding-Protein-Rho-Regulates.

C3 was also microinjected into HeLa cells that are very weakly adherent to the substrate and their neighbors (Figure 4). A large fraction of injected HeLa cells either detached from the dish during microinjection or failed to initiate anaphase. Of the limited number of cells injected with C3 that proceeded into anaphase, which we took as an indication of proper injection, all failed cytokinesis. Some of them showed clear signs of the initiation of cleavage, but never made further progress. HeLa cells that were microinjected with buffer alone were able to cleave successfully.

The opposite manipulation, artificial activation of Rho, can be achieved by microinjecting either excess wild-type or GTP-independent, dominant positive mutant forms. In the present study this approach was limited to one isoform, RhoA. Neither excess wild-type nor dominant positive mutants of RhoA caused visible effects on dividing NRK cells (not shown), even though the same proteins were found to induce rounding of interphase macrophages as reported by Allen et al. In addition, no effect was observed when cells were treated with lysophosphatidic acid, which activates Rho proteins in interphase cells.

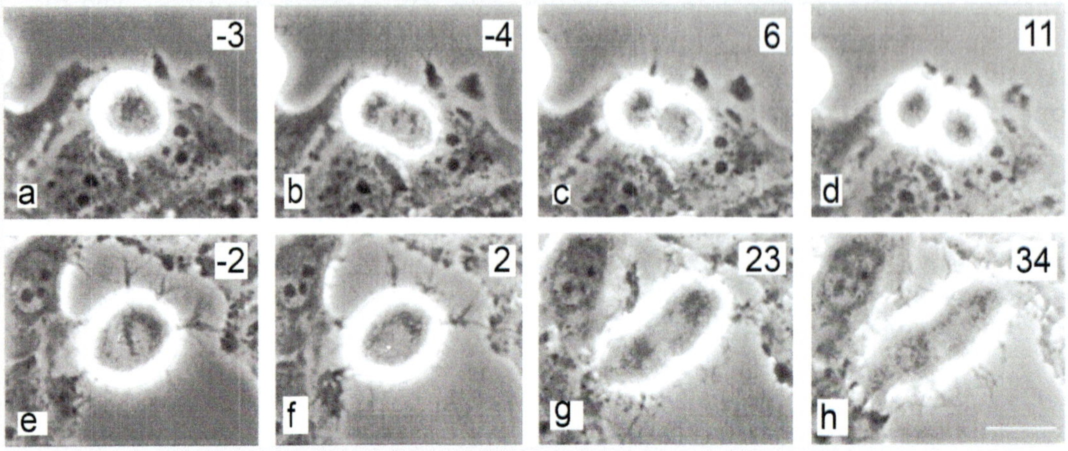

Figure 4. Time lapse images of control (a–d) and C3- injected HeLa cells (e–h).

Source: https://rupress.org/jcb/article/144/2/305/16023/The-Small-GTP-binding-Protein-Rho-Regulates.

Inhibition of Rho Causes Premature Cortical Flow in Dividing Cells

The reorganization of cortical actin during normal cell division is known to involve directional movements toward the equator, which can be easily followed by tracking associated surface receptors using charged fluorescent latex beads. In control NRK cells, such movement occurs only after anaphase onset, and is limited in a region roughly defined by the spindle interzone.

When cells were injected with C3 transferase at pro-metaphase or metaphase, surface-bound beads immediately started directional movement toward a central region of the cell, near the congregated chromosomes. In addition, the movements were observed over most of the, if not the entire, top surface. During anaphase, the cluster of beads either followed the movement of one set of chromosomes, or split into two and moved with each set, as if they were tethered to the chromosomes.

C3-induced Cortical Ingression Persists upon the Depolymerization of Microtubules

The above observations suggest that C3 may disrupt the coordination between mitosis and cytokinesis. In control NRK cells, depolymerization of microtubules before anaphase onset causes arrest of mitosis and inhibition of cytokinesis. Depolymerization after the initiation of the cleavage furrow causes it to regress or to follow an irregular path. However, all C3-injected NRK cells proceeded with cleavage when treated with the microtubule depolymerizing drug nocodazole after anaphase onset. Even when nocodazole was added at metaphase, broad ingressions developed despite the inhibition of mitosis. Unlike NRK cells, 3T3 cells showed no C3-induced cleavage activity in the presence of nocodazole.

C3 Transferase Causes No Direct Effect on Mitosis

Contrary to the dramatic effects on the cortex and actin organization, no apparent effect was observed in C3-injected cells with regard to chromosomal separation or nuclear envelope reformation, confirming that the effects on cortex were not caused by the disruption of mitosis or cell cycle. Moreover, imaging of microinjected fluorescent tubulin showed no change in microtubule organization during metaphase or anaphase, nor was there apparent effect on the elongation of astral microtubules. During telophase, wavy interzonal microtubules formed along the equator, most likely due to irregular cortical ingressions. In addition, the elongated furrows were probably responsible for the formation of an extended bundle of microtubules flanking the midbody. However, no prominent microtubule bundle was detected near ectopic furrows.

TD60 antigen, a chromosomal passenger protein that relocates to the equatorial cortex during telophase, has been implicated in signaling cytokinesis. The pattern of TD60 redistribution was unaffected by the microinjection of C3, and no concentration of TD60 was found near ectopic furrows. Similarly, there appeared to be no disruption to spindle poles, as visualized by staining with anti–γ-tubulin antibodies.

Discussion

All observations to date indicate that the mitotic apparatus directs the formation, initiation, and progression of the cleavage furrow in tissue culture cells. There is evidence that at least part of the signal originates from the chromosomal region at anaphase onset. However, except for several novel proteins such as TD60 and INCENP, few molecules have been identified that may mediate the coordination between mitosis and cytokinesis.

The Rho family of small GTP-binding proteins represents attractive candidates for regulating cortical activities during cell division. In smooth muscle cells, Rho is believed to promote contraction through the phosphorylation of myosin II regulatory light chains. In dividing echinoderm and frog embryos, inhibition of Rho prevented cytokinesis from taking place and caused preformed furrows to regress. In one simple model, Rho is deactivated as cells enter mitosis, causing the disassembly of focal adhesions and stress fibers. During anaphase and telophase, localized reactivation of Rho along the equator triggers actin–myosin II interactions and activates cytokinesis.

However, our observations suggest a considerably more complicated picture. We found that adherent NRK cells respond immediately to C3 injection during prometaphase and metaphase, suggesting that some Rho activities are not only maintained during mitosis but are required for a normal cell morphology. In addition, unlike embryos, most cultured cells cleaved successfully upon the injection of C3. Thus, either some actin–myosin II interactions can take place in the presence of high concentrations of C3, or a Rho-independent motor is involved in the cleavage of these adherent cells.

Inhibition of Rho also induces several striking effects on dividing NRK cells. First, cortical actin moved immediately toward the central region during metaphase, accompanied by a transient spreading of cells that adhered to their neighbors. Second, all injected cells showed cleavage or cortical ingressions, without an apparent concentration of actin or myosin filaments along the furrow. Third, the cleavage furrow became elongated, and ectopic furrows appeared in many cells, creating multiple anuclear fragments. Fourth, cleavage persisted even when mitosis was blocked with nocodazole. Since these effects were never observed in control cells, they cannot be easily explained as incomplete inhibition of an existing activity by suboptimal concentrations of C3. Moreover, identical results were obtained at increasing concentrations of C3 up to 1.0 mg/ml, which readily causes cell retraction and stress fiber disassembly in interphase cells.

The effects of C3 appeared to be highly dependent on cell adhesion. Ectopic cortical ingressions were observed before anaphase in well adherent NRK cells, but were not apparent in poorly adherent HeLa cells. Such effects took place in 3T3 cells only when the cells started to respread. In addition, while both NRK cells and respreading 3T3 cells developed elongated furrows, cleavage never even passed beyond an early stage in HeLa cells. Our results suggest that Rho is required for cytokinesis in the absence of traction forces, as in embryos and HeLa cells. However, Rho is involved in the spatial and temporal regulation of cortical ingression in the presence of adhesion forces, as in adherent epithelial cells.

Taken together, our observations cannot be easily explained with a model of Rho-activated cortical contraction, which would predict an inhibition of cytokinesis in all cell types by C3 as when cells were treated with cytochalasins. They indicate that Rho plays a more delicate role in coordinating the spatial and temporal pattern of cleavage with mitosis. One plausible explanation is that Rho regulates the mechanical integrity and strength of the cortex. Weakening of the cortex by C3 during metaphase causes the collapse of cortical actin toward the cell center, as well as an apparent spreading of the cell due to outward mechanical forces exerted by the neighbors. Subsequent interplay of cleavage forces and cell–cell, cell–substrate adhesive forces then leads to irregular ingressions and ectopic furrowing throughout a weakened cortex. This model suggests that cytokinesis of adherent cells is normally driven not only by localized contractions along the equator, but by localized weakening or "solation" of the cortex coupled to traction or adhesion forces.

The involvement of cortical disintegration in cytokinesis was first suggested based on electron microscopic observation of the thickness of the cortex. The cortex maintained a constant thickness despite the decrease in the diameter of the furrow, suggesting that materials are being removed continuously during the contraction. There is now evidence that cleavage can take place when such cortical disassembly is coupled to traction forces. In Dictyostelium, myosin II–null mutants cannot divide in suspension, but can divide when cells are adhered to the substrate, through a "traction-mediated cytofission" or "attachment-assisted mitotic cleavage". As in the present case, cleavage took place without an apparent concentration of actin or myosin II along the equator. It is important to determine the spatial and temporal distribution of Rho activities during cell division. We cannot rule out the possibility that high activities of Rho are present along the equator during the early stage of cytokinesis, as suggested by the localization of RhoA. However, the lack of effects with excess wild-type RhoA, constitutively active mutant forms of RhoA, or lysophosphatidic acid in the present experiments suggests that RhoA activity may normally be maintained at a high level in cultured cells, and that any localization of activities involves redistribution rather than net activation. Equally important is whether and how Rho activities decrease during subsequent stages of cytokinesis. Our observation that late injections of C3 resulted in the largely normal cleavage is consistent with the idea that Rho activities may decrease normally after furrow initiation, through either a GDP- or proteolysis-mediated pathway. This may lead to the solation of the cortex and allow the cleavage to proceed to completion.

Finally, it is important to determine the relationship between Rho proteins and signals that emanate from the mitotic spindle. Our observations indicate that the effects of C3 did not involve spindle microtubules, since both the mitotic spindle and an associated chromosomal passenger protein, TD60, appear to be unperturbed by the injection of C3 toxin. In addition, while normal cytokinesis requires the continuous presence of interzone microtubules, cortical responses to C3 transferase can occur without the proximity of these microtubules and can proceed after microtubule depolymerization. Therefore, Rho-mediated signals are likely to lie downstream from spindle signals. By manipulating directly the activities of Rho, we have likely caused a "short circuit" of the regulatory mechanism and induced cleavage independent of signals from the mitotic spindle.

CLASS ACTIVITY

- Ask students to create a 10–20 question quiz to test their understanding of concepts related to cell biology. Tests could be in the form of:
    - Multiple choice
    - Short answer
    - Essay
    - Blank diagram to label
- Remind students to create an answer key for their quizzes.
- Instruct them to swap back and grade each other's work.

SUMMARY

- The cell is the basic structural and functional unit of all forms of life. Every cell consists of cytoplasm enclosed within a membrane; many cells contain organelles, each with a specific function.
- Tissue is an assembly of similar cells and their extracellular matrix from the same embryonic origin that together carry out a specific function.
- Embryonic stem cells (ESCs) are pluripotent stem cells derived from the inner cell mass of a blastocyst, an early-stage pre-implantation embryo.
- Embryonic stem cells (ESCs), derived from the blastocyst stage of early mammalian embryos, are distinguished by their ability to differentiate into any embryonic cell type and by their ability to self-renew.
- Embryonic stem cells are derived from the inner cell mass of the early embryo, which are harvested from the donor mother animal.
- The cytoskeleton is a complex, dynamic network of interlinking protein filaments present in the cytoplasm of all cells, including those of bacteria and archaea.
- Eukaryotic cells contain three main kinds of cytoskeletal filaments: microfilaments, microtubules, and intermediate filaments. In neurons the intermediate filaments are known as neurofilaments.
- Microfilaments, also known as actin filaments, are composed of linear polymers of G-actin proteins, and generate force when the growing (plus) end of the filament pushes against a barrier, such as the cell membrane.
- The cytoskeleton is a highly anisotropic and dynamic network, constantly remodeling itself in response to the changing cellular microenvironment.
- Small GTP-binding proteins (G proteins) are monomeric G proteins with molecular masses of 20–40 kDa. The Ha-Ras and Ki-Ras genes were first discovered as the v-Ha-Ras and v-Ki-Ras oncogenes of sarcoma viruses around 1980.
- In biology, the extracellular matrix (ECM), also called intercellular matrix (ICM), is a network consisting of extracellular macromolecules and minerals, such as collagen, enzymes, glycoproteins and hydroxyapatite that provide structural and biochemical support to surrounding cells.
- Cell movement or motility is a highly dynamic phenomenon that is essential to a variety of biological processes such as the development of an organism (morphogenesis), wound healing, cancer metastasis and immune response.
- Morphogenesis is the biological process that causes a cell, tissue or organism to develop its shape. It is one of three fundamental aspects of developmental biology along with the control of tissue growth and patterning of cellular differentiation.

REVIEW QUESTIONS

1. What are embryonic stem cells used for?
2. How are embryonic stem cells different from other stem cells?
3. Differentiate between eukaryotic cytoskeleton between prokaryotic cytoskeleton.
4. Discuss the common features and differences between prokaryotes and eukaryotes
5. What do small GTP-binding proteins do?
6. Discuss the process of cell movement.
7. Describe the forces of cell movement.

REFERENCES

1. Aktories, K., Rosener, S., Blaschke, U., & Chhatwal, G. S. (1988). Botulinum ADP-ribosyl transferase C3. Purification of the enzyme and characterization of the ADP-ribosylation reaction in platelet membranes. *European Journal of Biochemistry, 172*, 445–450.
2. Block, S. M., Asbury, C. L., Shaevitz, J. W., et al. (2003). Probing the kinesin reaction cycle with a 2D optical force clamp. *Proceedings of the National Academy of Sciences USA, 100*, 2351–2356.
3. Cai, Y., Biais, N., Giannone, G., et al. (2006). Nonmuscle myosin IIA-dependent force inhibits cell spreading and drives F-actin flow. *Biophysical Journal, 91*(10), 3907–3920.
4. Dogterom, M., & Yurke, B. (1997). Measurement of the force-velocity relation for growing microtubules. *Science, 278*(5339), 856–860.
5. Euteneuer, U., & Schliwa, M. (1986). The function of microtubules in directional cell movement. *Annals of the New York Academy of Sciences, 466*(1), 867–886.
6. Fletcher, D. A., & Mullins, R. D. (2010). Cell mechanics and the cytoskeleton. *Nature, 463*(7280), 485–492.
7. Gittes, F., Mickey, B., Nettleton, J., et al. (1993). Flexural rigidity of microtubules and actin filaments measured from thermal fluctuations in shape. *Journal of Cell Biology, 120*(4), 923–934.
8. Helfand, B. T., Chang, L., & Goldman, R. D. (2004). Intermediate filaments are dynamic and motile elements of cellular architecture. *Journal of Cell Science, 117*(Pt 2), 133–141.
9. Herberts, C. A., Kwa, M. S. G., & Hermsen, H. P. (2011). Risk factors in the development of stem cell therapy. *Journal of Translational Medicine, 9*(29), 29.
10. Jurado, C., Haserick, J. R., & Lee, J. (2005). Slipping or gripping? Fluorescent speckle microscopy in fish keratocytes reveals two different mechanisms for generating a retrograde flow of actin. *Molecular Biology of the Cell, 16*(2), 507–518.

11. Kouros-Mehr, H., & Werb, Z. (2006). Candidate regulators of mammary branching morphogenesis identified by genome-wide transcript analysis. *Developmental Dynamics, 235*(12), 3404–3412. https://doi.org/10.1002/dvdy.20978

12. Lee, J., Go, Y., Kang, I., Han, Y. M., & Kim, J. (2010). Oct-4 controls cell-cycle progression of embryonic stem cells. *The Biochemical Journal, 426*(2), 171–181.

13. Lin, C. H., Espreafico, E. M., Mooseker, M. S., et al. (1996). Myosin drives retrograde F-actin flow in neuronal growth cones. *Neuron, 16*(4), 769–782.

14. Pampaloni, F., Lattanzi, G., Jonas, A., et al. (2006). Thermal fluctuations of grafted microtubules provide evidence of a length-dependent persistence length. *Proceedings of the National Academy of Sciences USA, 103*(27), 10248–10253.

15. Thoumine, O., Kocian, P., Kottelat, A., et al. (2000). Short-term binding of fibroblasts to fibronectin: Optical tweezers experiments and probabilistic analysis. *European Biophysics Journal, 29*(6), 398–408.

CHAPTER

6 Human Early Development

LEARNING OBJECTIVES

After studying this chapter, you will be able to:

- Discuss on human reproduction;
- Know about the human embryonic stem cells;
- Give an overview on human post-implantation development.

KEY TERMS FROM THIS CHAPTER

Brain development
Fetal development
Maternal health
Reproductive technology
Cognitive development
Human development
Motor skills

INTRODUCTION

Early human development is the crucial time from conception to the first few years of life, during which complex and swift changes in cognition, emotion, and physiology occur. This stage is fundamental because it establishes the conditions for subsequent development, health, and growth. The fusion of sperm and egg cells results in the formation of a zygote, which is the first stage of human development from conception. Rapid division and differentiation of this single cell results in the blastocyst, which implants in the uterine wall and matures into an embryo. Major organs and structures begin to form during the first eight weeks of pregnancy, known as the embryonic stage. Since the foundations of the heart, brain, and other essential organs are being formed during this crucial time, it is especially vulnerable to external influences and possible disruptions. From the ninth week of pregnancy until delivery, the fetal stage is defined by the development and maturation of already formed structures. The development of neurons and synapses, which are essential for cognitive function, occurs in the brain. By the end of the fetal stage, the fetus can respond to outside stimuli as its senses start to develop. Prenatal nutrition, the health of the mother, and environmental influences all remain important in influencing the development of the fetus and consequently the birth weight, brain development, and general health.

The neonatal period, which lasts for the first 28 days after birth, is a time of quick adaptation to life outside the womb. Crucial physiological changes occur in newborns, including the start of breathing, control of body temperature, and independent feeding. The brain continues to grow at an astonishing rate, supporting the emergence of sensory and motor skills and improving neural connections through significant synaptic growth and myelination. Rapid physical growth and motor development are characteristics of infancy, which lasts from one month to one year. Babies learn to regulate their own movements, develop voluntary behaviors from reflexive ones, and start investigating their surroundings. This is a crucial time for cognitive development because it is during this time that babies learn to identify faces, react to their names, and begin to solve simple problems. Infants form attachments to caregivers, which are essential for their emotional security and social competence, as part of the process of emotional and social development. This trajectory of learning and growth is maintained during the toddler years, which span from one to three years. Toddlers show enhanced motor skills, greater autonomy, and expanded language abilities. Symbolic thinking, pretend play, and the start of logical reasoning are characteristics of cognitive development. Socially, toddlers begin to comprehend and participate in cooperative play, grow in empathy, and learn basic social norms.

Development is shaped during these formative years by the interaction of environmental factors like nutrition, caregiver interactions, and socioeconomic status with genetic predispositions. Adverse experiences, like neglect or trauma, can impede developmental trajectories and have an impact on long-term outcomes. Conversely, positive early experiences and environments support optimal brain development and overall well-being.

6.1. HUMAN REPRODUCTION

Sexual reproduction in humans is the process by which a human is fertilized to create an offspring. Usually, it involves a sexually mature human male and female engaging in sexual activity. The male and female reproductive systems interact during sexual activity, causing the sperm to fertilize the ovum and create a zygote. Normal cells have 46 chromosomes (23 pairs), whereas gamete cells only have 23 single chromosomes. Genetic recombination takes place when these two cells combine to form a single zygote cell, giving the new zygote 46 chromosomes (23 pairs). After that, the zygote goes through a specific developmental process called human embryogenesis, which marks the beginning of the nine-month gestation period that ends with childbirth. Artificial insemination techniques are another way to fertilize an ovum without having a sexual encounter. There is also technology for assisted reproduction.

6.1.1. Process of Fertilization

Although artificial insemination can be used, copulation is the typical starting point for human reproduction. This is followed by a nine-month pregnancy that ends with childbirth. Contraceptives, such as intrauterine devices and condoms, can prevent pregnancy.

6.1.1.1. Copulation

Sexual contact results in internal fertilization, which is the natural process of human reproduction. The man enters the woman's vagina with his erect penis inserted, and both partners then begin rhythmic pelvic thrusts until the man experiences an orgasm, which causes him to ejaculate sperm-filled semen into the vaginal canal. The term "gametes" refers to the sperm and the ovum, which are produced through meiosis and each contain half of the genetic material of the parent. The sperm enters the uterus or fallopian tubes through the vagina and cervix, where it is one of the 250 million sperm that are produced during a normal ejaculation. Of the sperm ejaculated, only 1 in 14 million will pass through the fallopian tube. The egg simultaneously exits the ovary and travels through the fallopian tube. When a sperm comes into contact with an ovum, it penetrates it and fertilizes it, producing a zygote. The fetus then undergoes internal gestation in the uterus following fertilization and implantation. Sexual activity during the menstrual cycle, from about 5 days before to 1 to 2 days after ovulation, is when pregnancy rates are highest. Sexual activity should be done either every one or two days or every two or three days for the best chance of becoming pregnant. Research has indicated that there is no noteworthy distinction between the incidence of pregnancy and various sexual positions, provided that the ejaculation occurs within the vagina.

6.1.1.2. Alternative Methods

Artificial insemination, in which sperm are injected into the female reproductive system without the penis being inserted, is an alternative to spontaneous sexual activity. A variety of assisted reproductive technologies are also available, such as *in vitro* fertilization, which involves removing one or more egg cells from a woman's ovaries and co-incubating them outside the body with sperm. After that, the resultant embryo can be transferred back into the woman's womb.

6.1.1.3. Pregnancy

The time a fetus develops and divides inside the uterus through mitosis is known as pregnancy. During this period, the placenta, which is connected to the fetus' abdomen by an umbilical cord, filters the mother's oxygenated blood and provides the fetus with all of its nourishment. The mother, who must consume slightly more calories, may find this nutritional drain to be quite taxing. Furthermore, some vitamins and other nutrients must be consumed in larger amounts than usual, which frequently results in strange eating patterns. In humans, the gestation period lasts roughly 266 days. The baby passes through three stages while inside the uterus: the short zygote stage, the eight-week-long embryonic stage (during which the major organs develop), and the fetal stage (during which the fetus continues to grow in size and develops new bone cells). According to estimates, between three and five percent of couples are infertile, and each menstrual cycle, roughly thirty percent of couples are fecund.

6.1.1.4. Labor and Birth

There are 4 phases to labor. The cervix dilates by 6 to 10 cm to separate the latent and active phases in the first stage. The pushing stage is the second. The placenta is delivered during the third stage. The uterus's contraction is the final phase. Chemical cues trigger the birthing process, which starts with the fetus being forced out of the birth canal, once the fetus is fully developed. Soon after birth, the newborn—referred to as an infant in human language—should normally start breathing on its own. The placenta eventually separates on its own not too long after. The umbilical cord may also be severed by the person aiding with the birth.

6.1.2. Biological and Legal Requirements

For human reproduction to occur, a person has to have reached puberty, which means that ovulation for females and spermarche for males must have happened before engaging in sexual activity or becoming pregnant naturally. Humans are infertile prior to puberty because their genitals are only capable of releasing urine. Legal considerations are also essential to the success of human reproduction. In the former scenario, an older party may be charged with statutory rape in some jurisdictions if the minor does not give legal consent to sexual activity or artificial reproductive substitutes. Comprehensive sex education counsels both consenting parties to use contraception in order to prevent STDs and early, unwanted, or unplanned pregnancies, even for minors who are of legal consent. Girls under the age of 15 are especially discouraged from getting pregnant because their reproductive systems are still developing.

6.1.3. Development of the Reproductive Organs

A child's sex is decided at the moment

the spermatozoon fertilizes the ovum. The chromosomes that each possesses in the cell nuclei of a male and female determine their genetic differences. Following the identification of the genetic sex, a series of alterations typically transpire, culminating in the maturation of an adult male or female. Nonetheless, during the first eight weeks of an embryo's existence inside the uterus, there is no outward sign of the sex of the creature. During this neutral or indifferent stage, an embryo's sex can only be determined by looking at its cell's chromosomes. The next stage, called differentiation, starts a week or so later in gonads that will eventually become ovaries and first in those that will become testes. The two sexes' embryos are similar at first because they have similar external genitalia, which are represented by three simple protuberances, and similar duct systems connecting the undifferentiated gonads with the exterior. Each embryo has four ducts, and what happens to these later on greatly influences the final anatomical differences between men and women. The term "mesonephric," or "wolffian," ducts refers to two ducts that are closely associated with the growing urinary system. Each mesonephric duct in males differentiates into four related structures: the seminal vesicle, the ductus deferens, the ejaculatory duct, and the duct of the epididymis. The mesonephric ducts are mostly suppressed in females. The other two ducts, known as the paramesonephric or müllerian ducts, are mostly suppressed in males but continue to develop into the uterus, fallopian tubes, and a portion of the vagina in females. Additionally, differentiation takes place in the primitive external genitalia, which develop into the vulva (the clitoris, labia, and vestibule of the vagina) in females and the penis and scrotum in males. The organs that correspond to each sex are developed and in their adult positions at birth, but they are not functional. A number of anomalies can arise when the sex organs develop in embryos, resulting in conditions such as pseudohermaphroditism and hermaphroditism as well as other chromosomally induced disorders. All reproductive organs grow steadily throughout childhood and gradually develop activity until puberty. The onset of increased sex gland activity and the gradual development of secondary sexual characteristics coincide with puberty. Male puberty is characterized by the development of the ejaculatory ability, enlargement and activation of the testes, and enlargement of the external genitalia. With an increase in testicular hormone secretion, there are noticeable changes in weight and height. The voice becomes deeper as a result of the larynx, also known as the voice box, expanding. Specific skeletal characteristics, such as the pelvic bones and skull, become more prominent. The pubic hair and underarm hair grow thicker and more numerous. Along with hair on the chest, abdomen, and limbs, facial hair also grows. At the temples, hair recedes. Apocrine glands, a type of sweat gland located in the groin, armpits, and area surrounding the anus, become more active than other skin glands.

When a girl reaches puberty, her external genitalia enlarge and her uterus begins to function on a regular basis when she menstruates. The development of the breasts and the accumulation of body fat conform to the typical features of the mature female. Pubic and axillary (underarm) hair growth is increased, and the hair thickens.

6.1.4. The Male Reproductive System

The internal components of the male reproductive system are the testes, epididymis, vas deferens, and prostate, while the external components are the scrotum and penis (Figure 6.1). In addition to producing vital androgens for male development, these well-vascularized structures have numerous glands and ducts that aid in the formation, storing, and ejaculation of sperm for fertilization. Testosterone, the primary androgen in men, is generated in the testes from Leydig cells. A more active form of testosterone, dihydrotestosterone, can be produced peripherally by 5-alpha-reductase or aromatase, respectively. The Sertoli cells in the testes produce two other important hormones: inhibin B and Mullerian Inhibiting Substance (MIS) hormone. Follicle-stimulating hormone (FSH) and luteinizing hormone (LH) are significant hormones that modulate these. FSH and LH are released from the anterior pituitary gland, and gonadotropin-releasing hormone (GnRH) is produced by the hypothalamus. The hypothalamic-pituitary-gonadal axis, which is made up of these hormones together, supports and preserves male sexual development and function.

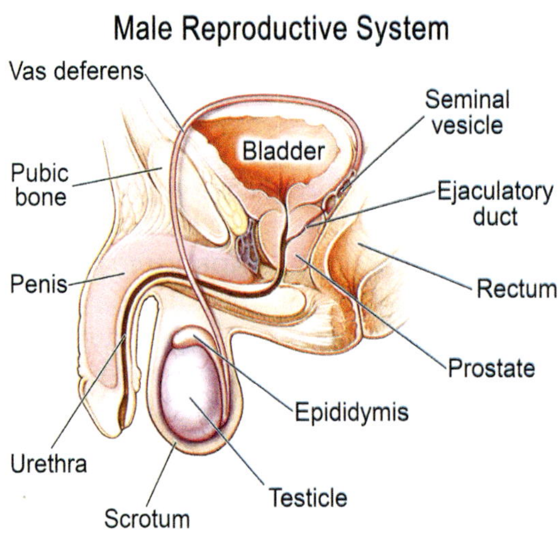

Figure 6.1. *The male reproductive system consists of internal and external organs. These organs help you have sexual intercourse and pee.*

Source: https://my.clevelandclinic.org/-/scassets/images/org/health/articles/9117-male-reproductive-system.

6.1.4.1. Cellular Level

Leydig and Sertoli cells, which are located in the testes, are the main functional cells of the male reproductive system. Adjacent to the seminiferous tubules in the testicular interstitium are Leydig cells. They have pink cytoplasm on histology, and pink Reinke crystals help to identify them. They create testosterone, a steroid hormone that acts by binding to various tissues' intracellular receptors and controlling the expression of certain proteins. The seminiferous tubules' periphery contains sertoli cells. They encourage the process of spermatogenesis, which starts at the tubules' periphery. They join forces to create the blood-testis barrier, which serves to confine germ cells within the seminiferous tubules. Tight junctions allow them to communicate with one another. The way these cells relate to primitive spermatogonia or germ cells defines them. Compared to nearby germ cells, sertoli cells are substantially larger and have less pronounced nuclei. As they develop, germ cells move from the inside of the seminiferous tubules toward the lumen. The nuclei of these cells are noticeable, dense, and dark.

6.1.4.2. Development

The default sexual differentiation in human embryos is female. On the other hand, the Y chromosome indicates differentiation into the male reproductive system and phenotype. The sex-determining region (SRY) gene, which codes for testicular development, is located on the Y chromosome. During development, the testes descend from the posterior abdominal wall and eventually land in the scrotum. The female reproductive tract is formed by the Mullerian ducts, which recede when the testes produce Sertoli cells, which in turn produce MIS. The primary regulator of male reproductive development, testosterone, is also produced by Leydig cells that grow in the testes.

The male fetus's Wolffian ducts, which eventually develop into the testes, epididymis, vas deferens, and seminal vesicles, are stimulated in part by testosterone. Testosterone is also responsible for erythropoiesis, pubertal growth spurt, bone density, closure of epiphyseal plates, deepening of the voice, increase in muscle mass, male physique development and libido. Additionally, 5-alpha-reductase, an enzyme produced peripherally by the prostate, can convert testosterone to dihydrotestosterone (DHT). The same androgen receptors are intracellularly bound by both DHT and testosterone, although DHT has a greater affinity. The development of the penis, scrotum, and prostate is stimulated by DHT. Acne, elevated sebaceous gland secretion, and male pattern balding are all caused by DHT. It is also the cause of male hair pattern (facial, axial, and pubic hair). These hormones work together to support male reproductive system maintenance after puberty.

In males, the growth of the testes usually signifies the start of puberty, which happens between the ages of 11 and 13. An abrupt increase in GnRH from the hypothalamus causes the anterior pituitary to release more FSH and LH, which in turn causes this. The growth and pigmentation of the penis and scrotum are caused by testosterone, which is increased by LH stimulating Leydig cells. Next come secondary sexual traits like deepening of the voice, growth spurt, and growth of hair on the face, axilla, chest, and pubic area. The onset of fertile ejaculations at this point indicates mature reproductive function. Pubertal development ends when the epiphyseal growth plates eventually close. Male pubertal development can last well into his twenties.

6.1.4.3. Organ Systems Involved

The development of secondary sex traits, sperm production, and sexual maturity are all significantly aided by the hypothalamic-pituitary-gonadal axis. Throughout a man's life, it sustains spermatogenesis and sexual function. To activate the anterior pituitary, the hypothalamus secretes GnRH into the hypothalamo-hypophyzeal portal system. The peptide hormone known as GnRH is pulsatilely released by hypothalamic neurons. By binding to and activating a G protein receptor, it affects the anterior pituitary's gonadotrophs. This stimulates the anterior pituitary to release FSH and LH by increasing intracellular calcium through the activation of inositol 1,4,5-triphosphate (IP3). Prolactin, estrogen, testosterone, and estradiol all inhibit GnRH.

The anterior pituitary releases LH and FSH into the bloodstream in retaliation. These gonadotropic hormones affect the Leydig and Sertoli cells of the testes, respectively, by acting on membrane receptors. The beta subunits in each hormone differ, but they both belong to the same glycoprotein family and have identical alpha subunits, which allows them to perform different functions. Both stimulate Sertoli and Leydig cells by binding to and activating a G protein receptor, which in turn activates adenylyl cyclase and raises cellular cAMP levels. Leydig cells in the testicular interstitium are stimulated by LH to convert cholesterol into testosterone. Desmolase is the first enzyme whose rate of conversion from cholesterol to pregnenolone is limited, and it is promoted by LH. Dehydroepiandrosterone (DHEA) and androstenedione are the two main weak androgen intermediates that are produced as a result. The process of converting androstenedione to testosterone is finished by the enzyme 17-beta-hydroxysteroid dehydrogenase. Testosterone reduces the release of LH and FSH by acting through negative feedback on the anterior pituitary and hypothalamus. Testosterone can also have some effect on Sertoli cells, which are located on the outside of the testes' seminiferous tubules. ABP, or androgen-binding protein, is released by Sertoli cells in response to FSH and testosterone. During spermatogenesis, ABP gives germ cells testosterone. FSH causes the release of MIS and inhibin B from Sertoli cells, which in turn increases sperm production. Sertoli cells use inhibitor as a negative feedback mechanism to reduce the release of FSH from the hypothalamic-pituitary system.

The levels of gonadotropins and androgens usually stay low and steady before puberty. When a person reaches puberty, the hypothalamus releases GnRH pulsatile every one to two hours to maintain levels of plasma testosterone, LH, and FSH. These levels are all regulated by one another to maintain hormonal balance. It has been observed that testosterone levels decrease in the third decade of life. The adrenal cortex produces some testosterone, but the Leydig cells in the testes produce the majority of a man's testosterone. The hypothalamus and anterior pituitary work together to form the hypothalamic-pituitary-adrenal axis, which regulates the adrenal glands similarly to the hypothalamic-pituitary-gonadal axis. Adrenocorticotropic hormone (ACTH) is released from the anterior pituitary gland in response to stimulation by corticotrophin-releasing hormone (CRH) released from the hypothalamus. Similar to how testosterone is synthesized in the testes, ACTH stimulates the enzyme desmolase in the adrenals to convert cholesterol into pregnenolone. Particularly, the adrenal medulla's zona reticularis produces the weak androgens DHEA and androstenedione, which are subsequently peripherally converted to testosterone or estradiol.

6.1.4.4. Function

In order to sustain male reproductive function and to facilitate spermatogenesis and its transfer into the female reproductive system for fertilization, the male reproductive system produces androgens like testosterone. Due to their roles in the production and transportation of sperm as well as androgen, the testes function as both endocrine and exocrine organs.

6.1.4.5. Mechanism

The germ cells in the basement membrane of the seminiferous tubules of the testes are where spermatogenesis begins during puberty. Spermatogenesis is regulated by FSH-stimulated sertoli cells. About every 13 days, spermatogenesis starts a new cycle; however, not all seminiferous tubules experience spermatogenesis at the same time. When diploid spermatogonia undergo mitosis to become primary spermatocytes, the first stage of spermatogenesis begins. Meiosis I is carried out by these spermatocytes to create haploid secondary spermatocytes, which subsequently go through Meiosis II to become haploid spermatids. The earliest spermatocytes are located in the periphery of the seminiferous tubules and develop by moving in the direction of the lumen. Spermatids undergo cytoplasmic reduction to become spermatozoa. These immature spermatozoa are released into the tubules where they will eventually mature and travel to the epididymis. The head, body, and tail make up the coiled structure known as the epididymis. Eventually, the tail unites with the vas deferens, giving mature sperm a place to ejaculate. It takes the sperm twelve days to mature and become motile in the epididymis. After that, they are kept in the epididymis' tail until ejaculation. A fully developed sperm has a head, midpiece, and tail. The nucleus and very little cytoplasm are located in the head. Lysosomes are found within an acrosome, or cap, which covers the head and facilitates fertilization. The sperm's flagellum or tail receives energy from the numerous mitochondria found in the midpiece.

Vasodilation brings blood to the penis during physical or psychological sexual arousal. Blood flows along the corpus spongiosum and corpora cavernosa in the penis to erect and enlarge it. Blood flow to the genitalia is sustained during sexual stimulation, and the testes grow larger in anticipation of ejaculation. During ejaculation, sperm are forced into the ductus deferens (vas deferens), which is located in the spermatic cord, by smooth muscle contractions of the epididymis. Through its junction with the seminal vesicle duct close to the prostate, the ductus deferens transports sperm to the ejaculatory duct. Fructose, which is produced by the seminal vesicles, gives sperm their energy for motility. It is discharged into a liquid, which combines with the sperm to create semen. The prostate secretes an alkaline fluid that thickens semen and aids in sperm retention within the female reproductive system once it enters the ejaculatory duct. The bulborethral glands, also known as Cowper's glands, release a thick fluid after the semen passes through them, lubricating the urethral opening and clearing the urethra of any leftover urine. The sperm can then travel to and fertilize a potential egg within the female reproductive system once the semen enters the female vaginal canal.

6.1.5. The Female Reproductive System

Menstruation and reproduction are made easier by the internal and external organs that make up the female reproductive system. This organ system is in charge of creating gametes, also known as eggs or ova, controlling sex hormones, and caring for fertilized eggs as they grow into fully formed fetuses that are ready to be delivered. The first menstrual cycle, or menarche, and the menopause, which is the cessation of menses for a period of 12 consecutive months, mark a woman's reproductive years. During this time, ova are periodically expelled from the ovary, with the possibility of fertilization by sperm or male gametes. This periodic discharge of ovum is a typical aspect of the menstrual cycle. The process of gestation, also known as pregnancy, is facilitated by the internal and external female genitalia acting independently to enable fertilization from the unison of ova and sperm. One or more fetuses that are referred to as infants upon birth may result from the gestation, depending on how the conception develops, usually within nine months or three trimesters.

The internal and external sex organs of the female reproductive system are responsible for producing new offspring. The female reproductive system in humans is immature at birth and matures to produce gametes and carry a fetus to term during puberty. The uterus, fallopian tubes, ovaries, and vagina are the internal sex organs. The uterus, fallopian tubes, and vagina make up the female reproductive tract, which is prone to infection. The vagina, which is joined to the uterus at the cervix, facilitates childbirth and sexual activity. The embryo, which grows into the fetus, is housed in the uterus, also known as the womb. Additionally, the uterus secretes substances that facilitate sperm transportation to the fallopian tubes, where sperm fertilize ova (egg cells) generated by the ovaries. The term "genitals" also refers to the vulva's external sex organs, which include the clitoris, labia, and vaginal opening.

An ovum is released by the ovaries during the menstrual cycle and enters the uterus through the fallopian tube. One sperm cell may enter the uterus and combine with an egg cell to fertilize it into a zygote if the two cells come into contact while the egg cell is traveling there. The process of fertilization, which typically takes place in the fallopian tubes, initiates embryogenesis. After dividing over a sufficient number of cell generations, the zygote forms a blastocyst that implants itself in the uterine wall. The embryo will continue to develop until full term from this point on during the gestational period. The cervix dilates and the uterus contracts to force the baby through the birth canal (the vagina) once the fetus has grown sufficiently to survive outside the uterus. The male reproductive system is the analogous system in males.

6.1.5.1. Internal genitalia

1. **Vagina:** The vagina is a fibromuscular canal that connects the uterus's cervix to the exterior of the body. Fibrous and muscular tissue make up this canal. In the context of pregnancy, it is also called the birth canal. During sexual activity, a penis is accommodated by the vagina. During an orgasmic moment, the penis ejaculates spermatozoa-containing semen into the vagina, which may facilitate ovum fertilization. Figure 6.2 shows the external female genitalia; a human vulva.

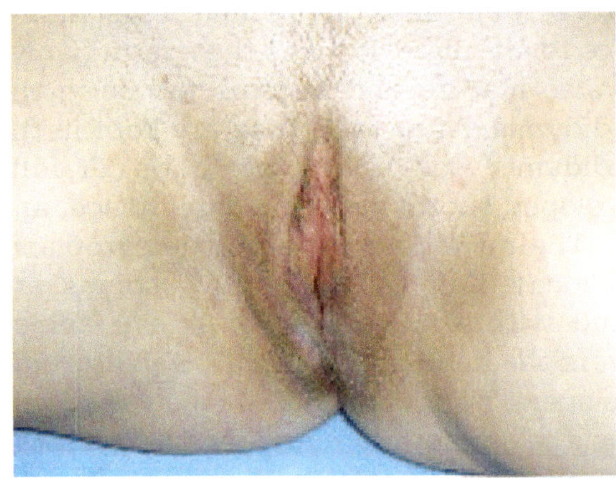

Figure 6.2. *External female genitalia; a human vulva.*

Source: https://upload.wikimedia.org/wikipedia/commons/thumb/7/75/Michaela.JPG/220px-Michaela.JPG.

2. **Cervix:** The narrow, lower part of the uterus where it joins the upper part of the vagina is called the cervix. Shaped like a cylinder or a cone, it emerges through the upper anterior vaginal wall. About

half of its length is visible; the remaining portion is hidden above the vagina. The opening where the fetus emerges during delivery is the vagina, which has a thick layer covering it externally.

3. **Uterus:** The main reproductive organ in women is the uterus, also known as the womb. The uterus offers the developing embryo (weeks 1 to 8) and fetus (week 9 until delivery) mechanical protection, nutritional support, and waste removal. Furthermore, the uterine wall's contractions play a crucial role in forcing the fetus out during delivery. Three suspensory ligaments in the uterus help to limit the uterus's range of motion and stabilize its position. The body is prevented from moving anteriorly and inferiorly by the uterosacral ligaments. The uterus's posterior movement is restricted by the round ligaments. Additionally, the cardinal ligaments stop the uterus from moving inferiorly.

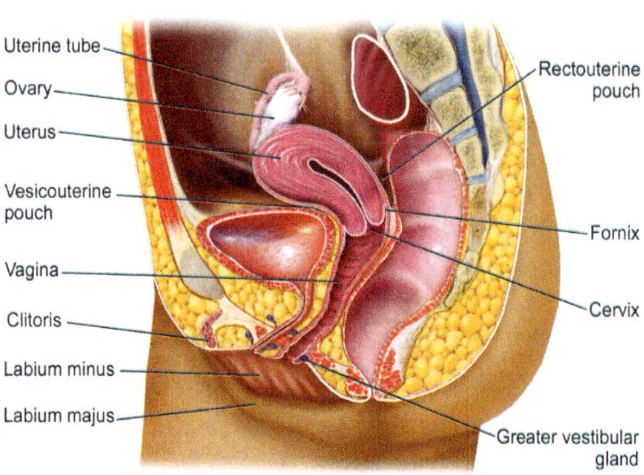

Figure 6.3. Depicts a healthy female reproductive system.

Source: https://upload.wikimedia.org/wikipedia/commons/thumb/8/8e/Blausen_0400_FemaleReproSystem_02.png/800px-Blausen_0400_FemaleReproSystem_02.png.

The uterus is a muscular organ with a pear shape. Its primary job is to receive a fertilized ovum, which is inserted into the endometrium and needs blood vessels that grow specifically for this purpose to nourish it. The fertilized ovum develops into an embryo, grows into a fetus, and carries a child until delivery. The female starts her menstrual cycle if the egg does not embed in the uterine wall.

4. **Fallopian Tubes:** Two tubes that connect the ovaries to the uterus are called fallopian tubes. An ovum can escape and enter the fallopian tube when it reaches maturity because of the rupture of the ovary's wall and follicle. There, cilia on the inner lining of the tubes push it in the direction of the uterus. Days or hours pass during this trip. When the fertilized ovum reaches the uterus, it typically implants in the endometrium, signifying the start of pregnancy, if fertilization occurred while the ovum was still in the fallopian tube.

Figure 6.4 depicts the frontal view of reproductive organs.

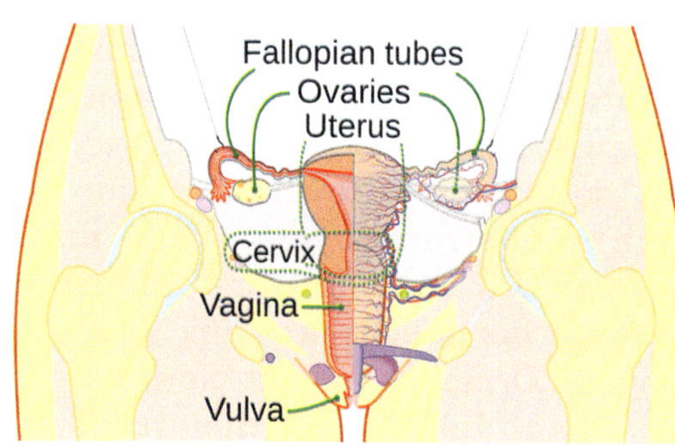

Figure 6.4. *Frontal view as scheme of reproductive organs.*

Source: https://upload.wikimedia.org/wikipedia/commons/thumb/6/68/Scheme_female_reproductive_system-en.svg/1024px-Scheme_female_reproductive_system-en.svg.png.

5. **Ovaries:** The ovaries are tiny, paired gonads that are situated close to the pelvic cavity's lateral walls. These organs are in charge of secreting hormones and producing ova, or egg cells. Ovulation is the process through which the egg cell, or ovum, is released. Periodic ovulation has a direct influence on how long a menstrual cycle lasts. The egg cell leaves the ovum and passes through the fallopian tube on its way to the uterus. If fertilization takes place, it usually takes place in the fallopian tube, after which the fertilized egg tries to implant on the lining of the uterus. The egg cell participates in fertilization by releasing specific molecules that help guide the sperm and enable the egg's surface to adhere to the sperm's surface. After the sperm has been absorbed by the egg, fertilization can start.

6.1.5.2. Function

A woman's first menstrual cycle, or menarche, is indicated by her first menstrual bleeding episode. During puberty, pubic and axillary hair growth, a growth spurt, and breast growth all occur prior to menarche. Several primordial follicles in the woman's ovaries continue to develop at the start of each menstrual cycle. One follicle develops into a dominant one and grows, while the other follicle atrophy and stops growing. A Graafian follicle emerges from the dominant follicle. At this stage, the ovum is no longer in prophase I arrest and meiosis I is complete. The corpus luteum, the surrounding tissue around the Graafian follicle, is where the ovum is expelled during ovulation. Menstrual bleeding is the result of the egg and secretory endometrial lining being released from the uterus if fertilization is unsuccessful, caused by a decrease in progesterone levels. If fertilization takes place, the egg implants in the uterus and is kept alive by the corpus luteum's progesterone until the placenta takes over.

6.1.5.3. Mechanism

Ovulation takes place in between the luteal and follicular phases of the normal menstrual cycle. Menstrual bleeding marks the start of the follicular phase, which concludes prior to the LH surge. The LH surge marks the start of the luteal phase, which concludes with the onset of menses. A normal cycle has a duration of roughly 28 days. The follicular phase is more erratic, and the luteal phase lasts for 14 days. The follicular phase starts when progesterone and estradiol levels in the blood are low. Absence of inhibitory feedback raises pulsatile levels of gonadotropin-releasing hormone (GnRH), which in turn causes FSH and LH to rise. A certain number of follicles continue to grow as a result of the stimulation of follicular maturation caused by this increase in FSH levels. Estradiol and FSH levels rise as a result of these follicles growing. The dominant follicle has grown to a size of about 20 to 25 mm by the end of the follicular phase. The endometrium thickens as a result of the rise in estradiol to prepare for the possible implantation of a fertilized egg.

The negative feedback effect of estradiol on LH is positive when levels reach a critical point, causing a sharp increase in LH concentration (and a less dramatic increase in FSH levels). The oocyte is released from the dominant follicle and enters the uterus through the fallopian tube about 36 hours after the LH surge. Progesterone is released by the corpus luteum, which is the follicular tissue that remains after the oocyte is expeled. This process inhibits the release of LH and FSH and promotes the development of the secretory endometrium. Declining LH levels lead to a decline in progesterone and estradiol levels in the event that fertilization is unsuccessful. The corpus luteum and progesterone level are maintained if fertilization takes place, as the oocyte implants into the endometrium and releases chorionic gonadotropin.

6.1.5.4. Pathophysiology

1. **Amenorrhea:** When there is no menstruation by the age of 13 in the absence of normal development and secondary sexual development, or when there is no menstruation by the age of 15 in the case of individuals with female genitalia, this is known as primary amenorrhea. In women who have previously been menstruating, secondary amenorrhea is defined as the absence of menses for longer than six months. A healthy hypothalamic-pituitary-ovarian endocrine axis is necessary for regular menstruation. Steroid hormone stimulation and an internal-to-external outflow tract elicit responses from this functional endometrium. Many investigations have determined that gonadal dysgenesis (e.g., constitutional delay of puberty, Mullerian agenesis, and Turner Syndrome.)
2. **Infertility:** The capacity to conceive is known as fecundability. A complex series of events, including ovulation, fertilization, and implantation into a viable uterine cavity, are necessary for a successful pregnancy. The likelihood of becoming pregnant during ovulation, or timed sexual activity, ranges from 16 to 28%, with slight variations observed across different studies. The inability to conceive after a year (12 months) of consistent, unprotected sexual activity is known as infertility. Those who have never conceived are classified as having primary infertility, whereas those who have conceived in the past but are currently facing infertility are classified as having secondary infertility. Infertility can result from male, female, or combined factors. In decreasing order, ovulatory disorders, endometriosis, pelvic adhesions, tubal blockage, other uterine/tubal abnormalities, and hyperprolactinemia are the most common female factors that cause infertility.

6.2. HUMAN POST-IMPLANTATION DEVELOPMENT

A crucial developmental stage that includes significant morphogenetic modification of embryonic and extra-embryonic tissues, axis formation, and gastrulation events begins with the implantation of the human embryo. Due to ethical and technical restrictions on accessing *in vivo* samples, our understanding of the mechanisms underlying this period of human life is still restricted. Furthermore, there are no human stem cell models of early post-implantation development that include the morphogenesis of both extra-embryonic and embryonic tissues. Here, we introduce iDiscoid, which is created using a synthetic gene circuit that has been engineered from human induced pluripotent stem cells. In a model of human post-implantation, iDiscoids display reciprocal co-development of engineered extra-embryonic niche and human embryonic tissue. Unexpectedly, they show tissue boundary formation and self-organization that recapitulates the characteristics of the yolk sac-like tissue specification, including extra-embryonic mesoderm and hematopoietic traits. They also develop an amniotic-like cavity, a bilaminar disc-like embryonic morphology, an anterior-like hypoblast pole, and a posterior-like axis. iDiscoids provide a high-throughput, repeatable, scalable, and user-friendly platform for investigating various facets of human early post-implantation development. As a result, they might offer a manageable human model for disease modeling, developmental toxicology, and drug testing (Oldak et al., 2023).

6.2.1. Engineering Co-development of Embryonic and Extra-Embryonic Endoderm Tissues

In order to express human GATA6, a crucial transcription factor for extra-embryonic endodermal fate, we engineered a hiPSC cell line with an inducible gene circuit (henceforth referred to as iGATA6). When the small molecule Doxycycline (Dox) is added, these iGATA6 cells express varying levels of GATA6, which can be distinguished by the expression of the reporter enhanced green fluorescent protein (EGFP). Similar to what happens during the implantation period *in vivo*, EGFP+ extra-embryonic endoderm cells segregated to the outer layer when iGATA6 and wild-type (WT) hiPSCs were co-assembled in 3D. The general arrangement of these aggregates, however, did not replicate the structural layout of the yolk sac and bilaminar disc tissues. This method produced aggregates lacking the parietal tissue, which forms apart from the epiblast cells, and in which the entire EGFP+ extra-embryonic endoderm is in close contact with wild type epiblast-like cells. Furthermore, we discovered that WT cells were unable to reliably cause symmetry breaking to amnion-like and epiblast-like domains in these 3D aggregates. Furthermore, it has been demonstrated that organizational instability leads to tissue organization that is unable to represent the developmental stages that are close to gastrulation when synthetic and real human blastocysts are reattached *in vitro* to a dish to simulate implantation. It is technically challenging to control the stable co-development of extra-embryonic and embryonic cell types from an initial 3D state through post-implantation embryonic stages.

Figure 6.5 illustrates engineering co-development of embryonic and extra-embryonic endoderm tissues in iDiscoids.

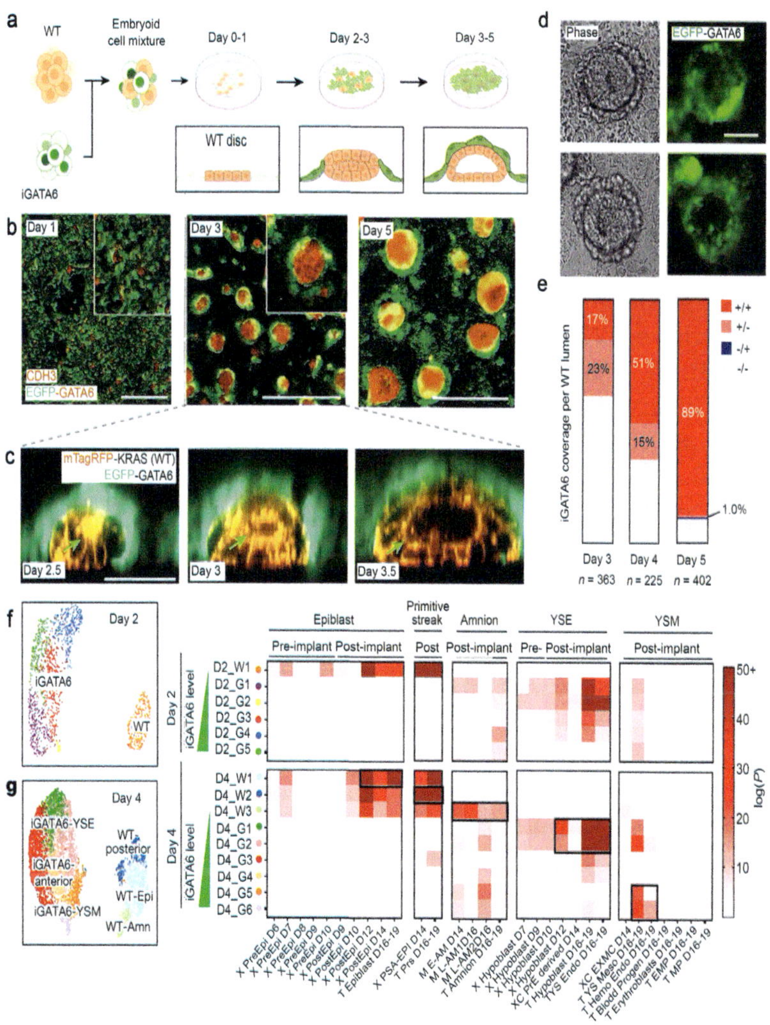

Figure 6.5. *Engineering co-development of embryonic and extra-embryonic endoderm tissues in iDiscoids.*

Source: https://www.biorxiv.org/content/biorxiv/early/2023/06/15/2023.06.15.545118/F1.medium.gif.

In order to overcome these obstacles, our goal was to create a different method of creating structures outside of cell culture plates by causing hiPSCs to self-organize from two dimensions to three dimensions. The WT and iGATA6-hiPSCs were combined, and the resulting cell population was seeded at a specific ratio and cell density onto standard culture plates (Figure 6.5A). Following Dox treatment, the iGATA6 cells lose the pluripotency markers NANOG and as expected, upregulate GATA6 and EGFP. Induced iGATA6 cells and WT cells multiply during the first 48 hours from an initial state of random distribution, and WT cells are arranged into disc-shaped clusters that are surrounded by iGATA6 cells. Then, together with 3D growth inside the WT disc, comes the migration of the iGATA6 cells across the upper surface of these WT clusters. Over a multilayer WT structure, these processes produce a membrane of extra-embryonic endoderm-like cells that express GATA4 and SOX17 but do not have OCT4 (Figure 6.5 A). The upward migration of iGATA6 occurs concurrently with the laminin membrane surrounding the WT clusters being deposited. The iGATA6 cells' laminin deposition can then cause cell polarization in the WT cluster's cells, which encourages the transformation of cellular rosettes into lumens (Figure 6.5 C. Expression of ZO-1, an apical tight junction protein, and PODXL, an apically expressed antiadhesive surface protein implicated in embryonic lumen formation, facilitate this. Under similar conditions, laminin deposition and lumen formation are not seen in WT-only cultures.

We found that lumen formation was absent from WT clusters for which an iGATA6 layer was not present around the cluster (Figure 6.5 E). As the iGATA6 cell migration proceeds, the number of clusters lacking iGATA6 layer coverage steadily declines by day 5 (Figure 6.5 E). Finally, after the lumen has expanded, the two cell monolayers of endoderm-like GATA6+ cells and epiblast-like WT (NANOG+) cells are positioned on either side of a laminin membrane (Figure 6.5 A, 6.5 C), creating a cellular configuration that resembles the bilaminar disc found in human embryos shortly after implantation. High levels of secreted AFP and APOA1, two proteins known to be produced by yolk sac and primitive endoderm, were supported by analysis of iDiscoid culture media. In our first analysis, we found that cavity formation happened via multiple lumens for WT clusters at day 5 with higher area and often lower circularity. With the aim of achieving cluster area most consistently within a size range corresponding to the E9-E17 human bilaminar disc, we tested a range of seeding ratio and densities for iGATA6 and WT cells. This was done in order to minimize the occurrence of large WT clusters with multiple lumens and control the cluster size and number. In order to achieve this, we chose an 81:5 iGATA6:WT seeding ratio at a density of 54k cells per cm^2. Using this optimal seeding density, we were able to identify 74% of the iDiscoids within the predicted physiological size range. Crucially, we also saw that most iDiscoids in this size range (70:9%) only produced one lumen. Therefore, we could control the size, number, circularity, and lumenogenesis of iDiscoids by modifying the initial parameters.

6.2.2. Single Cell RNA Transcriptomics Reveals Embryonic and Extra-Embryonic Fate Trajectories

From the first to the fifth day of iDiscoid cultures, we used single cell RNA sequencing (scRNA-seq) to examine the cellular identities of the developed population. An unsupervised analysis of the transcriptomes of the E16–19 human embryo, the E6-E14 human embryo samples, and the E18–20 cynomolgus embryo demonstrated that the identified iDiscoid subpopulations acquired different post-implantation cellular fates (Figure 6.5 (F, G)). iDiscoids revealed the appearance of cell populations (D1_G1, G2, G3) with transcriptomic resemblance to human hypoblasts, as well as a corresponding epiblast-like population (D1_W1), as early as 24 hours (D1) following induction. iGATA6 cells statistically resemble the human E12 hypoblast lineage by 36 hours (D1.5), according to Xiang et al., and through day five of culture, this similarity gets stronger. Significant similarities to yolk sac endoderm and hypoblast appear at day 1 and strengthen through day 5 ($p = 6.00\times10{-}^{54}$ to $2.84\times10{-}^{93}$ similarity significance to YS Endoderm at D5) (Figure 6.5F) when compared to the E16–19 human and cynomolgus embryo.

Within the iGATA6 population (Clusters D2_G1, D2_G2, D2_G3, D2_G4; $p = 1.3\times10^{-17}$ to 9.5×10^{-69}), there is statistically significant similarity to yolk sac endoderm on day 2. The highest GATA6-expressing cells (Cluster D2-G5) diverge from this fate towards a more yolk sac mesoderm-like state ($p = 6.8\times10^{-11}$). By day 4, we saw that the iGATA6 subpopulations' identities were becoming stronger and that the WT compartment's fate was diverging. Key yolk sac marker genes (GATA4, PDGFRA, LAMA1, CUBN, AMN, and NODAL) were expressed more frequently in the lower-GATA6 clusters (D4_G1 and D4_G2), which also displayed statistical similarities to human yolk sac endoderm. A "medium" GATA6 population (D4_G3 and D4_G4) exhibited a specific upregulation of anterior markers (LHX1, HHEX, FZD5, CER1, LEFTY1) and a strong anterior hypoblast identity (homologous to mouse Anterior Visceral Endoderm, AVE). Despite this population showed less statistical similarity to either of these fates. D4_G5 and D4_G6, the highest-GATA6 clusters, showed statistical similarities to human yolk sac mesoderm (Figure 6.5G). When combined, these findings show that there is a differentiation in the fates associated with the yolk sac within iGATA6 cells, which is connected to varying levels of GATA6 expression. Specifically, populations expressing low, medium, and high average levels of GATA6 are categorized by the presence of yolk sac endoderm, anterior hypoblast, and mesoderm-like cells in the yolk sac. Zhai et al., have recently made observations in cynomolgus monkeys imply a comparable GATA6 level stratification across a range of tissues, most notably between populations of mesoderm (higher GATA6) and yolk sac endoderm (lower GATA6).

The absence of GATA6 or EGFP transcripts within the WT clusters indicates the divergence of three distinct populations from the epiblast-like compartment. The first and largest cluster (D4_W1) expressed the pluripotency markers POU5F1 (OCT4), SOX2, and NANOG ($pW1 = 1.15\times10^{-49}$), maintaining similarities to human epiblast. The human primitive streak ($pW2 = 3.6\times10^{-95}$ to D16 PrS) and human epiblast ($pW2 = 3.4\times10^{-20}$ to $D16^{-19}$ Epi) were more similar to the second (D4_W2) WT cluster statistically (Figure 6.5G). Based on our analysis, the third wild type cluster (D4_W3) at this timepoint ($pW3 = 8.0\times10^{-15}$ and 1.6×10^{-25}, respectively) had a distinct transcriptomic similarity to human and cynomolgus amnion (Figure 6.5G). No resemblance to the trophoblast lineages was noticed by us. The projection space showed that the WT and iGATA6 lineages occupied different compartments when we integrated all clusters over various days. We did, however, notice population spatial positioning that matched the time of sampling. Similar outcomes can be seen in populations sampled at different times (i.e., iGATA6 yolk sac endoderm cells (D3 versus D5). The most highly upregulated genes exhibit unique patterns in various cell fates, with genes that are shared by similar fates sampled at various times.

6.2.3. Specification of Amniotic Ectoderm in iDiscoids

Since mouse amniogenesis differs from primate amniogenesis in both structure and timing, human model systems are essential for shedding light on the dynamics of human development. *In vivo*, the cells of the epiblast go through dorsal-ventral (D-V) patterning to separate the epiblast from amniotic ectoderm after lumenogenesis within the epiblast of the primate post-implantation embryo. Our single-cell transcriptome analysis in wild-type cells revealed a subset of cells that emerges around day 3 and has transcriptomic similarities to human and cynomolgus amnion (Figure 6.6 (A-B). These cells express amnion markers, such as ISL1, TFAP2A (AP-2α), and GATA3, and they continue to express OCT4, but they significantly downregulate NANOG expression (Figure 6.6 (A-B). This result was confirmed by our analysis using immunofluorescent staining, which showed an ISL1+/AP-2α+ layer in iDiscoids (Figure 6.6 C-E). The amnion-like lineage is positioned away from the bilaminar iGATA6/WT disc, forming a membrane against the tissue culture dish. We observe that these cells are spatially segregated within the cavitated sac-like structure. NANOG expression is seen in the WT cells above within the bilaminar iGATA6/WT disc (Figure 6.6 C-E).

Additional pSMAD1, pSMAD5, and pSMAD8/9 (BMP4 effector) immunofluorescence staining demonstrated BMP4 signal transduction in a ring pattern centered at the iDiscoid's edges (Figure 6.6 F). On day 4, scRNA-seq revealed that the amnion-like population specifically upregulated BMP4 targets and related genes. It has been proposed recently that primate amniotic tissue actively stimulates BMP4 signaling by acting downstream of ISL1. Additionally, it was noted that in a human embryo model, BMP4 can encourage the development of ISL1+ amniotic tissue. In order to ascertain the significance of amnion-like tissue formation in iDiscoid, we thus set out to block BMP signaling by administering the inhibitor Noggin. Effective suppression of the phosphorylation of SMAD1/5/8 was observed (Figure 6.6 G), and a significant reduction in the acquisition of amnion fate as determined by ISL1-expression (Figure 6.6 H-I). Nonetheless, NANOG expression was unaltered in cells that resembled epiblasts (Figure 6.6 H). In order to achieve this, we demonstrated that, on day 4 of iDiscoid development, 98.8% of WT clusters expressed the amnion marker ISL1, a finding that held true for all experiments. We demonstrate that ISL1 expression is higher in all (n = 87) observed WT clusters than in the noggin treatment group, with the exception of one (n = 88). All things considered, we find that iDiscoids exhibit robust amniogenesis, including the formation of a dorsal-ventral (D-V) axis that is reliant on BMP4 signaling. This lends support to the use of iDiscoids in research on this crucial stage of human development.

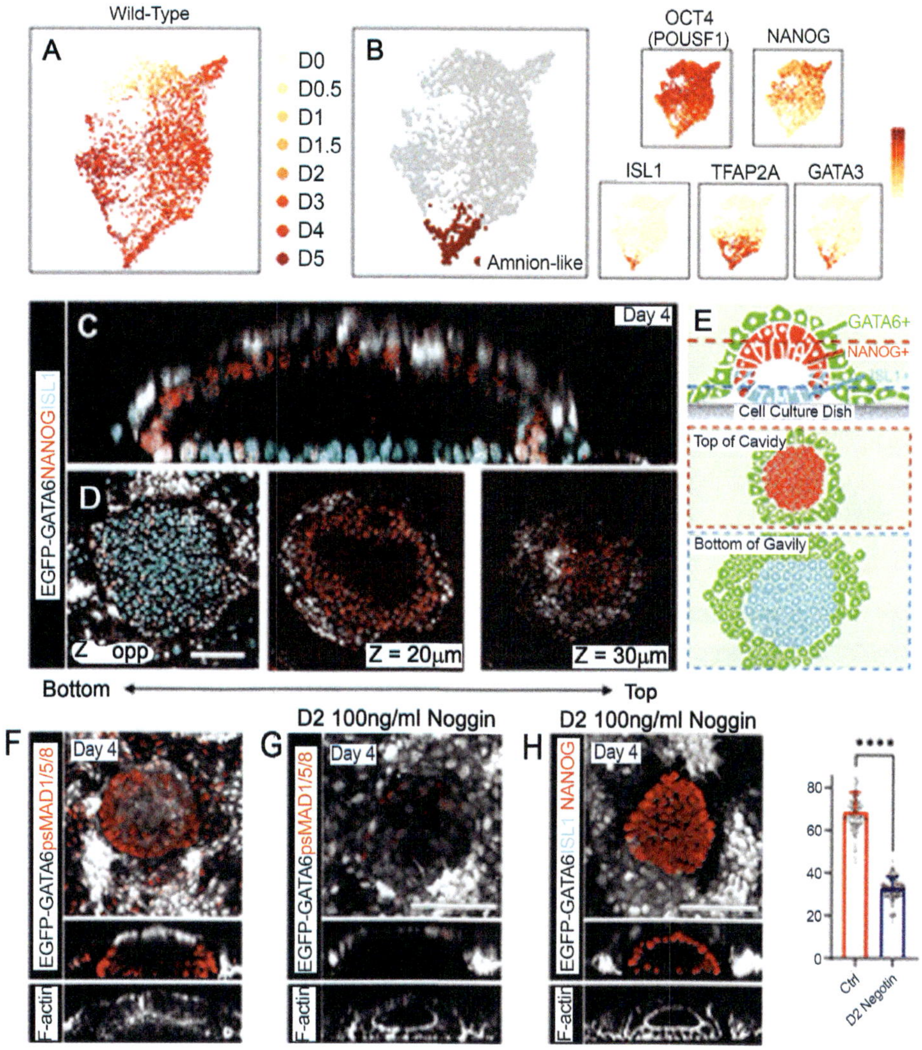

Figure 6.6. *Amniotic-like cavity formation and expansion.*

Scale bars = 100μm. n represents separate biological samples from a single experiment harvested on the corresponding day.

Source: https://www.biorxiv.org/content/biorxiv/early/2023/06/15/2023.06.15.545118/F2.medium.gif.

6.2.4. Development of Anterior Hypoblast Poles and Posterior Axis in iDiscoids

Anterior hypoblasts are a group of cells with anterior visceral endoderm characteristics found in both human and non-human primates, according to *in vivo* evidence. It has been demonstrated that in mice, this tissue patterns the posterior axis specification through the expression of signaling inhibitors, such as CER1. Nonetheless, popular human embryonic stem cell models *in vitro* (i.e., blastoids) have not been able to fully embody this requirement. We questioned whether iDiscoids could display characteristics linked to the formation of the anterior-posterior (A-P) axis (Figure 6.7). As early as day 2, our scRNA-seq analysis of iGATA6 cells identified a cluster of cells exhibiting anterior hypoblast-associated markers. These cells continued to exhibit CER1, LHX1, HHEX, LEFTY1, and LEFTY2 markers in the days that followed (Figure 6.7A). Furthermore, on days three and four, we also noticed the appearance of a unique cluster of cells exhibiting TBXT (Brachyury), MIXL1, GDF3, and EOMES, important indicators linked to the development of the embryo's posterior pole (Figure 6.7 B).

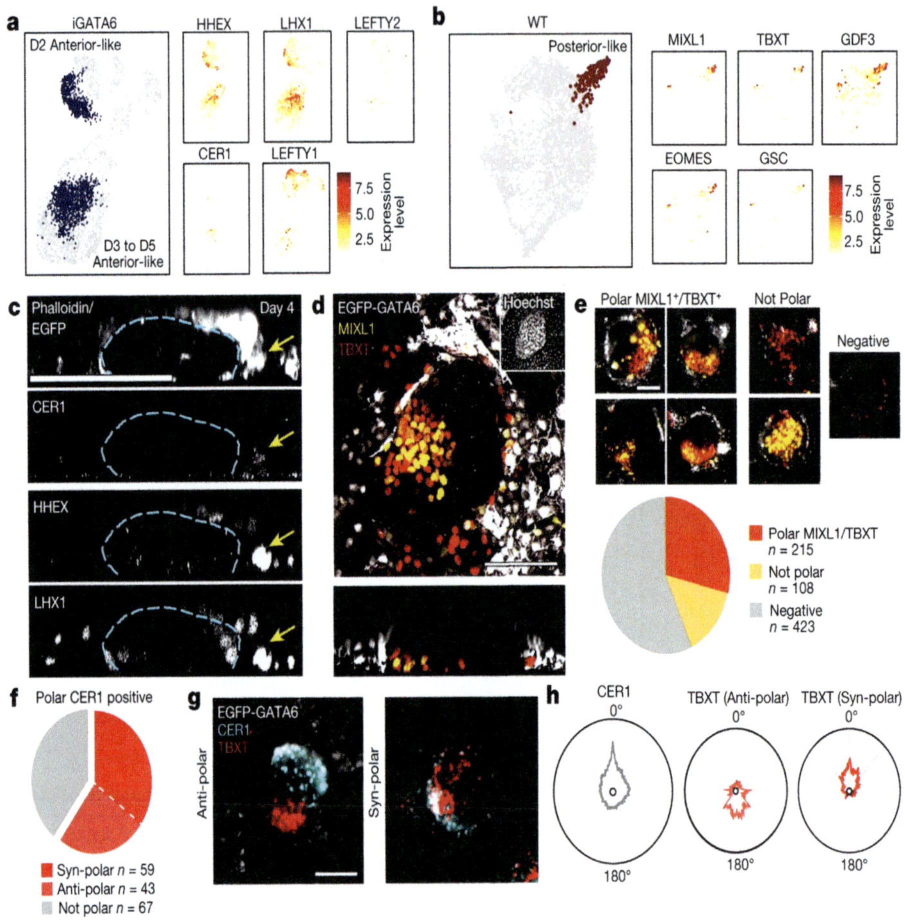

Figure 6.7. *Anterior hypoblast-like domain and posterior axis in iDiscoids.*

Scale bars = 100 µm. n represents separate biological samples from a 3–4 separate experiments harvested on day 4 after iDiscoid induction.

Source: https://www.biorxiv.org/content/biorxiv/early/2023/06/15/2023.06.15.545118/F3.medium.gif.

In a subset of extra-embryonic cells surrounding the WT clusters, immunostaining revealed domains of cells expressing the anterior hypoblast-related proteins (CER1, LHX1, HHEX) (Figure 6.7C). On day 4, we found 42.4±4.1% (169) discs with polar CER1-expressing domains out of the 396 iDiscoids that were assessed (Figure 6.7C), demonstrating an efficiency comparable to current *ex vivo* human blastocyst cultures36. Parallel to this, we observed the formation of TBXT+ domains in 46.4±9.0% (323 out of 746) WT clusters in iDiscoids on day 4, of which 30.1±6.3% (215) of these clusters exhibited asymmetric (polar) expression (Figure 6.7(D-E). Noggin, a tiny molecule, has the ability to remove TBXT+ poles without affecting the growth of CER1-expressing cells, indicating that these posterior domains are dependent on local BMP4 signaling. This is consistent with previous research conducted in these tissues in iDiscoids and suggests a role for BMP4 from amnion in posterior axis specification. We then assessed A-P axis positionings by examining iDiscoid cultures. On day 4, we found that in CER1-expressing cells, 60.4±5.3% (102 of 169) of iDiscoids with polar configuration had a TBXT-expressing pole (Figure 6.7F). We saw two different patterns in iDiscoids with polarity in both TBXT and CER1 (Figure 6.7F, G). Comparable to the A-P axis positioning in mice at E6.5, 42.2±3.2% show anti-polar configuration, where the TBXT+ pole occupies the opposite radial region as the CER1+ domain (Figure 6.7 F-H). We observed high expression of CER1 near $TBXT^{+}$ domains in a second set of polar iDiscoids (Figure 6.7 F-H). Unlike previously published literature, this set of iDiscoids displays CER1+ and TBXT+ poles with the same radial region (syn-polar). In certain instances, we observed that a subset of $TBXT^{+}$ cells expressed CER1, and scRNA-seq analysis revealed the existence of a subpopulation of CER1+/TBXT+ cells in the cluster that denotes posterior identity.

When combined with an examination of the available *in vivo* data, our analyzes of iDiscoids imply that the expression and positioning of $TBXT^{+}$ and $CER1^{+}$ cells in iDiscoids can be observed *in vivo* and reveal different developmental stages. Syn-polar positioning and co-expression of CER1 and TBXT is indicative of more advanced stages post-E14, where CER1 expression acts to either prevent posterior domain from spreading or represent early mesendoderm commitment derived from a primitive streak. Anti-polar positioning of CER1 and TBXT domains reflect positioning of anterior hypoblast-like and posterior domains during early primitive streak formation around E14. We suggest using iDiscoids as a model to investigate the biology of the A-P axis, the development of anterior hypoblast-like cells, the posterior pole, and the cellular offspring of each in a human setting.

6.2.5. Specification of Yolk Sac Mesoderm-like Cells with Hematopoietic Potential

The yolk sac is an example of a multipurpose extra-embryonic tissue that develops from E16–18 in human development and serves as a center for early blood production in addition to providing nutritional support for the developing embryo. The lack of multicellular models of human yolk sac morphogenesis that include subsets of cells derived from mesoderm and possess *in vivo*-like structural features presents a challenge for the dissection of the molecular mechanisms underlying the process. In the iGATA6 compartment of iDiscoids, we observed progressive developmental maturation in line with yolk sac cellular fates (Figure 6.5G). We also investigated whether the development of mesodermal subsets within the growing yolk sac endoderm could be mimicked by iGATA6 cells in iDiscoids. When comparing the E16–19 human embryo to the clusters exhibiting the highest expression of GATA6 on days 4 (D4_G6) and 5 (D5_G3), hypergeometric statistical analysis demonstrates a strong resemblance to human yolk sac mesoderm, hematopoietic lineage progenitors, and hemogenic endothelial cells. The D5_G3 cluster exhibits increased expression of ECM proteins and the yolk sac mesoderm marker cell surface marker BST2. The markers of human yolk sac mesoderm (CREB3L1, NR2F2, PLAGL1, ANXA1, NID2), endothelial cell markers (CD34, PECAM1 (CD31), TEK, ICAM1, ETV2), and human hematopoiesis markers (GATA1, GATA2, KLF1, KLF2, ZEB1, ZEB2, GYPA (CD235a), and GYPB (CD235b)) are also more commonly expressed in subpopulations of cells within this cluster.

Within the tissues of the yolk sac, hematopoietic processes manifest as blood islands during *in vivo* development. Immunofluorescence analysis of iDiscoids first identifies cells with spindle shape that are positive for TAL1 (scl), a crucial regulator of blood development. Subsequent examination reveals that a portion of these cells also exhibited positive signals for RUNX1, a master transcription factor of hematopoiesis, and CD34 and ERG, markers linked to hematopoietic and endothelial fates. Z confocal section image analysis reveals that ERG+ cells regularly emerged beneath an iGATA6 layer. This morphology may resemble the initial *in vivo* localization of human yolk sac mesoderm, in which this tissue is positioned against extracellular matrix in the region beneath the yolk sac endoderm. $EGFP^+$ was also present in the observed $CD34^+$/$TAL1^+$ cells, suggesting that they originate from a parental population that is derived from iGATA6 and supporting a hypoblast origin (Figure 6.8C). Staining for KDR (VEGFR2) and CDH5, two additional endothelial markers, verified a similar localization in day 5 iDiscoids. After starting the colony-forming unit (CFU) assay with $CD34^+$ cells and analyzing the cultures using flow cytometry, the cultures produced CD45+ cells, erythroid cells ($CD71^+$, CD235a—(early stage) or $CD235a^+$ (late stage)), and $CD11b^+$ myeloid cells.

After noticing early hematopoietic potential in the iDiscoid on day 5, we continued to follow iDiscoids to gain a deeper understanding of the system's hematopoietic potential. A small amount of CD34+ area expansion was seen when the culture was kept on mTeSR. Therefore, after day five, we used basal media (IMDM without Dox) to aid in the differentiation of progenitor cells. Day 5 and Day 12 showed a significant expansion of the $CD34^+$ endothelial-like cell area, with the cell area covering 1.2±0.46% of the culture area on Day 5 and 14.7±6.0% on Day 12, which was further supported by flow cytometry analysis. During this period, we saw the formation of condensed regions with spherical cells inside the compartment that resembled a yolk sac (Figure 6.8D). Immunofluorescence staining of these regions demonstrated the presence of spherical cells expressing Leukosialin (CD43), a marker of hematopoietic progenitors, encircled by a CD31+ endothelial layer (Figure 6.8E. Numerous of these spherical cells were found to express hemoglobin and CD235a, two erythropoietic marker genes, upon closer examination. We further supplemented iDiscoid cultures at the time of seeding with high-GATA6 expressing cells and observed significantly increased generation of these $CD43^+$ hematopoietic foci, demonstrating predictable and controllable fate augmentation at the time of seeding. We had previously observed that yolk sac mesodermal and hemogenic progenitors of this cell type primarily correlating within the GATA6- hi clusters in scRNA-seq.

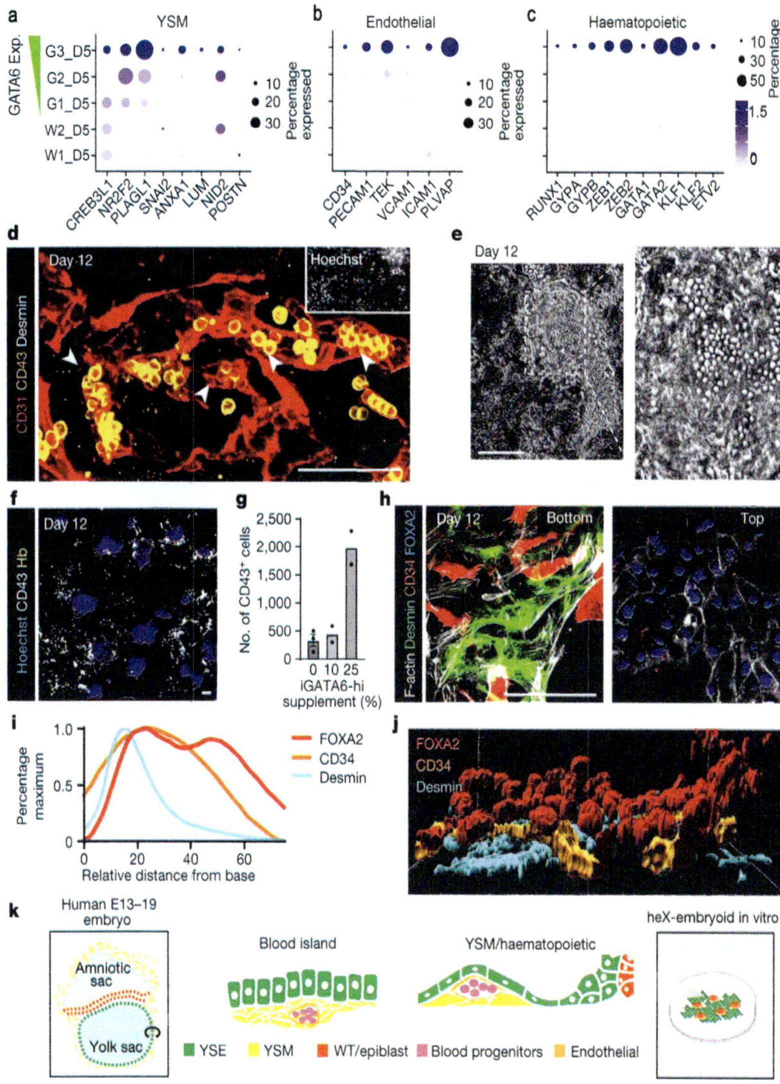

Figure 6.8. *iDiscoid hematopoietic lineages and blood island-like structures.*

Source: https://www.biorxiv.org/content/biorxiv/early/2023/06/15/2023.06.15.545118/F4.medium.gif.

CASE STUDY

CASE STUDY: TRYING TO CONCEIVE

Isabella, 28, and Omar, 30, have been together for three years. A year ago, they decided they wanted to have a baby, and they stopped using birth control. At first, they did not pay attention to the timing of their sexual activity in relation to Isabella's menstrual cycle, but after six months passed without Isabella becoming pregnant, they decided to try to maximize their efforts.

Figure 6.9. *Couples holding hands.*

Source: https://bio.libretexts.org/Bookshelves/Human_Biology/Human_Biology_(Wakim_and_Grewal)/22%3A_Reproductive_System/22.01%3A__Case_Study%3A_Making_Babies.

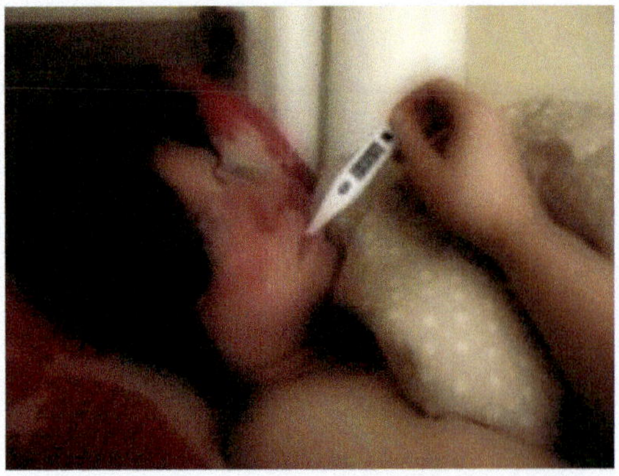

Figure 6.10. *Monitoring body temperature before getting out of bed in the morning can often help tell if and when a woman is ovulating. Usually, a special, highly sensitive thermometer is used.*

Source: https://bio.libretexts.org/Bookshelves/Human_Biology/Human_Biology_(Wakim_and_Grewal)/22%3A_Reproductive_System/22.01%3A__Case_Study%3A_Making_Babies.

They knew that in order for a woman to become pregnant, the man's sperm must encounter the woman's egg, which is typically released once a month through a process called ovulation. They had also heard that for the average woman, ovulation occurs around day 14 of the menstrual cycle. To maximize their chances of conception, they tried to have sexual intercourse on day 14 of Isabella's menstrual cycle each month.

After several months of trying this method, Isabella is still not pregnant. She is concerned that she may not be ovulating on a regular basis because her menstrual cycles are irregular and often longer than the average 28 days. Omar is also concerned about his own fertility. He had some injuries to his testicles (testes) when he was younger, and wonders if that may have caused a problem with his sperm.

Isabella calls her doctor for advice. Dr. Bashir recommends that she try taking her temperature each morning before she gets out of bed. This temperature is called basal body temperature (BBT), and recording BBT throughout a woman's menstrual cycle can sometimes help identify if and when they ovulate. Additionally, Dr. Bashir recommends she try using a home ovulation predictor kit, which predicts ovulation by measuring the level of luteinizing hormone (LH) in the urine. In the meantime, Dr. Bashir sets up an appointment for Omar to give a semen sample, so that his sperm may be examined with a microscope.

Here you will learn about the male and female reproductive systems, how sperm and eggs are produced, and how they meet each other to ultimately produce a baby. You will learn how these complex processes are regulated, and how they can be susceptible to problems along the way. Problems in either the male or female reproductive systems can result in infertility, or difficulty in achieving a successful pregnancy. As you read the chapter, you will understand exactly how BBT and LH relate to ovulation, why Dr. Bashir recommended that Isabella monitor these variables, and the types of problems she will look for in Omar's semen. At the end of the chapter, you will find out the results of Isabella and Omar's fertility assessments, steps they can take to increase their chances of conception, and whether they are ultimately able to get pregnant.

SUMMARY

- Human early development encompasses the critical period from conception through the first few years of life, characterized by rapid and complex physiological, cognitive, and emotional changes. This phase is foundational, as it sets the stage for future growth, health, and developmental outcomes.
- The fetal stage follows, from the ninth week of gestation until birth, characterized by the growth and maturation of already formed structures. The brain undergoes significant development, with the formation of neurons and synapses, which are crucial for cognitive function.
- Human reproduction is sexual reproduction that results in human fertilization to produce a human offspring. It typically involves sexual intercourse between a sexually mature human male and female.
- In order for human reproduction to be achieved, an individual must have undergone puberty first, requiring ovulation in females and the spermarche in males to have occurred prior to engaging in sexual intercourse or achieving pregnancy through non-penetrative means. Before puberty, humans are infertile, as their genitals lack reproductive function (only being able to discharge urine).
- The sex of a child is determined at the time of fertilization of the ovum by the spermatozoon. The differences between a male and a female are genetically determined by the chromosomes that each possesses in the nuclei of the cells. Once the genetic sex has been determined, there normally follows a succession of changes that will result, finally, in the development of an adult male or female.
- In females at puberty, the external genitalia enlarge and the uterus commences its periodic activity with menstruation. The breasts develop, and there is a deposition of body fat in accordance with the usual contours of the mature female. Growth of axillary (armpit) and pubic hair is more abundant, and the hair becomes thicker.
- In human embryos, the default sexual differentiation is female. However, having the Y chromosome defines differentiation into the male phenotype and the male reproductive system. The Y chromosome contains the sex-determining region (SRY) gene, which encodes for development of the testes. The testes descend from the posterior abdominal wall during development to lie in the scrotum at maturity.
- The hypothalamic-pituitary-gonadal axis plays a major role in promoting sexual maturity, sperm production and the development of secondary sex characteristics. It maintains spermatogenesis and sexual function throughout the male's lifetime. The hypothalamus secretes GnRH into the hypothalamo-hypophyzeal portal system to stimulate the anterior pituitary.
- The function of the male reproductive system is to produce androgens such as testosterone that maintain male reproductive function and to promote spermatogenesis and transport into the female reproductive system for fertilization. The testes act as both endocrine and exocrine organs in that they

are responsible for androgen production and sperm production and transport.

- During sexual arousal (physical or psychological), vasodilation brings blood to the penis. The penis contains corpora cavernosa and a corpus spongiosum where blood flows along to enlarge and erect the penis. As sexual stimulation continues, blood continues to flow to the genitals, and the testes enlarge in preparation of ejaculation.
- The female reproductive system comprises internal and external organs that facilitate menstruation and procreation. This organ system is responsible for producing gametes (termed eggs or ova), regulating sex hormones, and maintaining fertilized eggs as they develop into mature fetuses ready for delivery. A woman's reproductive years are between menarche (the first menstrual cycle) and menopause (cessation of menses for 12 consecutive months).
- The uterus is a pear-shaped muscular organ. Its major function is to accept a fertilized ovum, which becomes implanted into the endometrium, and derives nourishment from blood vessels, which develop exclusively for this purpose. The fertilized ovum becomes an embryo, develops into a fetus and gestates until childbirth. If the egg does not embed in the wall of the uterus, the female begins menstruation.

REVIEW QUESTIONS

1. What is the process of fertilization?
2. How do reproductive organs develop?
3. What are the key components and functions of the female reproductive system?
4. What are cell replacement therapies and how are they utilized?
5. What techniques and conditions are essential for derivation and culture in cell biology?
6. How is the co-development of embryonic and extra-embryonic endoderm tissues engineered?

REFERENCES

1. Bayerl, J., et al. (2021). Principles of signaling pathway modulation for enhancing human naïve pluripotency induction. *Cell Stem Cell, 28*, 1549–1565.
2. Carlson, B. M. (1999). Formation of germ layers and initial derivatives. In *Human Embryology & Developmental Biology* (pp. 62–68). Mosby, Inc. [First published 1997].
3. Delemarre-van de Waal, H. A. (1993). Environmental factors influencing growth and pubertal development. *Environmental Health Perspectives, 101*(Suppl 2), 39–44.
4. Jelenkovic, A., Ortega-Alonso, A., Rose, R. J., Kaprio, J., Rebato, E., & Silventoinen, K. (2011). Genetic and environmental influences on growth from late childhood to adulthood: A longitudinal study of two Finnish twin cohorts. *American Journal of Human Biology, 23*(6), 764–773.

5. Lampl, M., Veldhuis, J. D., & Johnson, M. L. (1992). Saltation and stasis: A model of human growth. *Science, 258*(5083), 801–803.

6. Merrick, J. (2013). Child health and human development over the lifespan. *Frontiers in Public Health, 1*, 1.

7. Oldak, B., et al. (2023). Transgene-free ex utero derivation of a human post-implantation embryo model solely from genetically unmodified naive PSCs. *bioRxiv*. https://doi.org/10.1101/2023.06.14.544922.

8. Rivron, N. C., Arias, A. M., Pera, M. F., Moris, N., & M'hamdi, H. I. (2023). An ethical framework for human embryology with embryo models. *Cell, 186*, 3548–3557.

9. Rossant, J., & Tam, P. P. L. (2021). Opportunities and challenges with stem cell-based embryo models. *Stem Cell Reports, 16*, 1031–1038.

10. Smith Agreda, V., Ferrés Torres, E., & Montesinos Castro-Girona, M. (1992). Organización del desarrollo: Fase de germinación. In *Manual De Embriología Y Anatomía General* (pp. 72–85). Universitat de València.

11. Tarazi, S., et al. (2022). Post-gastrulation synthetic embryos generated ex utero from mouse naive ESCs. *Cell, 185*, 3290–3306.e25.

CHAPTER

7

Techniques for Studying Organogenesis and Postnatal Development

LEARNING OBJECTIVES

After studying this chapter, you will be able to:

- Focus on genetics;
- Understand clonal analysis;
- Explain tissue and organ culture;
- Know cell analysis and separation.

KEY TERMS FROM THIS CHAPTER

Bioinformatics
Chromosomes
Cross-pollination
Heredity
Pedigrees
Tissue-culture
Cell separation
Clonal selection theory
Enzymes
Inheritance
Phenylketonuria
Chlorophyll
Codons
Genetics
Organ culture
Replication

INTRODUCTION

The process by which an organism develops its internal organs and particular structures, like its limbs, is known as organogenesis. The coordination of several developmental processes, including pattern formation, morphogenesis, cell proliferation, differentiation, and embryonic induction, is known as organogenesis. Puberty, childhood, and infancy are the three main stages of postnatal growth, each of which is impacted by a unique set of growth factors. Infant growth is mostly growth hormone independent, mostly determined by nutrition, and mostly a continuation of growth that occurs in utero.

7.1. GENETICS

Genetics is the study of genes in particular and heredity in general. One of the main tenets of biology is genetics, which has connections to many other fields including biotechnology, medicine, and agriculture. Ever since the beginning of human civilization, people have understood the importance of heredity and have used its principles to improve domesticated animals and crops. For instance, a more than 6,000-year-old Babylonian tablet displays horse pedigrees and suggests potential inherited traits. Date palm trees cross-pollinating is depicted in other ancient carvings. However, until the 19th century, when genetics emerged as a systematic science, the majority of the mechanisms underlying heredity remained unknown (Griffiths et al., 2000).

The discovery of genes, the basic building blocks of heredity, gave rise to the field of genetics. The study of genes at all levels, including how they function within cells and how they are passed down from parents to children, is known as genetics. Deoxyribonucleic acid, or DNA, is the chemical building block of genes. Modern genetics focuses on how DNA influences the chemical reactions that make up a cell's biological processes. Interaction with the environment is necessary for gene action. For instance, green plants have genes that carry the instructions needed to produce the chlorophyll, a pigment used in photosynthetic reactions that gives them their green color. Because the gene encoding chlorophyll is only expressed when it interacts with light, chlorophyll is synthesized in environments that receive light. Chlorophyll synthesis ceases in a plant in a dark environment because the gene is no longer expressed.

The mid-1800s work of Gregor Mendel laid the foundation for the scientific field of genetics. Despite his ignorance of the physical or chemical properties of genes at the time, Mendel's suspicion that traits were inherited as discrete units laid the groundwork for the current understanding of heredity. Mendel's discovery of the rules governing trait inheritance is the source of all current genetics research. English biologist William Bateson, who supported Mendel's theories of inheritance and was one of the people who first discovered Mendel's work, coined the term genetics in 1905 (Poczai & Santiago-Blay, 2022).

7.1.1. Historical Background

7.1.1.1. Ancient Theories of Pangenesis and Blood in Heredity

Even though Mendel's research produced the first scientific proof of genetic inheritance patterns, history demonstrates that curiosity about heredity must have existed among humans long before civilization emerged. First, curiosity had to have been sparked by similarities within the human family, including similarities in voice, body type, gait, and gestures. These ideas played a crucial role in the formation of families and royal dynasties. The characteristics of the animals that the early nomadic tribes domesticated, herded, and probably selectively bred interested them. It seems that the earliest human settlements to engage in farming chose crops with advantageous traits. Racehorse breeding pedigrees with distinct physical trait inheritance clearly depicted in ancient tomb paintings of the horses. Despite this curiosity, the earliest known theories about heredity date back to the ancient Greeks, some of whose concepts are still relevant today.

Hippocrates (*c.* 460–*c.* 375 BCE), regarded as the founding father of medicine, postulated pangenesis as a theory to explain his belief in the inheritance of acquired traits. According to his theory, every parent's organs released invisible "seeds" during sex that reassembled themselves in the mother's womb to form an infant. These seeds resembled tiny building blocks.

Aristotle (384–322 BCE) highlighted the role that blood plays in inheritance. He reasoned that blood served as the foundation for transmitting this generative power to the following generation and that blood supplied the generative material needed to build every part of the adult body. Actually, he thought that a woman's menstrual blood was her equal to semen and that a man's semen was pure blood. Together, the contributions of the male and female made a baby in the womb. Although he thought the baby would develop under the influence of these essences rather than being built from the essences themselves, the blood did contain some sort of hereditary essences.

The widely held belief that blood plays a role in heredity can likely be traced back to Aristotle's theories regarding the role of blood in reproduction. People still refer to some characteristics as being "in the blood," "blood lines," and "blood ties" in modern society. The Mendelian model of inheritance was not the same as the Greek model, which invoked a veritable rainbow of substances. Mendel proposed that unique individual differences are caused by variations in a single, potent hereditary factor. Genes were found to be the sole hereditary component. Gene copies are passed from sperm to egg and direct the development of the progeny. The unique characteristics of both parents that are evident in their offspring are also inherited by their offspring due to genetics.

7.1.1.2. Preformation and Natural Selection

There were not many new theories about the nature of heredity discovered in the two millennia that separated Aristotle and Gregor Mendel. The concept of preformation was first introduced in the 17th and 18th centuries. With the aid of recently developed microscopes, scientists conjectured that they might observe microscopic human duplicates inside sperm heads. The theory of "the inheritance of acquired characters," as proposed by French biologist Jean-Baptiste Lamarck, served as a model for evolution rather than an explanation for heredity. Although the fixity of species was taken for granted during the time he lived, he insisted that fixity could only exist in a stable environment. He established the law of use and disuse, which says that when certain organs develop uniquely in response to a need in the environment, that development is inherited and can be passed on to offspring. According to his theory, giraffes descended from deer-like creatures that had to constantly extend their necks in order to reach the tall leaves on trees over many generations (Griffiths et al., 2000).

Naturalist Alfred Russell Wallace of Britain is credited with developing the theory of evolution through natural selection. Nonetheless, natural selection was supported by Charles Darwin's observations made during his globe-circumnavigating voyage on the HMS Beagle (1831–1836), which also suggested a common ancestor between humans and animals. Mendel's experiments led to the theory of heredity, and Darwin's theories did not seem to fit with the hereditary mechanism that many scientists at the time believed in. This mechanism was a version of the ancient Greek concept of pangenesis.

7.1.1.3. The Work of Mendel

Prior to Gregor Mendel, most theories regarding a hereditary mechanism were grounded in conjecture and reasoning rather than scientific research. Mendel obtained pure-breeding lines of garden pea variants and conducted numerous cross-pollination experiments between them in his monastery garden. He noticed that all of the progeny seeds (the first generation, or F_1) were yellow when he crossed peas with yellow seeds to those with green seeds. The F_1 individuals' offspring (F_2) displayed a 3:1 ratio (3/4 yellow and 1/4 green) when they self-pollinated or crossed with one another. Because yellow is more prevalent than green, he reasoned that even though the F_2 generation contained some green individuals, the determinants of greenness had to have been present in the F_1 generation as well. He inferred not only the existence of discrete hereditary units (genes), but also that the units were present in pairs in the pea plant and that the pairs split during gamete formation from the exact mathematical 3:1 ratio (of which he discovered several other examples). Therefore, it was suggested that the two original lines of pea plants would be YY (yellow) and YY (green). These produced an F_1 generation of Yy that were yellow in color due to Y's dominance. The gametes from these were Y and y. The F_2 generation was created by random mating of 1/4 Yy, 1/2 YY, and 1/4 YY, which explains the 3:1 ratio. In the F1 generation, half of the gametes were Y and the other half were Y. Alleles are the variations of the pea color genes, Y and y (Szabó & Poczai, 2019).

Mendel also examined pure lines that varied between characters in pairs, like the shape and color of seeds (round versus wrinkled) and yellow versus green. The dominance of the yellow and round traits was demonstrated by the cross between yellow round seeds and green wrinkled seeds, which produced an F1 generation that was exclusively yellow and round. A 9:3:3:1 ratio, on the other hand, was displayed by the F2 generation that resulted from the self-pollination of F1 plants (9/16 yellow round, 3/16 yellow wrinkled, 3/16 green round, and 1/16 green wrinkled; keep in mind that a 9:3:3:1 ratio is just the sum of two 3:1 ratios). He inferred the independent assortment of distinct gene pairs during gamete formation from this and similar results.

Mendel's traditional experimental methodology is partly responsible for his success. He conducted numerous controlled experiments to gather data, making a wise choice for his experimental organism. He created astute explanatory theories based on his findings, which he then put to the test through experiments. Mendel's approach created a genetics prototype that is being utilized today to find new genes and comprehend the hereditary characteristics of inheritance.

7.1.1.4. How the Gene Idea Became Reality

Mendel's genes were merely postulated elements that one could assume to be real in order to account for his findings. The understanding of the nature and function of genes has advanced significantly during the 20th century. Mendel's works were not included in the scientific literature until 1900, when multiple other researchers came to the same conclusions. Hundreds of studies that demonstrated Mendelian inheritance in a broad range of plants and animals, including humans, came after that. Mendel's theories appeared to have broad applicability. Numerous biologists noticed that during nuclear divisions, or meiosis, which take place in cell divisions just before gamete formation, the inheritance of genes closely resembled the inheritance of chromosomes.

1. **The Discovery of Linked Genes:** Genes appeared to be components of chromosomes. The proof of the parallel inheritance of some Drosophila (a kind of fruit fly) genes on sex-determining chromosomes by American zoologist and geneticist Thomas Hunt Morgan in 1910 reinforced this theory. In addition to demonstrating that some genes appeared to be connected on the same chromosome, Morgan and Alfred Henry Sturtevant also demonstrated that the frequency at which new chromosomal combinations appeared could be used to determine the distance between genes on the same chromosome (these were proposed to be caused by chromosomal breakage and reunion, also known as crossing over). Calvin Bridges, a different pupil of Morgan's, utilized fruit flies with an extra chromosome in 1916 to establish beyond a shadow of a doubt that the presence of the extra chromosome was the only explanation for the aberrant inheritance of specific genes. Hermann Joseph Müller, an American geneticist, demonstrated for the first time that exposure to X-rays could cause cells to produce new alleles, or mutations, at high frequencies. (mutations can also arise spontaneously). American scientists Barbara McClintock and botanist Harriet Creighton showed in 1931 that newly discovered allelic combinations of linked genes were associated with physically exchanged chromosomal segments.
2. **Early Molecular Genetics:** The important theory that alkaptonuria, a human disease, and some other hereditary diseases are caused by inborn errors of metabolism was put forth by British physician Archibald Garrod in 1908; this was the first time that linked genes were suggested to have molecular action at the cell level. In 1941, American geneticist George Beadle and American biochemist Edward Tatum demonstrated that the genes they were examining in the fungus *Neurospora crassa* functioned by coding for catalytic proteins known as enzymes, marking the beginning of molecular genetics in earnest. This theory was expanded upon by research on various organisms, which demonstrated that genes typically code for proteins. The discovery that DNA makes up bacterial genes was soon demonstrated by American biologist Maclyn McCarty, Canadian American geneticist Colin M. MacLeod, and American bacteriologist Oswald Avery. This discovery was later extended to all organisms.

3. **DNA and The Genetic Code:** The development of the double helix model for DNA structure by American geneticist and biophysicist James D. Watson and British biophysicists Francis Crick and Maurice Wilkins in 1953 marked a significant milestone. Rosalind Franklin, a British scientist, contributed to their breakthrough by shedding light on the helical structure of the DNA molecule through X-ray diffraction studies. The double helix model demonstrated that complementary strands in DNA can be separated and used as templates to synthesize new DNA molecules, demonstrating the ability of DNA to replicate itself. It was suggested that each of the entwined strands of DNA was a chain of chemical groups known as nucleotides, of which there are four known varieties. Since amino acids make up proteins, it was suggested that a particular DNA nucleotide sequence may encode the amino acid sequence and, consequently, the structure of a protein. By expanding on previous research in Drosophila, American molecular biologist Seymour Benzer demonstrated in 1955 that the mutant sites within a gene could be mapped in relation to one another. According to his linear map, a gene's structure is linear.

   American geneticist Franklin W. and molecular biologist Matthew Meselson first demonstrated the strand-separation method (also known as the semiconservative method) for DNA replication experimentally in 1958. Stahl. It was demonstrated in 1961 by South African biologist Sydney Brenner and scientist Francis Crick that the genetic code needs to be read in triplets of nucleotides, or codons.

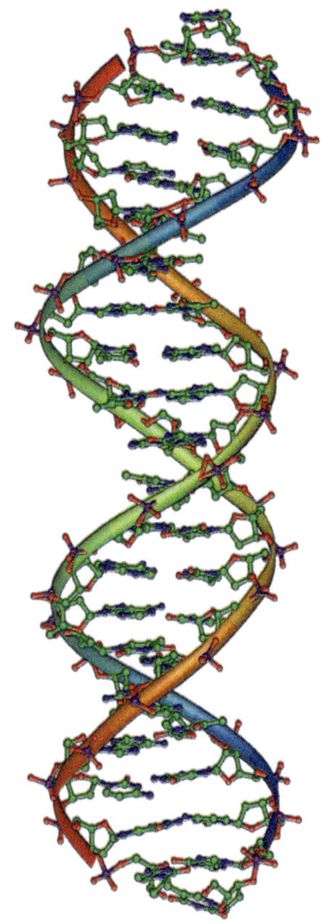

Figure 7.1. *DNA, the molecular basis for biological inheritance. Each strand of DNA is a chain of nucleotides, matching each other in the center to form what look like rungs on a twisted ladder.*

It was demonstrated by American geneticist Charles Yanofsky that the locations of mutant sites within a gene exactly matched the locations of changed amino acid sequences in the corresponding protein. The precise amino acids that each of the 64 potential triplet coding units (codons) codes for were determined in 1966 by American biochemists Marshall Nirenberg and Har Gobind Khorana. Further research on a wide range of organisms revealed that almost all living things, including fungi, bacteria, viruses, plants, and animals, share the same genetic code, replication mechanism, and double helical structure of DNA. The classic model for gene regulation was established in 1961 by the French biologists François Jacob and Jacques Monod, who demonstrated that regulatory proteins can bind to a region just upstream of a gene's coding region to turn on and off bacterial genes, thereby initiating transcription into RNA and protein synthesis.

4. **Recombinant DNA Technology and the Polymerase Chain Reaction**: The advancement of genetic understanding has been significantly aided by technological developments. The specialized class of enzymes known as restriction enzymes was discovered in 1970 by American microbiologists Daniel Nathans and Hamilton Othanel Smith. These enzymes cut DNA at particular nucleotide target sequences. The first artificial recombinant DNA molecule was created in the early 1970s by American biochemist Paul Berg thanks to this discovery. He isolated, cut, and joined DNA molecules from various sources in a test tube. The American biochemists Herbert W. and Katherine M. Stanley N. and Boyer. Cohen developed techniques for creating recombinant plasmids, or extragenomic circular DNA elements, that, when introduced into bacterial cells, underwent spontaneous replication. By splicing individual genes into self-replicating DNA molecules, like plasmids or viruses, and inserting these into living bacterial cells, these developments made it possible to clone (amplified to a high copy number) individual genes. These techniques gave rise to the recombinant DNA technology field, which eventually took over the field of molecular genetics. The English biochemist Fred Sanger and the American molecular biologists Allan Maxam and Walter Gilbert developed two distinct techniques in 1977 for figuring out the nucleotide sequence of DNA. Numerous conclusions regarding genes that were previously made indirectly have been confirmed as a result of these technologies, which enable direct examination of the structure of genes through nucleotide sequencing.

Site-directed mutagenesis, a method of inducing precisely tailored mutations at specific sites within a gene, was developed in the 1970s by Canadian biochemist Michael Smith, who transformed the field of gene redesigning. American biochemist Kary B. was born in 1983. Mullis invented the polymerase chain reaction, a method for rapidly detecting and amplifying a specific DNA sequence without cloning it. Complete DNA sequences of several viruses, bacteria, plants, and animals were determined in the final decade of the 20th century thanks to advancements in recombinant DNA technology and automated sequencing equipment. The full human DNA sequence, comprising roughly three billion nucleotide pairs, was released to the public in 2001.

7.1.2. Methods in Genetics

7.1.2.1. Experimental Breeding

It is possible to cross genetically distinct lines of organisms to create various allele combinations within a single line. For instance, crossing parental lines results in the F_1 generation, which is subsequently permitted to mate at random to create offspring with genotypes specific to (i.e., *AA*, *bb*, *cc*, or *DD*). New plant and animal lines are created through this kind of experimental breeding, and they are crucial for creating laboratory stocks for fundamental research. Genetically modified organisms (GMOs) are transgenic commercial lines created through experimentation that are used in commerce. Many of the plants and animals used by humans today (e.g., cows, pigs, chickens, sheep, wheat, corn (maize), potatoes, and rice) have been bred in this way.

7.1.2.2. Cytogenetic Techniques

The field of cytogenetics is concerned with the microscopic analysis of the genetic elements of cells, such as genes, chromosomes, and gene products. Earlier methods of cytogenetics include embedding cells in paraffin wax, cutting them into thin slices, and getting them ready for microscopic examination. Studying the contents of entire squashed cells is the newer and faster squash technique. They use dyes that specifically stain different areas of the cell; the genes, for instance, can be found by specifically staining the DNA that makes up each gene. The location of different genes and gene products within the cell can be ascertained with the use of radioactive and fluorescent tags. Before squashing, cells can be grown using tissue-culture techniques; for example, white blood cells can be grown from human blood samples and analyzed using the squash technique. The diagnosis of aberrant chromosomal complements, such as Down syndrome (resulting from an extra copy of chromosome 21) and Klinefelter syndrome (affecting males with an extra X chromosome), is one of the main uses of cytogenetics in humans. Certain prenatal diagnoses are made using placenta or amniotic fluid cell samples.

7.1.2.3. Biochemical Techniques

Cell extracts are typically used in biochemistry experiments conducted at the cellular or subcellular level. The primary chemical components of genetics, namely DNA, RNA, and protein, are studied using biochemical techniques. The activities of genes within cells and the substrates and products of gene-controlled reactions are both determined by biochemical techniques. One method involves grinding up the cells and fractionating the substitute chemicals for additional examination. Particular methods (e.g., chromatography and electrophoresis) are employed to disentangle the constituents of proteins, thereby unveiling hereditary variations in their structural makeup. For instance, it has been possible to identify more than 100 distinct types of human hemoglobin molecules. Compounds with radioactive tags are useful for researching the biochemistry of entire cells. One such molecule is thymine, which is unique to DNA. When radioactive thymine is added to a tissue-culture medium containing growing cells, genes utilize it to replicate. Analyzes of cells containing radioactive thymine reveal that the DNA molecule splits in half during duplication, with each half synthesizing the missing pieces.

Chemical tests are employed to differentiate between specific inherited human conditions, such as certain inherited abnormalities, such as phenylketonuria

(PKU), cystinuria, alkaptonuria, gout, and galactosemia, are detected by urinalysis and blood analysis. A series of diagnostic tests that can be performed on an individual's DNA have been made possible by genomics. A few of these tests are applicable to developing fetuses.

7.1.2.4. Physiological Techniques

Genetic investigations also employ physiological techniques aimed at investigating functional properties or organisms. The majority of genetic variations in microorganisms involve some crucial cell function. For example, certain strains of Escherichia coli can produce the vitamin thiamin from basic compounds; other strains, on the other hand, are dependent on the presence of thiamin to survive because they lack the enzyme required for this synthesis. By putting the two strains on a thiamin-free mixture, it is possible to distinguish between them: the strains that grow have the enzyme gene, while the strains that do not grow do not. Since many hereditary human abnormalities are brought on by a defective gene that is unable to produce an essential enzyme, the technique is also applied to human cells. Albinism, for example, is a condition in which a person lacks the ability to produce the pigment melanin, which is found in the skin, hair, or iris of the eyes.

7.1.2.5. Molecular Techniques

The direct study of DNA is at the heart of molecular genetics techniques, despite their overlap with biochemical techniques. The development of recombinant DNA technology has completely changed this field. To create recombinant DNA, a chromosome can be removed from a donor organism (like a human) and its DNA inserted into a vector. This recombinant DNA can then be amplified, altered, researched, or used to transgenically alter the genomes of other organisms. Amplification is a critical stage in the recombinant DNA technology process. This is accomplished by introducing the recombinant DNA molecule into a bacterial cell, which multiplies and creates numerous copies of both the recombinant DNA molecule and the bacterial genome (forming a DNA clone). A genomic library is a collection of many clones of donor DNA molecules that have undergone recombinant modification. These kinds of libraries serve as the foundation for sequencing whole genomes, like the human genome. Single nucleotide polymorphisms, or SNPs, are small molecular variants that function as chromosomal tags to associate specific regions of DNA that have a property of interest and may be involved in a human disease or disorder. These variants can now be found in genomes through scanning.

7.1.2.6. Immunological Techniques

Many substances (e.g., proteins) are antigenic; i.e., when introduced into a vertebrate body, they stimulate the production of specific proteins called antibodies. Red blood cells contain a variety of antigens, including those that comprise the major blood groups found in humans (A, B, AB, O). Immunogenetics is the study of these and other genetically determined antigens. Human blood antigens come in inherited variations, and an individual's specific antigen combination is nearly as unique as their fingerprints. This makes it useful for applications like paternity testing, though DNA-based methods have largely replaced this method.

Immunological methods are applied in organ transplants, blood transfusions, and the determination of Rhesus incompatibility during childbirth. The human leukocyte antigen (HLA) genes contain specific antigens that are linked to human diseases and disease predispositions. The seemingly limitless capacity of antibodies to match any antigen that is presented is based on unique DNA shuffling processes between antibody genes. Antibodies also have a genetic basis. Immunology is also helpful in locating particular recombinant DNA clones that produce a particular target protein.

7.1.2.7. Mathematical Techniques

Genetics uses a lot of mathematical techniques because a lot of its findings are based on quantitative data. When it comes to crossbreeding, the laws of probability can be applied to forecast the frequencies of particular genetic constitutions in progeny. In order to assess the significance of variations from expected outcomes in experimental analyzes, geneticists also employ statistical techniques. Furthermore, a lot of the foundation of population genetics is mathematical reasoning, such as the Hardy-Weinberg equilibrium and its derivatives.

Bioinformatics handles and analyzes the massive amounts of data generated by genome sequencing projects using computer-centered statistical techniques. The computer program searches the DNA for genes, inferring their likely function from other genes that are similar to it, and comparing various DNA molecules to examine how they have evolved over time. The field of systems biology, which views and analyzes a cell's genes and gene products as a whole integrated system, is made possible by bioinformatics.

7.2. CLONAL ANALYSIS

The immunological theory of clonal selection explains how lymphocytes, which are immune system cells, react to particular antigens that invade the body. Australian physician Frank Macfarlane Burnet first proposed the idea in 1957 in an effort to explain the wide range of antibodies that are produced when the immune response is first activated (Cohn et al., 2007). The theory has gained widespread acceptance as the model explaining how the human immune system reacts to infection and how particular B and T cell types are chosen for the purpose of destroying particular antigens (Rajewsky & Klaus, 1996).

The hypothesis posits that within an already-existing population of lymphocytes (both B and T cells), a particular antigen triggers (i.e., selects) only its counter-specific cell, which prompts the growth of that specific cell to create identical clones for the synthesis of antibodies. The spleen and lymph nodes are examples of secondary lymphoid organs where this activation takes place. To put it briefly, the theory explains how different antibody specificities are generated (Jordan, Margaret Baxter, & Alan, 2007). The first experimental proof was presented by Gustav Nossal and Joshua Lederberg in 1958, who demonstrated that a single B cell can only ever produce one antibody (Nossal, Lederberg, & Joshua, 1958). In particular, the concept proved to be the cornerstone of adaptive immunity in molecular immunology (Medzhitov, 2013).

7.2.1. Postulates

The clonal selection theory can be summarised with the following four tenets:

- A single type of receptor with a distinct specificity—produced by V(D)J recombination—is present on every lymphocyte.
- Receptor occupation is required for cell activation.
- The differentiated effector cells derived from an activated lymphocyte bear receptors of identical specificity as the parent cell.
- Those lymphocytes bearing receptors for self-molecules (i.e., endogenous antigens produced within the body) are destroyed at an early stage.

7.2.2. Early Work

Paul Ehrlich postulated the "side chain theory" of antibody production in 1900. It states that certain cells display distinct "side chains" (i.e., membrane-bound antibodies) with the capacity to respond to various antigens. An antigen attaches itself to a side chain that matches when it is present. After that, the cell stops making any more side chains and begins synthesizing and secreting the antigen-binding side chain in large quantities to form a soluble antibody. Ehrlich's theory was a selection theory that was far more accurate than the illuminating theories that dominated immunology in the following decades, even though it was different from clonal selection.

Niels Jerne, a Danish immunologist, proposed in 1955 that the serum contains a wide variety of soluble antibodies even before an infection. There is only one kind of antibody that is selected to match an antigen once it enters the body. According to theory, this happened as a result of some cells phagocytosing the immune complexes and in some way duplicating the structure of the antibody to make more of it (Burnet, 1976).

In 1957, David W. Talmage hypothesized that antigens bind to antibodies on the surface of antibody-producing cells and "only those cells are selected for multiplication whose synthesized product has affinity for the antigen". Ehrlich's theory differed significantly from this one in that each cell was assumed to synthesize a single type of antibody. Following antigen binding, the cell divides and produces clones that have the same antibodies.

7.2.2.1. Burnet's Clonal Selection Theory

A modification of Jerne's theory of antibody production using the concept of clonal selection was the title of a paper that Australian immunologist Frank Macfarlane Burnet published later in 1957 in the rather obscure Australian Journal of Science. Burnet developed Talmage's concepts and called the resulting theory "clonal selection theory" in it. In his 1959 book *The Clonal Selection Theory of Acquired Immunity*, he further codified the theory. He described immune memory as the result of two different lymphocyte types being cloned. While one clone fights infection right away, the other takes longer to work, staying in the immune system and building immunity to that antigen. Burnet's hypothesis states that antibodies contain molecules that can likely, albeit to differing degrees of accuracy, correspond to all or almost all of the antigenic determinants found in biological material aside from those that are unique to the body. The core of the theory is that every cell naturally has representative reactive sites similar to those of the globulin they produce on its surface. Each type of pattern is a particular byproduct of a clone of lymphocytes.

It is believed that an antigen will adhere to the surface of any lymphocyte containing reactive sites that match one of its antigenic determinants when it enters the bloodstream or tissue fluids. Subsequently, the cell is triggered and experiences growth to generate an array of progeny. Thus, all clones whose reactive sites match the antigenic determinants on the body's antigens are initiated into preferential proliferation (Hodgkin et al., 2007). As with the parent forms, the descendants can actively release soluble antibodies and lymphocytes (Jordan, Margaret, Baxter, & Alan, 2007).

The first concrete proof in favor of the clonal selection theory was provided in 1958 when Gustav Nossal and Joshua Lederberg demonstrated that a single B cell can only ever produce one antibody (Nossal, Lederberg, & Joshua, 1958).

7.2.3. Theories Supported by Clonal Selection

Together, Burnet and Peter Medawar worked to understand immunological tolerance, a phenomenon that clonal selection also explains. As long as the introduction of cells happens early in the organism's development, this is the organism's capacity to withstand their introduction before the emergence of an immune response. The immune system contains a wide variety of lymphocytes, including both self-terrorizing and non-self-terrorizing cells. Only self-tissue-tolerant cells, though, make it through the embryonic stage. The lymphocytes that develop are the ones that incorporate the non-self-tissues as self-tissue if non-self-tissue is introduced.

Burnet (1959) postulated that foreign recipients could receive successful tissue transplants in specific situations. Both the immune system and tissue transplantation have greatly advanced as a result of this work. In 1960, Burnet and Medawar were jointly awarded the Nobel Prize in Physiology or Medicine. According to Niels Kaj Jerne's 1974 theory, interactions between the variable components of lymphocytes and the molecules they secrete regulate how the immune system operates as a network. Clonal selection is a fundamental tenet of immune network theory. Jerne's contributions to immune network theory were a major factor in his 1984 Nobel Prize in Physiology or Medicine victory.

7.3. TISSUE AND ORGAN CULTURE

Organ culture is the process of growing organs or portions of organs *in vitro* while preserving the anatomical relationships and functions of the various tissue components—such as parenchyma and stroma—so that the transplanted tissue closely resembles the parent tissue *in vivo*. Proper culture conditions discourage and minimize the outgrowth of isolated cells from the periphery of the explants, and the resulting "new" growth is made up of differentiated structures. As a result, new small bronchi form at the edge of the explant in lung tissue, or new glandular structures form in glands. They are made up of alveoli lined by bronchial epithelium or secretory, cuboidal, or columnar glandular epithelium. The epithelium follows a similar pattern of differentiation as in the organs *in vivo* in tissues lined with squamous epithelium, such as skin or the esophagus, or in bladder lined with transitional epithelium. Endocrine organs continue to secrete particular hormones, and hormone-dependent tissues remain responsive and sensitive to hormones. In conclusion, morphogenesis observed in fetal tissues *in vitro* is similar to that observed *in vivo* (Al-Lamki et al., 2017).

7.3.1. Methodology

Preserving the tissue's architecture and guiding it toward a normal development path are two important goals of organ culture. It is crucial to this technique that there be no disruption or damage to the tissue. It must therefore be handled carefully. Generally speaking, the media used for tissue culture are also used for growing organ cultures. Organ culture methods can be divided into two categories: (i) solid medium methods and (ii) liquid medium methods.

7.3.1.1. In vitro Culture

An easier option to regular organ culture derived from adult animals is embryonic organ culture. The four methods used for embryonic organ culture are as follows:

1. Plasma Clot Method: The general procedures for organ culture on plasma clots are as follows.

- In a watch glass, combine 15 drops of plasma and 5 drops of embryo extract to create a plasma clot.
- In order to prevent excessive evaporation from the dish, place a watch glass on top of a pad of cotton wool in a petri dish.
- Cover the plasma clots in watch glass with a tiny, finely cut piece of tissue.

The method has now been changed, and the tissue is placed on a raft made of rayon net or lens paper. The tissue can then be transferred with ease using a raft. After removing extra fluid, the net containing the tissue is repositioned on the newly created medium pool.

2. Agar Gel Method: Organ culture also uses agar-solidified media, which are made up of seven parts horse serum, three parts chick embryo extract, and one part 1% agar in BSS. Agar is also used with defined media, either with or without serum. The mechanical support for organ culture is provided by the agar-containing medium. It is not liquefiable. Agar generally promotes the growth of embryonic organs, but adult organ culture cannot be sustained on it. Because adult animals require more oxygen to thrive, it is more challenging to culture their organs or parts. A variety of adult organs (e.g., the liver) have been cultured with special equipment (Towell's II culture chamber) and special media. Serum was discovered to be toxic, so serum-free media were employed instead, and a special device allowed for the use of 95% oxygen.

3. Raft Methods: This method involves placing the explant onto a raft made of rayon acetate or lens paper that is floating on serum inside a watch glass. By applying silicone to each of the four corners of rayon acetate rafts, they are able to float on the serum. In a similar vein, applying silicone to lens paper improves its floatability. Four or more explants are typically placed on each raft. When using a combination of raft and clot techniques, the explants are first stored on a plasma clot and then placed on an appropriate raft. This modification keeps explants from sinking into liquefied plasma and facilitates simple media changes.

4. Grid Method: The grid method was first developed by Trowell in 1954. It makes use of 25 mm by 25 mm pieces of appropriate wire mesh or perforated stainless steel sheet, the edges of which are bent to form four legs that are approximately 4 mm tall. Softer tissues, such as glands or skin, are first placed on rafts, which are then maintained on the grids; skeletal tissues are typically placed directly on the grid. In order to meet the high O_2 requirements of adult mammalian organs, the grids themselves are placed in a culture chamber that is filled with fluid medium up to the grid. The chamber is then supplied with a mixture of O_2 and CO_2. A popular approach to studying the development and differentiation of adult and embryonic tissues is a modification of the original grid method.

7.3.2. Uses

An alternative to organs from other (dead or living) people is cultured organs. This is helpful because the number of transplantable organs (organs taken from living individuals) is decreasing in developed nations. An additional benefit is that cultured organs, which are made from the patient's own stem cells, enable organ transplants in which the immunosuppressive medication is no longer needed for the patient.

7.3.3. Limitations

Findings from studies conducted *in vivo* are frequently not transferable to *in vitro* organ cultures (e.g., research on medication action) because medications are not metabolized *in vitro* but rather *in vivo*.

7.4. CELL ANALYSIS AND SEPARATION

Cell separation was first achieved in 1964 by S. H. Seal, who separated large tumor cells from the smaller blood cells using filtration. Two years later, Pohl and Hawk conducted research on the various ways that different types of cells react to a nonuniform electric field. These discoveries led to the development of cell separation techniques, which enable the isolation of a particular cell type from a heterogeneous sample (Atajanov, Zhbanov, & Yang, 2018).

7.4.1. Cell Isolation Techniques

Cell isolation techniques are ways to take specific cells out of a complex mixture of cells and transfer them to other cells to form single cells or to arrange the cells in a desired way to create a homogenous cell population.

7.4.1.1. Circulating Cells

The methods needed to obtain isolated cells differ according to the type of cell that is needed. By drawing blood, circulating cells like blood cells or some tumor cells can be separated (Bhagwat & Carpenter, 2017). Blood samples comprise a diverse range of cell types, so a technique for sorting cells into different types is required. Flow cytometry is the most widely used technique for this, where a narrow stream of cells is inspected by an automated analyzer. One variation of this method involves shining a light on the stream of cells, and then quickly moving the cells of interest into a collection chamber using the analyzer's detection of the reflected light or fluorescence (Hu, Zhang, Xin, & Deng, 2016).

7.4.1.2. Solid Tissues

Obtaining tissue for cell isolation may be more difficult when working with solid tissues. Human tissue surplus may occasionally be obtained during planned surgery. For instance, during open heart procedures like coronary artery bypass surgery, specimens of the right atrial appendage are frequently removed and discarded (Voigt, Pearman, Dobrev, & Dibb, 2015). Samples of other tissues, such as bladder or pancreas, may be obtained for a biopsy. Alternatively, sacrificing the animal is a common way to obtain tissue from them. Once a tissue specimen has been acquired, it needs to be immersed or perfused with a solution that contains the nutrients and salts needed to maintain the cells' viability at the proper temperature. This can be done as simple as immersing the tissue in the solution, or it can entail more intricate setups like Langendorff perfusion. Typical solutions were variations of Krebs and Henseleit's solution or Tyrode's solution. These solutions have specific electrolyte concentrations, such as those of sodium, potassium, calcium, magnesium, phosphate, chloride, and glucose. Osmotic pressure must be taken into consideration when carefully balancing the concentrations of these electrolytes. It is necessary to control the solution's acidity, frequently with the use of a pH buffer like HEPES. Oxygenating the solution may help with tissue-specific cell isolation. Initially, it is helpful to perfuse the tissue with a calcium-free solution, especially when isolating cardiac myocytes, since the lack of calcium separates the intercalated disks. The solution can then be supplemented with proteolytic enzymes. When isolating cells from the heart or bladder, collagenases—enzymes that break down collagen—are frequently utilized (Kloskowski et al., 2014). Proteases, which are all-purpose enzymes that break down different types of protein, can also be employed. It might be necessary to use additional DNA-breaking enzymes, or DNAases, when separating cells from brain tissue. Apart from breaking down the extracellular matrix, these enzymes are also capable of breaking down other crucial proteins that are necessary for the targeted cells to operate. Cell death occurs when cells are exposed to these enzymes for an extended period of time, but incomplete extracellular matrix digestion occurs when cells are not exposed to the enzymes for an adequate amount of time. Cells can be mechanically separated, or dissociated, once the tissue has been perfused with a second solution devoid of enzymes to eliminate the enzymes from it. Cutting the tissue into small chunks and then agitating the chunks in a solution with a pipette is a basic method for dissociating cells.

7.4.2. Applications of Cell Separation and Isolation

In vitro cultivation of isolated cells can yield cell lines, which have a wide range of uses in fields like immunology, pharmacology, and stem cell therapy. One way to examine a drug's *in vitro* effects is on particular cell populations. Selecting the right B-cells for the synthesis of monoclonal antibodies also requires cell separation. Single-cell analysis also depends on isolating particular cells for investigation, as in the case of gene expression patterns and epigenetic effects. It is often necessary to separate specific tumor cells from the tissue in order to conduct oncological investigations that advance our knowledge of cancer cells. Therefore, practically all of the main disciplines of contemporary biology employ cell isolation and separation techniques.

7.4.3. Challenges of Cell Separation Methods

The challenge faced by cell separation techniques is a high noise-to-signal ratio, where a small number of target cells are present among the many different components that make up the tissue. The final isolate's purity is influenced by a number of factors, so standardizing the protocols for every target cell type and application is necessary. For instance, the yield of isolated cells is determined by how well the cells separate from the tissue. While incomplete digestion leads to cell type contamination, excessive digestion increases the amount of dead cells in the isolate. Affinity-based cell separation may also result in cell type contamination. Non-specific binding may happen when labeling a specific cell type with specific antibodies, introducing different cells into the preparation. Another limiting factor is the target cell type's population size in the tissue; low-abundance cells frequently require laborious protocols to produce enough numbers in the isolated cell populations.

CASE STUDY

COMATOSED NEWBORN: INBORN ERROR OF METABOLISM

A four day old male newborn was referred from a tertiary care NICU centre in Mumbai. The baby was born to a young couple who were related to each other by blood (first cousins). The child was born through normal vaginal delivery and weighed 3 kilograms at birth. After an initial uneventful period of two days, the baby suddenly had decreased feeding on the third day of life. This was followed by seizures and altered sensorium. He was comatose and unresponsive since the fourth day of life; necessitating mechanical ventilation and high-level intensive unit care. Inspite of the best medical care, the child succumbed on the fifth day. There was a significant family history. This couple had lost two previous children in an eerily similar fashion. Sensing that there could be a genetic basis to the given recurring disease, the physician had sought timely genetic opinion in this last baby's case. A detailed and accurate pedigree analysis was outlined. An inborn error of metabolism was suspected. Also, special genetic tests were routed in time, not losing the precious window of opportunity to test the child.

The initial biochemical tests also pointed towards an inborn error of metabolism; proximal urea cycle defect or a mitochondriopathy. However, a biochemical diagnosis is not sufficient to prevent recurrence of the disease in the next child. For prevention of recurrence, one needs to know the error 'at the level of the gene' (i.e., the mutation).

The child's DNA was analyzed after its death, by highly sophisticated technology, next-generation sequencing. This picked up the causative DNA error (mutation) which had resulted in three newborn deaths in the family. The mystery was unveiled; the cause was a genetic disease, a urea cycle defect. This disorder occurs, when each parent has one healthy copy of a gene and the other copy is a disease-causing one. When present in a single dose, the genetic error is insufficient to cause the disease, thus the parents remain asymptomatic throughout life. However, if the child inherits a disease-causing copy each from both the parents (homozygosity), it leaves no normal gene at the given position in the child. This precipitates the disease in such a child. However, by permutation and combination it is likely that the child may inherit both the healthy copies form both the parents or just one disease-causing gene, and the other one being healthy. In the latter two scenarios the child will remain asymptomatic. Having clinched the diagnosis at the level of the gene, it was possible to inform the couple an accurate risk of recurrence of the same disease in every subsequent pregnancy (25%).

With the genetic error outlined in this family, the couple was informed, that in the next pregnancy, they can select the embryos which do not have the given genetic mutation (preimplantation genetic diagnosis) and ensure the birth of an unaffected child. However, since this technology is relatively expensive, the couple was also counseled about the other option of prenatal diagnosis. In this method, when the mother is around

11–13 weeks pregnant, a small quantity of the fetal DNA is procured through a safe procedure (chorion villous biopsy). This fetal DNA can be checked for the disease-causing mutation in double dose (homozygosity). If present it predicts the same outcome as the previous pregnancies, in this baby too once it is born. In such a case, the parents can opt to discontinue the pregnancy medically and legally, before 20 weeks of pregnancy. On the other hand, if the baby inside is normal or carries just one copy of the disease-causing mutation, then the parents can continue with the pregnancy without stress and anxiety.

CLASS ACTIVITY

SEEING PEDIGREE SCIENCE: MAKING A FAMILY TREE OF TRAITS

Introduction

Has anyone ever told you that you look just like one of your parents or grandparents? Some characteristics, like the shape of your hairline or whether your earlobes are attached or detached, are inherited. In this activity you'll get to see how writing some characteristics onto a family tree can help you determine just how you inherited them. You will likely discover some characteristics that you got from your father, and for this Father's Day you can thank your father for giving them to you.

Background

When we look at members of a family it's easy to see that some physical characteristics, or traits, are shared. In the 1860s, Gregor Mendel discovered that in pea plants some traits are passed down in clear and predictable patterns. Today we know that offspring inherit half of their DNA from each parent. This results in two copies of every gene. Many genes come in different versions, called alleles. Alleles are changes in the actual DNA sequence of the gene. If you have two identical alleles, you are said to be homozygous for that gene, whereas if you have two different alleles you're heterozygous.

Some alleles are dominant, meaning that if you have one copy of that allele you will display that trait. Other alleles are recessive, meaning you need two copies of that allele to display the trait. For example, Mendel took pea plants that were homozygous for different traits and crossed them. When crossing homozygous purple flowered plants to homozygous white flowered plants, the offspring had purple flowers. The purple allele was dominant. When the heterozygous offspring were crossed to each other, they sometimes had white flowers. The white allele was recessive.

Materials

- Paper and a pencil or pen.
- Access to a photocopier or four sheets of paper total.
- At least three generations of people from the same family. The more members of the family that are available, the better the results will be.

Preparation

1. Draw a family tree, or pedigree, showing the different members of your family. Include all of the family members that you will be getting data from. You can designate the males by a square and the females by a circle.
2. Make three copies of your family tree (so you have four total).
3. Label each family tree one of the following: "Earlobes," "Widow's Peak," "Mid-digit Hair," and "Hitchhiker's Thumb."

Instructions

1. Gather the family members together that you put on your family tree. Alternatively you can do this activity with some members separately.
2. Determine whether each family member has attached earlobes or detached earlobes. (Attached earlobes are clearly attached to the side of the head at the bottom of the earlobe, whereas detached earlobes are not.) There can be a range of attachment – just do your best to determine how attached the earlobes are. Do many relatives have the same type of earlobe, or is there variation? Can you see how this trait was inherited? Write down your results in your "Earlobes" family tree.
3. Determine whether each family member has a widow's peak or not. (A widow's peak is where the hairline comes to a V-shaped point above the person's forehead.) Keep in mind that widow's peaks can vary considerably – when determining if a person has a widow's peak, count any sort of V-shaped hairline as a widow's peak. Do many relatives either have a widow's peak or don't have one? Can you see how this trait was inherited? Write your results in your "Widow's Peak" family tree.
4. Determine whether each family member has hair on their mid-digits (the middle joints on your fingers) or does not have mid-digit hair. You may need to look closely at each person's hands – if they have any hair on the mid-digit, even a tiny strand, then they have mid-digit hair. Do many relatives have hair on their mid-digits, or are they hairless? Can you see how this trait was inherited? Write your results in your "Mid-digit Hair" family tree.

5. Have each family member make a fist with their thumb sticking up and determine whether their thumb is straight or curved. (A curved thumb is also called a hitchhiker's thumb.) Keep in mind that thumbs come in a wide range of curvedness, from completely straight to completely curved – do your best to decide if a person's thumb looks curved or straight. Do many relatives have a hitchhiker's thumb, or are their thumbs mostly straight? Can you see how this trait was inherited? Write your results in your "Hitchhiker's Thumb" family tree.
6. Overall, can you see how different traits were passed down through your family? Are there traits that skipped a generation? Are there traits that were in every generation? Can you Figure out which traits might mostly be recessive and which might mostly be dominant?

Extra: There are several other genetic traits that you could investigate using this activity. How does it look like other traits are inherited, such as freckles, cleft chins, toe lengths (whether the big toe is longest or the second toe is longest), and which thumb is on top when you interlace your fingers? Extra: In this activity you looked at existing family members to investigate how different traits are inherited. Can you use your knowledge from this activity to predict what traits future offspring of different family members might have, if you know what traits their partner has? Extra: People have long debated whether handedness is a genetic trait. One scientific study showed a correlation between handedness and hair whorl direction, specifically suggesting that a greater percentage of left-handed people have counterclockwise hair whorls than right-handed people. Can you find a correlation between handedness and the direction of peoples' hair whorls in your family?

Observations and Results

Could you see how some traits were passed down? Did it look like having detached earlobes, a widow's peak, mid-digit hair, and a straight thumb were all dominant traits? While earlobe attachment, presence of a widow's peak, mid-digit hair, and having a hitchhiker's thumb are thought to be primarily genetically inherited traits, there is some controversy over exactly how they are inherited – in other words, they may be affected by more than one gene, more than two alleles, or factors other than genetics. That said, having detached earlobes, a widow's peak, mid-digit hair, and a straight thumb are generally considered to be dominant traits, although clearly their inheritance is complex. A dominant trait is one that only needs one copy of the dominant allele to be displayed. For a person to show a recessive trait, they generally need two alleles for the recessive trait. Because a person only needs one copy of the dominant allele to show the trait, they could be homozygous for the allele or heterozygous. This means that a person who shows a dominant trait could have a child with a person who shows the recessive trait (or even another person who shows the dominant trait) and the child may show either trait (since the parent with the dominant trait may be heterozygous). Two people who both show a recessive trait are most likely to have a child that also shows the recessive trait, but inheritance is often affected by other factors.

SUMMARY

- Genetics, study of heredity in general and of genes in particular. Genetics forms one of the central pillars of biology and overlaps with many other areas, such as agriculture, medicine, and biotechnology. Since the dawn of civilization, humankind has recognized the influence of heredity and applied its principles to the improvement of cultivated crops and domestic animals.
- Genetically diverse lines of organisms can be crossed in such a way to produce different combinations of alleles in one line. For example, parental lines are crossed, producing an F_1 generation, which is then allowed to undergo random mating to produce offspring that have purebreeding genotypes (i.e., *AA*, *bb*, *cc*, or *DD*).
- Cytogenetics focuses on the microscopic examination of genetic components of the cell, including chromosomes, genes, and gene products.
- Biochemistry is carried out at the cellular or subcellular level, generally on cell extracts. Biochemical methods are applied to the main chemical compounds of genetics—notably DNA, RNA, and protein.
- Physiological techniques, directed at exploring functional properties or organisms, are also used in genetic investigations.
- A fundamental step in recombinant DNA technology is amplification. This is carried out by inserting the recombinant DNA molecule into a bacterial cell, which replicates and produces many copies of the bacterial genome and the recombinant DNA molecule (constituting a DNA clone).
- In immunology, clonal selection theory explains the functions of cells of the immune system (lymphocytes) in response to specific antigens invading the body.
- Burnet and Peter Medawar worked together on understanding immunological tolerance, a phenomenon also explained by clonal selection. This is the organism's ability to tolerate the introduction of cells prior to the development of an immune response as long as it occurs early in the organism's development.
- Organ culture means the explanation and growth *in vitro* of organs or part of organs in which the various tissue components, such as parenchyma and stroma, and their anatomical relationship and function, are preserved in culture, so that the explanted tissue closely resembles its parent tissue *in vivo*.
- A key objective of organ culture is to maintain the architecture of the tissue and direct it towards normal development. In this technique, it is essential that the tissue is never disrupted or damaged. It thus requires careful handling.
- Cultured organs can be an alternative for organs from other (living or deceased) people. This is useful as the availability of transplantable organs (derived from other people) is declining in developed countries.
- Cell Isolation techniques are methods to separate and to transfer certain cells from a complex mixture of cells to obtain single cells or to sort the cells according to a property of choice and thus to generate a homogenous cell population.

- Isolated cells can be grown *in vitro* to produce cell lines, which have numerous applications in areas like pharmacology, immunology, and stem cell therapy. For example, the *in vitro* effects of drugs can be analyzed on specific cell populations.
- Cell separation techniques face a high noise-to-signal ratio, with a small number of specific cells (target) against a large number of varied components that form the tissue. Various factors affect the purity of the final isolate; hence, the protocols require standardization for each target cell type and application.

REVIEW QUESTIONS

1. What are the types of genetics?
2. Why is genetics important?
3. What does clonal mean in genetics?
4. What is meant by organ culture?
5. What are the techniques used in tissue and organ culture?
6. What is cell isolation?
7. What are the different cell isolation techniques and how do they work?

REFERENCES

1. Al-Lamki, R. S., Bradley, J. R., & Pober, J. S. (2017). Human organ culture: Updating the approach to bridge the gap from *in vitro* to *in vivo* in inflammation, cancer, and stem cell biology. *Frontiers in Medicine, 4*, 148. https://doi.org/10.3389/fmed.2017.00148.
2. Atajanov, A., Zhbanov, A., & Yang, S. (2018). Sorting and manipulation of biological cells and the prospects for using optical forces. *Micro and Nano Systems Letters, 6*(2). https://doi.org/10.1186/s40486-018-0064-3.
3. Bhagwat, N., & Carpenter, E. L. (2017). Flow cytometric methods for circulating tumor cell isolation and molecular analysis. In *Isolation and Molecular Characterization of Circulating Tumor Cells* (pp. 105–118). Advances in Experimental Medicine and Biology, Vol. 994. https://doi.org/10.1007/978-3-319-55947-6_5.
4. Burnet, F. M. (1976). A modification of Jerne's theory of antibody production using the concept of clonal selection. *CA: A Cancer Journal for Clinicians, 26*(2), 119–121. https://doi.org/10.3322/canjclin.26.2.119.
5. Cohn, M., Mitchison, N. A., Paul, W. E., Silverstein, A. M., Talmage, D. W., & Weigert, M. (2007). Reflections on the clonal-selection theory. *Nature Reviews Immunology, 7*(10), 823–830. https://doi.org/10.1038/nri2177.
6. Griffiths, A. J., Miller, J. H., Suzuki, D. T., Lewontin, R. C., & Gelbart, W. M. (Eds.). (2000). Introduction. In *An Introduction to Genetic Analysis* (7th ed.). New York: W. H. Freeman.

7. Hodgkin, P. D., Heath, W. R., & Baxter, A. G. (2007). The clonal selection theory: 50 years since the revolution. *Nature Immunology, 8*(10), 1019–1026. https://doi.org/10.1038/ni1007-1019.

8. Hu, P., Zhang, W., Xin, H., & Deng, G. (2016). Single cell isolation and analysis. *Frontiers in Cell and Developmental Biology, 4*, 116. https://doi.org/10.3389/fcell.2016.00116.

9. Jordan, M. A., & Baxter, A. G. (2007). Quantitative and qualitative approaches to GOD: The first 10 years of the clonal selection theory. *Immunology and Cell Biology, 86*(1), 72–79. https://doi.org/10.1038/sj.icb.7100140.

10. Kloskowski, T., Uzarska, M., Gurtowska, N., Olkowska, J., Joachimiak, R., Bajek, A., Gagat, M., Grzanka, A., Bodnar, M., Marszałek, A., & Drewa, T. (2014). How to isolate urothelial cells? Comparison of four different methods and literature review. *Human Cell, 27*(2), 85–93. https://doi.org/10.1007/s13577-013-0070-y.

11. Medzhitov, R. (2013). Pattern recognition theory and the launch of modern innate immunity. *The Journal of Immunology, 191*(9), 4473–4474. https://doi.org/10.4049/jimmunol.1302427.

12. Nossal, G. J. V., & Lederberg, J. (1958). Antibody production by single cells. *Nature, 181*(4620), 1419–1420. https://doi.org/10.1038/1811419a0.

13. Poczai, P., & Santiago-Blay, J. A. (2022). Themes of biological inheritance in early nineteenth century sheep breeding as revealed by J. M. Ehrenfels. *Genes, 13*(8), 1311. https://doi.org/10.3390/genes13081311.

14. Rajewsky, K. (1996). Clonal selection and learning in the antibody system. *Nature, 381*(6585), 751–758. https://doi.org/10.1038/381751a0.

15. Szabó, A. T., & Poczai, P. (2019). The emergence of genetics from Festetics' sheep through Mendel's peas to Bateson's chickens. *Journal of Genetics, 98*(2), 63. https://doi.org/10.1007/s12041-019-1108-z.

16. Voigt, N., Pearman, C. M., Dobrev, D., & Dibb, K. M. (2015). Methods for isolating atrial cells from large mammals and humans. *Journal of Molecular and Cellular Cardiology, 86*, 187–198. https://doi.org/10.1016/j.yjmcc.2015.07.006.

CHAPTER

8

Development of the Nervous System

LEARNING OBJECTIVES

After studying this chapter, you will be able to:

- Focus on overall structure and cell types.
- Discuss anteroposterior patterning of the neural plate.
- Define neurogenesis and gliogenesis.
- Understand the neural crest.
- Explain the development of neuronal connectivity.

KEY TERMS FROM THIS CHAPTER

Ependymal cells	Gliogenesis	Growth cone
Neural crest	Neural tube	Neuroblast
Neurogenesis	Neuroplasticity	Neurotrophins
Radial glia	Schwann cells	Synaptogenesis
Retinal development		

INTRODUCTION

The development of the nervous system, known as neurodevelopment, is a complex and dynamic process that begins early in embryonic life and continues into adulthood. It involves the formation, differentiation, and maturation of the nervous system's diverse cell types and structures. Neurodevelopment starts with the neural induction phase, where the neural plate forms from the ectoderm, the outermost layer of the embryo. This plate folds to create the neural tube, which will eventually develop into the brain and spinal cord. Proper closure of the neural tube is crucial, as defects can lead to conditions such as spina bifida or anencephaly.

As development proceeds, neurogenesis occurs, where neural stem cells proliferate and differentiate into various types of neurons and glial cells. Neurons then migrate to their destined locations, guided by a combination of genetic and environmental cues. This migration is essential for the establishment of functional neural circuits. Once in place, neurons extend axons and dendrites, forming intricate networks of connections through a process known as synaptogenesis. These connections are fine-tuned by synaptic pruning, where excess or improperly formed synapses are eliminated, ensuring efficient and precise neural communication. (Bally Cuif & Wassef, 1995).

Throughout neurodevelopment, the environment plays a critical role. Factors such as nutrition, exposure to toxins, and sensory experiences can significantly influence the development of the nervous system. Additionally, genetic factors contribute to the robustness and specificity of neural development. Disruptions in these processes can lead to a range of neurodevelopmental disorders, including autism spectrum disorders, intellectual disabilities, and schizophrenia.

8.1. OVERALL STRUCTURE AND CELL TYPES

The brain and spinal cord make up the vertebrate central nervous system (CNS), which develops from the neural plate of the early embryo. The cranial, spinal, and autonomic nerves, as well as the ganglia that support them, make up the peripheral nervous system (PNS). It is created by the neural crest, axons growing from CNS cell bodies, and epidermal placodes. The CNS's nerve cells, or neurons, typically consist of one axon that sends signals from the cell body and links to the dendritic trees of many other neurons, as well as many dendrites that take in input from other neurons (Figure 8.1a). The term "white matter" in the CNS refers to nerve axons, with the high lipid content of the myelin sheaths encasing the axons contributing to their white color. "Gray matter" in the CNS refers primarily to cells. Nuclei, which should not be confused with cell nuclei, are condensations of neuronal cell bodies in the central nervous system.

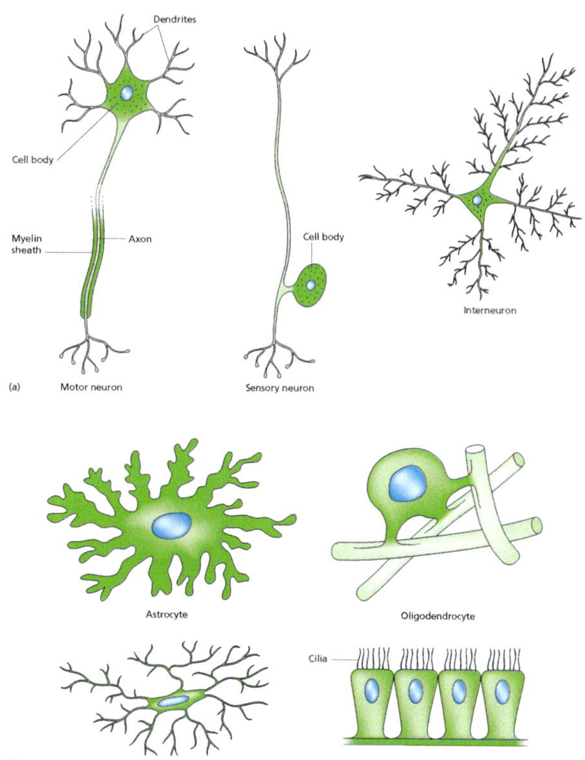

Figure 8.1. *Cell types found in the central nervous system: (a) neurons; (b) glia. Note that some cells with glial morphology are now considered to be neural stem cells.*

Source: https://dn790000.ca.archive.org/0/items/SlackJMWEssentialDevelopmentalBiology/Slack_J_M_W_Essential_Developmental_Biology.pdf.

Synapses are the connections that exist between neurons or between neurons and target cells that are not neurons. They function by causing a neurotransmitter to be released from the presynaptic terminal and a ligand-gated ion channel to open in the postsynaptic terminal, which causes the receiving cell to fire an action potential. The vertebrate nervous system is incredibly complex overall; the human brain has about 10^{11} neurons, each of which forms synapses with about 10^3 other neurons. The positioning and connectivity of the neurons, which are accomplished during embryonic development, determine the system's capabilities to a large extent. However, the vast complexity of the final product is built up by the sequential operation of several fairly simple processes.

Numerous neuronal types exist (Figure 8.1a). Axon and dendritic tree morphology, the neurotransmitter being used (e.g., acetylcholine, noradrenaline, or norepinephrine), and whether the target is another neuron, a muscle (motor neurons), a gland, or a sense organ (sensory neurons) can all be used to categorize them. The nervous system contains roughly ten times as many glial cells as neurons (Figure 8.1b). The most common type of cells in the central nervous system, astrocytes carry out a variety of structural and metabolic tasks. White matter contains more type 1 (= fibrous) astrocytes, whereas gray matter contains more type 2 (= protoplasmic) astrocytes. The CNS cells known as oligodendrocytes are responsible for creating the myelin sheaths that encase axons, while their PNS counterparts are known as Schwann cells. It is now understood that radial glial cells, which are only found during development, are a subset of neural stem cells. Their radial migration of neurons is guided by fine processes that extend from their cell bodies, which are located close to the neural tube's lumenal surface. Phagocytic cells derived from the bone marrow; microglia are not at all derived from the neural plate. Last but not least, the ependyma is the straightforward cuboidal epithelial lining of the spinal canal and ventricles. It serves as a storehouse for neural stem cells during the early phases of development.

The majority of protein synthesis in the cell body is done by neurons. Slow axonal transport or fast axonal transport in the case of neurotransmitter vesicles are the methods used to move materials along an axon. The retrograde transport of materials from the axon terminal to the cell body in a mature neuron maintains equilibrium in this material flow. This aspect of neuronal behavior is highly helpful to neuroanatomists because it allows them to follow the paths taken by axons as they navigate the intricate network of the central nervous system (CNS) by peripherally injecting a marker, like the enzyme horseradish peroxidase (HRP). The axons will pick this up and carry it back to the cell bodies, which can subsequently be identified using peroxidase histochemical staining. When HRP is injected into a single muscle, for instance, it will be transported back into the spinal cord's motor neurons, innervating the muscle and making it visible.

8.1.1. The Brain

The primary brain regions (Figure 8.2) emerge in the early stages of development from the neural tube. The telencephalon, which forms the paired cerebral hemispheres and functions as the olfactory organs' receptive area in lower vertebrates in particular, is formed by the anterior forebrain. The nasal placodes, thick patches of skin next to the telencephalon, are the source of the olfactory sensory cells. In higher vertebrates, where more complex associative and control behaviors are evident, the cerebral hemispheres are very large, and in mammals, the multilayered cerebral cortex is frequently referred to as the neocortex. The diencephalon, which creates the optic vesicles, is derived from the posterior forebrain. These expand and encroach to create the optic cups (Figure (8.3). The optic cup makes contact with the epidermis, which is where the lens of the eye originates. This results in the formation of a thicker lens placode, which rounds up and invaginates to become part of the eye. Each cup's inner layer develops into the sensory retina, its outer layer into the retinal pigment epithelium, and the optic nerve eventually travels through the tube that connects the brain to the eyecup. The retina differentiates into a multilayered structure that houses different types of neurons in addition to sensory photoreceptors. These use the optic stalks to transmit processes back to the brain. The optic nerve fibers terminate in the midbrain, not the diencephalon. The dorsal region of the midbrain enlarges into paired optic lobes, known as the optic tecta (singular tectum) in nonmammals, which serve as the main visual information reception centers. The auditory lobes are also formed by the midbrain. (Blumberg, 1997).

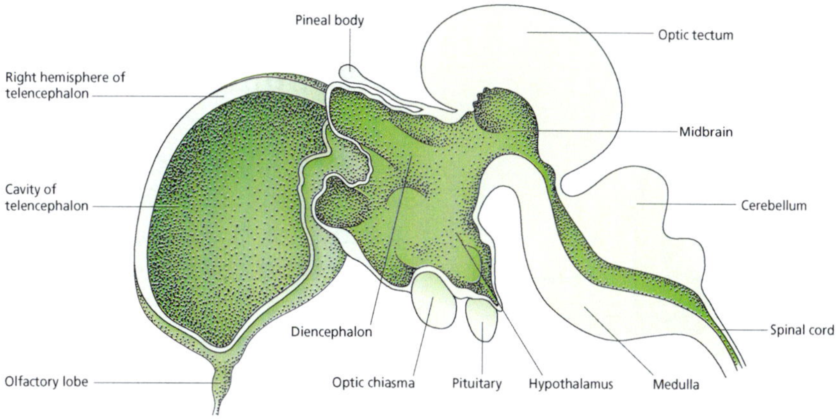

Figure 8.2. *Structure of the brain of an 8-day chick embryo. The mid- and hindbrain are sectioned in the median plane, while the right telencephalon and cerebellar swelling are shown in parasagittal section.*

Source: https://dn790000.ca.archive.org/0/items/SlackJMWEssentialDevelopmentalBiology/Slack_J_M_W_Essential_Developmental_Biology.pdf.

In addition to the eyes, the diencephalon also forms the pituitary (= hypophysis) ventrally, the pineal body (= epiphysis) dorsally, and the thalamus. The majority of the side walls are occupied by the thalamus, a crucial integration center. The pineal gland is later responsible for producing melatonin and is believed to be involved in circadian rhythms. With two distinct origins, the pituitary is a crucial endocrine organ that secretes a broad range of hormones. The placode that gives rise to the anterior part first forms just in front of the neural folds. It then invaginates through the oral aperture and develops into Rathke's pouch in the dorsal pharynx. The diencephalon's floor, the hypothalamus, is where the posterior portion originates and is permanently attached.

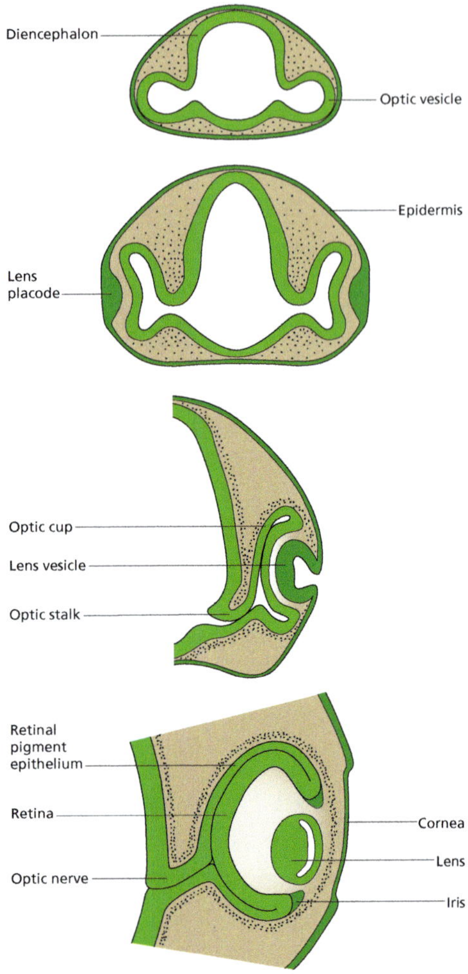

Figure 8.3. *Development of the eye. The optic vesicle grows out of the diencephalon while the lens develops from the epidermis. The inner layer of the cup becomes the retina, while the outer layer becomes a pigmented epithelium.*

Source: https://dn790000.ca.archive.org/0/items/SlackJMWEssentialDevelopmentalBiology/Slack_J_M_W_Essential_Developmental_Biology.pdf.

The cerebellum, which is involved in movement control, is located dorsoanteriorly in the hindbrain. More posteriorly, the medulla houses a number of cranial nerve nuclei. The cerebellum is formed by the first of seven rhombomeres that are formed from the segmentation of the early embryonic hindbrain. The otic placodes, located opposite rhombomeres 5 and 6, invade and grow to form the inner ear, which is made up of the semicircular canals, which control balance and perception of motion, and the cochlea, which is the organ of hearing. The neural tube's original lumen is still present in the brain's ventricles and the spinal cord's central canal. All vertebrates have the same basic set of ten cranial nerves: I is the olfactory nerve, II is the optic nerve, III, IV, and VI regulate the muscles of the eyes, V is the trigeminal nerve, VII is the facial nerve, VIII is the auditory nerve, IX is the glossopharyngeal nerve, and X is the vagus nerve.

8.1.2. Spinal Cord

The spinal cord is composed of an outer layer of white matter that is primarily made up of myelinated fibers, an inner layer of ependymal tissue, and a central region of gray matter that houses the majority of the cell bodies (Figure 8.4). The interneurons, which process sensory information, and the commissural neurons, which link the two sides of the spinal cord, are found dorsally in the gray matter, whereas the motor neurons are found ventrally.

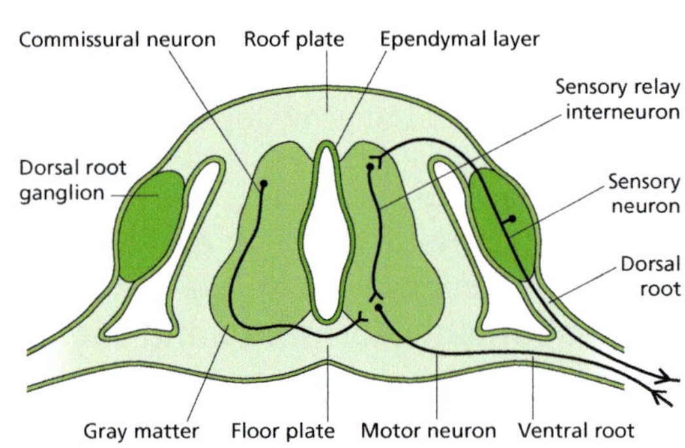

Figure 8.4. *Main structures of the developing spinal cord.*

Source: J.M.W. Slack; Essential Developmental Biology; ISBN-13: 978–1-4051–2216–0.

Each trunk segment gives rise to a single pair of spinal nerves, which are made up of the ventral and dorsal roots. The motor fibers from the spinal motor neurons that will innervate the muscles make up the ventral roots. Sensory fibers that link the interneurons of the dorsal spinal cord with the peripheral sense organs make up the dorsal roots. The dorsal root ganglia, which originate from the neural crest, house the sensory neurons (Figure 8.4). The spinal nerves unite at the levels of the forelimb buds and hindlimb buds, respectively, to form networks known as the brachial plexus and the lumbosacral plexus, from which the limb-supplying nerves emerge.

The neural crest also forms the autonomic system, which includes many small ganglia and connections to the central nervous system. It is separated into the sympathetic and parasympathetic nervous systems, and it also generates a large network of enteric neurons in the intestinal wall.

8.2. ANTEROPOSTERIOR PATTERNING OF THE NEURAL PLATE

The suppression of bone morphogenetic proteins (BMPs) by compounds like chordin that are released from the organizer is the cause of neural induction. As a result, a neuroepithelium with an anterior specification is induced, which enables it to express transcription factors like Otx and form forebrain structures in the absence of additional signals. The action of different posteriorizing factors is the first step towards the determination of more posterior structures. Among these, fibroblast growth factors (FGFs) are significant because they trigger Cdx transcription factors, which trigger the trunk and tail Hox genes. Wnt factors and retinoic acid are assumed to be significant constituents of the posterior signal.

8.2.1. Forebrain

A multitude of local interactions refine the CNS that is initially divided into a small number of large anteroposterior domains by these early inducing factors. Forebrain patterning functions similarly to other embryological interactions: stimulating factors activate distinct transcription factor combinations in various regions, and these regulate the ensuing pattern of differentiation (Figure 8.5). It was once believed that thalamic innervation influenced the cerebral cortex's patterning. However, mice lacking this innervation still display a typical initial pattern of transcription factor domains in the cortex. The anterior neural ridge's FGF is one of the inducing signals needed for forebrain regionalization. This was initially identified in zebrafish, where removal from a region at the extreme anterior of the neural plate caused the diencephalon to expand. When placed posteriorly, it may result in ectopic telencephalon formation. The mouse is a comparable source of FGF8 and FGF15. Numerous Wnts and BMPs are expressed in the cortical hem region, which is located at the posterior edge of the developing cerebral cortex. A Wnt signal is necessary in the posterior telencephalon, as evidenced by defects in the hippocampus and other posterior telencephalon defects in mice lacking wnt3A or Lef1. In loss of function mutants, a few transcription factors that are expressed in the forebrain exhibit gap phenotypes. This implies that in order for the later structures to form, they must exist in their normal expression domain. This is the case, for instance, with the expression of emx2 in the posteromedial cortex and pax6 in the anterolateral cortex (Cayouette & Raff, 2002).

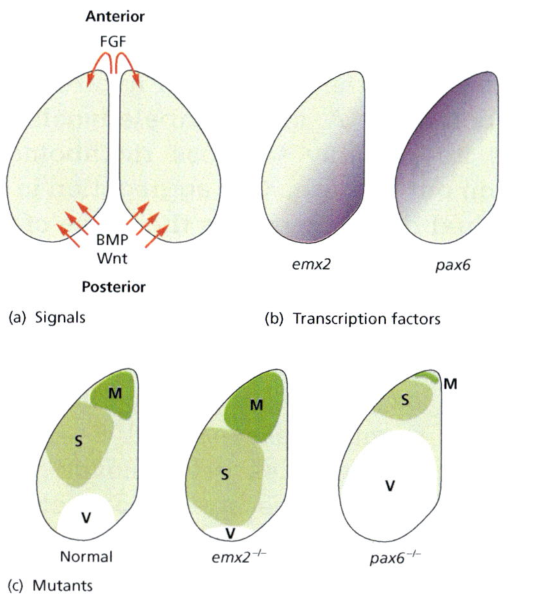

Figure 8.5. *Regionalization of the cerebral cortex of the mouse. (a) Source regions of inducing factors are the anterior rim and the cortical hem. (b) Various transcription factors show early regionally specific expression. (c) The phenotype of loss-of-function mutations suggests that these transcription factors control subsequent development. M, motor area; S, sensory area; V, visual area. For simplicity the diagrams represent the 11.5-day mouse embryo, but the signals operate at earlier stages.*

Source: J.M.W. Slack; Essential Developmental Biology; ISBN-13: 978–1-4051–2216–0.

8.2.2. Isthmic Organizer

The isthmic organizer is a crucial anteroposterior signaling hub. The isthmus is the narrow passage between the vesicles of the midbrain and the hindbrain (Figure 8.6). Transcription factor genes like lim1 and otx2 are expressed in the developing fore- and midbrain at the open neural plate stage. A strip of cells at their posterior boundary begins to produce the signaling factors Wnt1 and FGF8, just anterior to the future isthmus. The regional specification of the neural tissue on either side is managed by this organizing region. It induces the midbrain to the anterior and creates a gradient of engrailed 1 and 2 transcription factor expression. Rhombomere 1 is induced to the posterior, and the cerebellum follows. Transplanting the isthmic region, which can produce the cerebellum from the anterior region of the hindbrain and the midbrain from the prospective diencephalon, may demonstrate these effects. When applied on a bead, pure FGF8 protein shares these functions. Additionally, FGF8 will cause the expression of its own gene, engrailed, and wnt1, demonstrating the organizer's ability to sustain itself after initial establishment. Both wnt1 and engrailed1 knockout mice lack not only the isthmus region where they are expressed, but also sizable portions of the cerebellum and midbrain, demonstrating the critical role these components play in the isthmic organizer's signaling function. A significant posterior defect makes the fgf8 knockout mouse early lethal, but a hypomorphic mutant of fgf8 exhibits a midbrain–cerebellum defect and enough residual function to survive early development. The zebrafish mutant acerebellar, which has a fgf8 loss-of-function mutation but is viable for a longer amount of time than the corresponding mouse mutant, exhibits a similar phenotype.

The creation of a "Pax code," in which the expression of various combinations of the pax family transcription factor genes characterizes distinct brain regions, is a result of the isthmic organizer's activity. While pax2 and pax5 are expressed in the midbrain and rhombomere 1 region, the pax6 gene is expressed in the forebrain and hindbrain. The midbrain and cerebellum are absent in a mouse double knockout of pax2 and-5, demonstrating the requirement of these genes for the development of the corresponding structures.

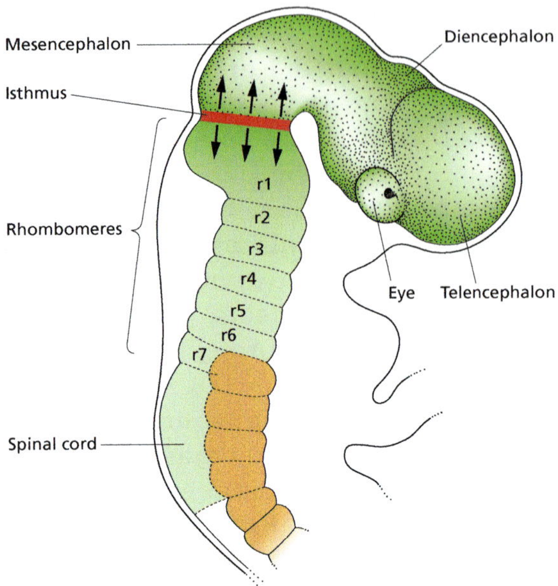

Figure 8.6. *Position of the isthmic organizer, in relation to main parts of the brain of the chick embryo.*

Source: J.M.W. Slack; Essential Developmental Biology; ISBN-13: 978-1-4051-2216-0.

8.2.3. Segmentation of the Hindbrain

In contrast to the more anterior regions of the brain, the hindbrain exhibits a clearly segmented pattern. Rhombomeres are a group of seven (sometimes eight) bulges that are visible following neural tube closure (Figures 8.6 and 8.7). The cerebellum develops from the first rhombomere. Every additional pair of rhombomeres adds fibers to one cranial nerve and has a reiterative set of motor nuclei. Therefore, rhombomeres 2–3 is the source of the trigeminal (V) nerve, rhombomeres 4–5 is the source of the facial (VII) nerve, and rhombomeres 6–7 is the source of the glossopharyngeal (IX) nerve. Neural crest cells are also contributed to one of the branchial arches by every other rhombomere. Thus, cells are contributed to the first arch by rhombomere 2, the second arch by rhombomere 4, and the third arch by rhombomere 6. Clonal analysis also clearly demonstrates the segmented nature of the rhombomeres. A single labeled cell will cause its progeny to cross rhombomeres early on but not later. Cell aggregation in the individual rhombomeres is the cause of this clonal restriction. The expression of ephrins and Eph-type receptors by alternating rhombomeres is at least partially responsible for this aggregation. Rhombomeres 1, 2, 4, and 6 express Ephrin B-type ligands, whereas rhombomeres 3 and 5 express Ephs. The boundaries are believed to be formed by a repulsive interaction between the cell groups expressing Ephs and Ephrins, since the boundaries do not form when a dominant negative Eph is overexpressed in the area. Since the rhombomeres' identities are still malleable during the open neural plate stage, potential hindbrain regions can be switched around in terms of location without changing the overall pattern. Nevertheless, transplantation is no longer an option for changing their character once the rhombomere boundaries start to show. The Notch system determines how the boundary cells actually form. It has been demonstrated that in zebrafish, cells that overexpress constitutive Notch are driven into boundary regions, but cells that overexpress a dominant negative Su(H) are kept out of the boundaries.

Hox genes are crucial for anteroposterior patterning from the hindbrain posteriorly, but they are not expressed in the fore- or midbrain. The hindbrain contains the anterior boundaries of the most anterior Hox genes, and the combinations of their activities result in a unique code carried by each rhombomere (Figure 8.7). It is believed that early gradients of retinoic acid and FGF activate the Hox genes, which are then activated by other transcription factors. One such intermediate transcription factor is the zinc-finger protein Krox 20, which is expressed in rhombomeres 3 and 5. This particular pair of rhombomeres does not release a neural crest stream and expresses Eph-type receptors. Rhombomeres 3 and 5 fail to form and 2/4/6 are fused together in a mouse knocked out of krox20.

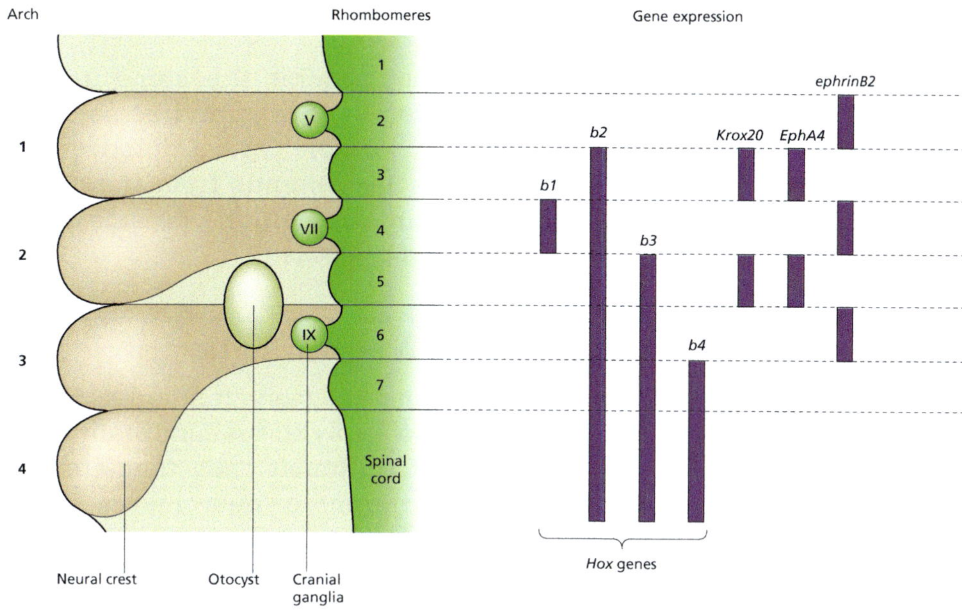

Figure 8.7. *Diagrammatic view of the rhombomeres of the mouse embryo showing gene expression domains and contributions to the cranial nerves and neural crest.*

Source: https://dn790000.ca.archive.org/0/items/SlackJMWEssentialDevelopmentalBiology/Slack_J_M_W_Essential_Developmental_Biology.pdf.

Adjusting the amount of retinoic acid has specific effects on the hindbrain. Dietary vitamin A is converted to retinaldehyde dehydrogenase, or RALDH2, an enzyme that is produced in the somites and lateral plate of the trunk region. The forebrain and midbrain contain the cytochrome enzyme Cyp26, which breaks it down. It is believed that retinoic acid forms a concentration gradient across the hindbrain region, with the source located in the posterior and the sink in the anterior. It is possible to almost completely eliminate retinoid from quail embryos by depriving their mothers of vitamin A. These embryos have tiny hindbrains that are entirely devoid of rhombomeres 5–7; however, if vitamin A is administered to the embryos prior to the early somite stage, the absent parts do develop. In the anterior spinal cord, ventral neuronal types are also disappearing. In addition to lacking rhombomeres 5–7, mouse embryos with RALDH2 knocked out die in the middle of gestation. When a dominant negative retinoic acid receptor is overexpressed in Xenopus, similar outcomes are observed. By dosing the mothers, an excess of retinoic acid can be administered to mouse embryos. In addition to the normal expression in rhombomere 4, treatment at 7.5 days results in the expression of hoxb1 in rhombomere 2. This results in the first arch becoming a duplicate of the second arch by changing the characteristics of the neural crest cells released by rhombomere 2. This is an excellent illustration of a teratogenic effect that has been thoroughly studied at the molecular level. Since retinoic acid can be taken orally to treat acne, it has actually shown a teratogenic risk to humans. This resulted in numerous abortions and a number of birth defects causing brain damage before the risk was recognized.

8.2.4. Spinal Cord

Compared to the brain, the spinal cord does not exhibit the same anteroposterior pattern. Hox boundaries do exist, though, at different levels, and some motor neuron groups are arranged anteroposteriorly in addition to their typical dorsoventral pattern. For instance, the column supplying the axial muscles extends from the cervical to the beginning of the lumbar region, while the lateral motor-neuron columns supplying the limbs are only present at the forelimb and hindlimb levels.

8.2.5. Dorsoventral Patterning of the Neural Tube

Throughout its entire length, the spinal cord maintains the same fundamental dorsoventral arrangement; in fact, many of its characteristics also apply to the hindbrain and midbrain. A floor plate located mid-ventrally is made up of columnar cells that are close to the notochord. The area that surrounds this is where the motor neurons are located. These are found in different concentrations; the lateral columns supply the limb muscles, while the medial column supplies the axial and body wall muscles. The set of muscles to which the motor neurons in each column eventually project is linked to the specific combination of Lim-type transcription factors that these motor neurons express. The pattern of motor neuron connectivity changes in response to experimental modifications in the combination of lim genes expressed. The dorsal sensory relay interneurons and the commissural neurons, which project to the opposite side of the spinal cord, are located in the dorsal half of the spinal cord. The roof plate is located in the dorsal midline.

Signals from the neural plate's surrounding tissues at first, and then from the floor plate and roof plate, regulate this dorsoventral pattern (Figure 8.8). A signal from the notochord activates the floor plate and, to a lesser extent, the motor neurons. Recombination experiments, where the notochord is grafted next to various dorsoventral levels of the neural tube, can demonstrate this. Nearby, an ectopic floor plate will form, and a little ways away, motorneurons. Three factors lead experts to conclude that the signal is a sonic hedgehog (Shh).

1. Shh is expressed in the notochord;
2. Shh will induce floor plate and motor neurons when applied to the neural tube;
3. The floor plate and motor neurons are lost in the mouse knockout of shh.

As fate-mapping studies in the chick have demonstrated, the lateral portions of the floor plate are induced by the surrounding neuroepithelium, whereas the medial portion is directly derived from Hensen's node during normal development.

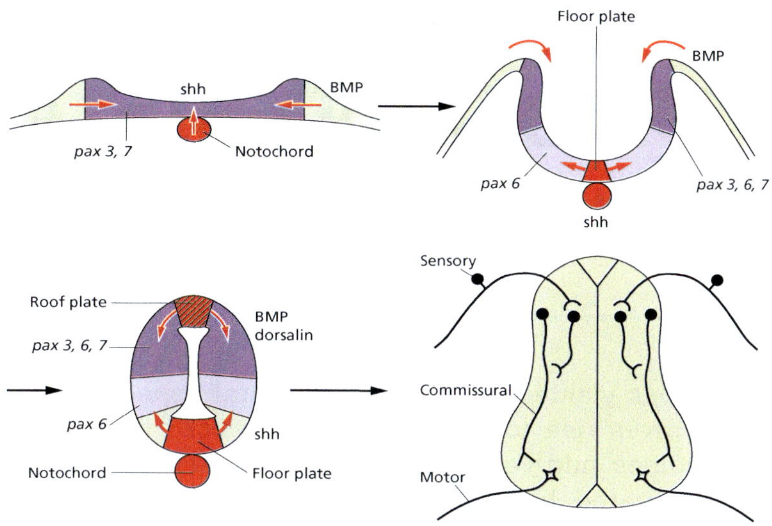

Figure 8.8. *Dorsoventral patterning of the neural tube. Sonic hedgehog from the notochord induces expression of its own gene in the floor plate. BMP from the epidermis induces the roof plate. The signals from the two centers induce a nested expression of pax genes that later brings about the pattern of neuronal differentiation.*

Source: J.M.W. Slack; Essential Developmental Biology; ISBN-13: 978–1-4051–2216–0.

Once formed, Shh assumes the role of the notochord in signaling because the floor plate itself makes Shh. The expression of several transcription factor genes, including pax3 and pax7, is suppressed by Shh. These are first expressed throughout the neural tube, but the signal causes them to be repressed ventrally. The formation of the ventral cell types will be hindered by the misexpression of these factors. At every dorsoventral level, later on, particular transcription factors start to express themselves. For instance, the premotor-neuron region expresses the bHLH transcription factors olig1 and olig2, which are essential for the development of oligodendrocytes and motor neurons.

The lateral edge of the neural plate, where it meets the epidermis, is where the dorsal midline of the neural tube originates. BMP4 and BMP7 are expressed in the epidermis, and when these BMP proteins are applied to the neural plate, they can increase the expression of pax3 and pax7. Following neural tube closure, the roof plate expresses the genes for dorsalin, a member of the TGF-β superfamily, as well as bmp4 and bmp7. All of these elements have the ability to stimulate dorsal cell types, including sensory relay neurons. The dorsal neuronal types do not form if the roof plate is not formed, as is the case, for example, in mice lacking the lim1a gene. For these reasons, it is believed that the gradients of BMPs and dorsalin from the roof plate, as well as of Shh from the floor plate, combine to produce the overall dorsoventral pattern. (Henrique & Schweisguth, 2003).

8.3. NEUROGENESIS AND GLIOGENESIS

8.3.1. Drosophila

Research on Drosophila has yielded insights into several aspects of neurogenesis. The neurogenic region that gives rise to the ventral nerve cord in Drosophila is lateral to the mesoderm and becomes mid-ventral after the mesoderm invaginates. The bHLH transcription factors generated by the achaete–scute complex (AS-C) are expressed in a segmentally repeated pattern of proneural clusters within the neurogenic region. Only one neuroblast and several epidermal cells are produced by each cluster, and the Notch lateral inhibition system regulates this process. The production of Delta is triggered by AS-C factors, which in turn stimulates Notch on neighboring cells. The Notch signal inhibits AS-C expression, which lowers the amount of Delta produced in the surrounding cells (Figure 8.9).

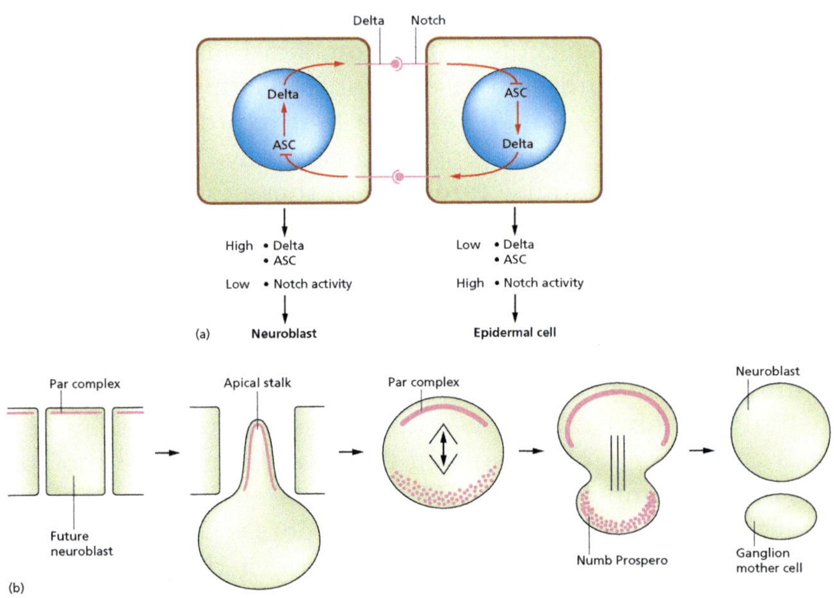

Figure 8.9. *Neurogenesis in Drosophila. (a) The Delta-Notch lateral inhibition system amplifies initial small differences between cells of a proneural cluster ensuring that only one ends up with high levels of products of the achaete–scute complex (AS-C) and becomes the neuroblast. (b) Asymmetrical division of neuroblast to generate another neuroblast and a ganglion mother cell.*

Source: *https://dn790000.ca.archive.org/0/items/SlackJMWEssentialDevelopmentalBiology/Slack_J_M_W_Essential_Developmental_Biology.pdf.*

The neuroblast is a single cell with a high level of AS-C, surrounded by cells with a low level of AS-C, which become epidermal, until the positive feedback system eventually magnifies tiny differences. This process clarifies why an excess of neuroblasts results from Notch loss of function. One instance of a symmetry-breaking system is the mechanism. The feedback that is a natural byproduct of interaction makes the initial uniform state unstable because there is a small variation in the levels of AS-C or Delta between cells.The neuroblasts separate from their neighboring cells once they have formed and descend into the interior, staying momentarily joined by an apical stalk in the epithelium's plane. Other proteins, such as the homeodomain transcription factor Prospero and the tyrosine phosphatase Numb, are repelled to the basal region by the phosphorylation of intracellular trafficking system components by the Par complex. A large neuroblast that inherits the apical stalk and a smaller ganglion mother cell that is enriched in Numb and Prospero are the two asymmetric sides of the next division. Prospero stimulates ganglion mother cell-specific genes and represses those required for additional cell division, whereas Numb further suppresses Notch signaling by causing endocytosis of the Notch protein from the cell surface. Much of the evidence supporting this mechanism comes from the behavior of mutants. For instance, embryos lacking Par complex components do not exhibit Numb and Prospero localization, and their neuroblasts divide symmetrically.

8.3.2. Vertebrate Primary Neurogenesis

The neuroepithelium of the open neural plate serves as the source of the first neurons in lower vertebrates, providing an excellent opportunity to study the process. The process that drives neuronal differentiation is mediated by multiple neurogenic transcription factors. These include the bHLH-type factors neurogenin and neuroD, which are members of the same biochemical family as AS-C in Drosophila. When neurogenin is overexpressed, it will activate neuroD. Neurogenin is expressed in the neural plate regions where the initial primary neurons form. Primary neurons from open neural plate stages also express neuroD. In Xenopus, overexpression of either factor will result in a higher percentage of neurons derived from the neuroepithelium. A mechanism of lateral inhibition that is strikingly similar to that observed in Drosophila determines the spacing between the primary neurons (Figure 8.10a). In a patch that expresses neurogenin, every cell has the potential to develop into a neuron. However, Notch signaling inhibits this tendency, and Delta on nearby cells stimulates Notch on one cell. In the Xenopus neural plate, overexpression of Delta, or a constitutive form of Notch, inhibits the development of primary neurons. On the other hand, a greater density of primary neurons results from the overexpression of a dominant negative Delta.

8.3.3. Later Neurogenesis

Neurogenesis persists throughout the development of the embryo and into the postnatal period. Despite the neuroepithelium's single cell thickness, its appearance is stratified due to the numerous levels at which the nuclei are present. When it comes to epithelial morphology, the apical side faces the neural tube's lumen, and the basal side faces the outside world. This means that, in contrast to other epithelia, apical is typically positioned at the bottom of a diagram and basal at the top. All of the cells are mitotic in the early stages, and during S phase and mitosis, the nuclei migrate to approach the lumenal surface and the exterior surface, respectively. Since neurons cannot divide once they are formed, the formation time of each class of neuron can be determined by measuring the pulse of BrdU or ^{3}HTdR. When tissue is examined after a pulse label, the cells that stopped dividing soon after incorporating the label will still have the label on them, while all the cells that have kept dividing will have lost it due to dilution with unlabeled precursors. The final S phase prior to neuronal differentiation is frequently referred to as the neuron's "birthday."

The newly formed neuron is divided from the neural tube's lumenal surface by this last, tangential division. It is connected to the Numb protein's localization on the lumenal side. Similar to Drosophila, the Par complex, which is connected to the intercellular junctions at the cell's lumenal ends, regulates this. In contrast to Drosophila, the Par complex in vertebrates appears to draw Numb to the same area of the cell rather than repel it to the other side. Following division, the postmitotic, internalized neuron loses its Numb content while the mitotic lumenal cell keeps its Numb protein (Figure 8.10 b, c). Although it is believed that this cytoplasmic location is a crucial component of neuronal determination, the destiny of the Numb-containing cell appears to vary with stage. Numb suppresses neurogenesis and promotes proliferation in early mouse embryos, but it may segregate to the neuronal daughter, stimulate neurogenesis, and suppress proliferation in later stages. Increased neurogenesis is seen in numb (and the related gene numblike) knockout mice, which is consistent with numb's early role in promoting proliferation. The fact that the Numb protein localizes to the basal rather than the apical side in the chick further implies that the effects of Numb may vary depending on the situation.

The neural tube stops being one cell thick when the tangential divisions occur (Figure 8.10d). The ventricular zone (and eventually the ependymal layer) are the names given to the mitotic lumenal layer after that. After the enveloping layer of connective tissue known as the pia mater (Latin for "tender mother"), the outer surface is referred to as the pial surface. An "inside-out" pattern of morphogenesis gives rise to the layered structure of the central nervous system. As cells move inward from the ventricular layer, they create a mantle layer that eventually becomes known as the gray matter and is home to both glia and neurons. A cell-poor marginal zone forms beneath the pial surface as axons from the neurons accumulate. This becomes the white matter after the axons are myelinated. The neurons must move from the ventricular zone, where they were formed, to their final locations as the neural tube thickens. The radial glial cells, which extend processes from the inner to the outer surface of the tube, are responsible for directing this migration. Early-born neurons typically migrate shorter distances, while late-born neurons migrate farther. The spinal cord retains the original three-layered structure of the ventricular, mantle, and marginal zones; however, later on, it becomes much more complex in certain areas of the brain. Neurons from the mantle zone invade the marginal zone in the mammalian cerebral cortex to form six more layers of neocortex. The majority of mouse embryos' cortical neurons develop between 12.5–17.5 days of gestation. Certain mitotic cells in the cerebellum move to the outer layer to form an external germinal layer, which is followed by "outside-in" neurogenesis.

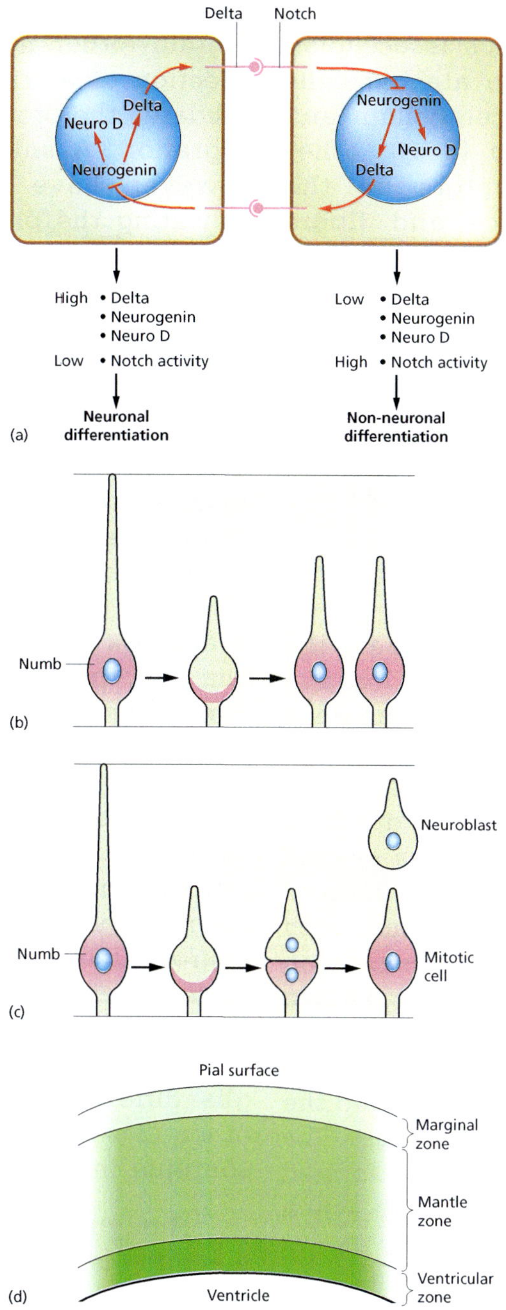

Figure: 8.10. *(right) (a) Lateral inhibition mediated by the Delta–Notch system in vertebrates ensures that only some cells from the neurogenic regions become neurons. (b, c) Symmetrical and asymmetrical divisions of cells at the lumenal surface (= ventricular zone; apical surface of neuroepithelium). In a symmetrical division Numb protein is partitioned equally. In an asymmetrical division the lumenal daughter keeps the Numb, and the internal daughter, which in early neurogenesis will become a neuroblast, loses it. (d) Three-layered structure of the later CNS.*

Source: *J.M.W. Slack; Essential Developmental Biology; ISBN-13: 978–1-4051–2216–0.*

Moreover, the mantle zone splits into multiple layers, one of which is occupied by Purkinje neurons, which are big cells that are exclusive to the cerebellum and can communicate with up to 10^5 other neurons. With photoreceptors on the original ventricular surface and fibers leading to the optic nerve on the original pial surface, the outer region of the optic cup becomes pigment epithelium and the inner layer becomes the retina in the eye. The retina is made up of six cell layers. There is some lateral neuronal migration in different regions of the brain in addition to the common radial migration.

8.3.4. Neural Stem Cells

The ventricular zone's mitotic cells have the capacity to both sustain themselves and produce offspring that develop into neurons and glia. This layer is still mitotic later on, when it is referred to as the ependyma, even though there aren't many cell divisions at that point. The central nervous system (CNS) is now believed to be structured similarly to other cell-renewal systems, with a higher concentration of transit amplifying cells and fewer stem cells. It is now known that, just as in lower vertebrates, there is continuous cell renewal in two areas of the mammalian brain: the hippocampus and the areas next to the lateral ventricles. The olfactory bulbs receive the newly formed neurons from the lateral ventricles when they migrate anteriorly.

There are two ways that the potency of individual CNS mitotic cells has been investigated. One method involves using a replication-incompetent retrovirus that expresses GFP or β-galactosidase to perform *in vivo* clonal analysis. The infected cells are kept apart and each positive cell group can be regarded as a single clone if this is carried out at a low multiplicity. This technique demonstrates that the majority of clones contain either neurons or glia, indicating infection of a transit amplifying cell, and validates the progeny's predominant radial migration. A small percentage of clones, however, have both neurons and glia, indicating the pluripotency of the original labeled cell. Individual cell culture *in vitro* is the alternative technique. About 7% of clones in cultures from the lateral ventricular zone of the rat embryo cerebral cortex form large clones containing neurons, astrocytes, and oligodendrocytes, while the majority of clones form small numbers of glia. It is assumed that in both kinds of experiments, neural stem cells give rise to the large mixed clones and transit amplifying cells give rise to the small single-type clones. It appears that the cells known as radial glia are, in fact, the neural stem cells during the time of maximal neurogenesis. Although their nuclei are located close to the ventricular surface, radial glia have fine processes that reach the pial surface. They repeatedly divide, and labeling by retroviral infection reveals that the progeny are neurons or neuronal precursors. These neurons follow the elongated radial glial mother cell towards the pial surface in order to reach their final cell layer. After neurogenesis is finished in mammals, radial glia differentiate into astrocytes, but in other classes of vertebrates, they might continue into adulthood.

Though it's unclear how much of these markers distinguish transit amplifying cells from actual stem cells, a number of markers have been proposed to describe the CNS stem cells. These consist of nucleostemin, an intermediate filament protein called nestin, a GTP-binding protein, and the complex carbohydrate SSEA-1, which is also present on embryonic stem cells. Cells from the CNS's stem cell regions can be cultured when FGF and epidermal growth factor

(EGF) are present. Grown in suspension, they aggregate into small structures called neurospheres that comprise a combination of transit amplifying cells and stem cells. Neurospheres plated on a laminin substrate will develop into neurons, oligodendrocytes, and astrocytes.

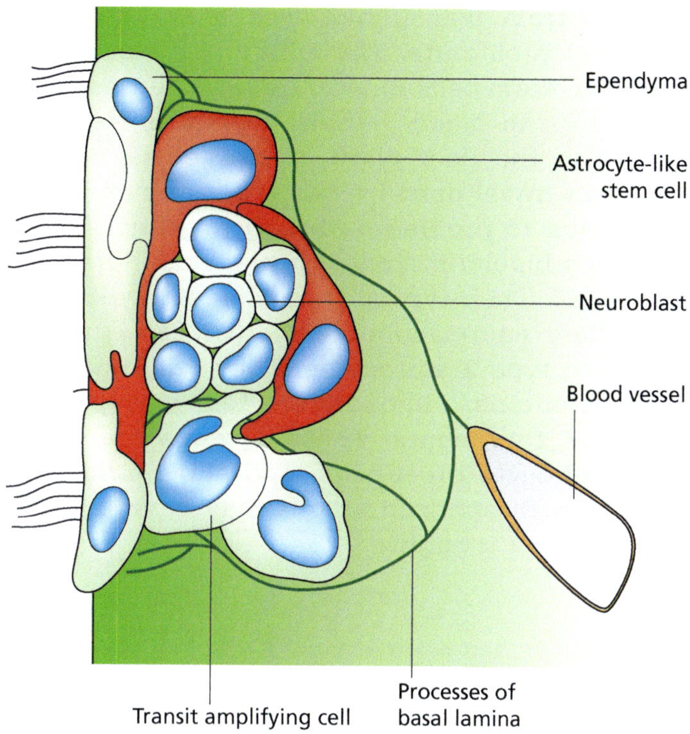

Figure 8.11. *A model for the stem cell niche in the subventricular zone of the mammalian lateral ventricle.*

Source: J.M.W. Slack; Essential Developmental Biology; ISBN-13: 978–1-4051–2216–0.

The subventricular zone, which is next to the ependyma, is home to the stem cells in the adult neurogenesis regions. The niche for stem cells (Figure 8.11) is believed to be created by blood vessels close by, which sequester growth factors through their basal laminae. Evidence that the stem cells are astrocyte-like cells is as follows:

1. The subventricular astrocytes are the only cells that show long-term retention of BrdU.
2. They are able to endure treatment with ara-C, an antimitotic medication that destroys the rapidly proliferating transit amplifying cells. Neurogenesis most likely starts from these survivors because it continues after treatment is stopped.
3. GFP-specific labeling of glial cells allows for the observation of their progeny, which include neurons migrating to the olfactory bulbs and dividing transit amplifying cells. This particular labeling is achieved by infecting transgenic mice with a virus that expresses an avian leukosis virus receptor and is driven by the

glial-specific (GFAP) promoter. In this transgenic line, the virus will specifically infect and label the glial cells, despite the fact that it normally cannot infect mice.

4. Some of these labeled cells can be grown into neurospheres. However, if every astrocyte is eliminated—which can be achieved by applying the medication ganciclovir to cells from transgenic mice that have the GFAP-TK gene—neurospheres will not proliferate. The ability of endothelial cells to sustain the self-renewal of neural stem cells *in vitro*—even in a trans filter culture setting—provides evidence for the blood vessels' role in preserve the niche. Glial cell clusters form close to the floor plate in the ventral region of the ventricular zone. These cells can travel dorsally and enter any area of the central nervous system. They continue to proliferate during migration, in contrast to neurons. One of these cells is a bipolar-morphological oligodendrocyte precursor cell, also known as the O2A cell. Their typical course is to develop into oligodendrocytes. However, when ciliary neurotrophic factor (CNTF) is present in culture, they can also develop into type 2 astrocytes. They will develop into neuronal stem cells and produce neurons, oligodendrocytes, or type 1 astrocytes if given high serum and FGF treatments. These recently identified characteristics of radial glia, oligodendrocyte precursors, and subventricular astrocytes suggest that certain cell types categorized as glia may have far greater developmental potential than previously thought.

8.4. THE NEURAL CREST

A group of cells called the neural crest originates from the neural folds. At the moment of closure, they are in the dorsal region of the neural tube, but they quickly move out of the tube and into the surrounding tissues. They form a variety of different cell types of which the following are the most significant:

1. Neurons and glia of the sensory and autonomic systems;
2. Adrenal medulla and calcitonin cells of thyroid;
3. Pigment cells (excluding those of the pigmented retina);
4. Skeletal tissues of the head;
5. Part of cardiac outflow tract.

There are two main methods that have been used to determine the crest's typical fate. One is the grafting of neural tube segments from quail to chick embryos, with the quail cells being localized afterwards. The second technique involves localized dye injections to label neural crest *in vivo*. DiI would be injected extracellularly to label a patch of cells, or fluorescent dextran would be injected intracellularly to label a single cell. After the label is applied, the embryo is given time to develop before the location and cell types of the labeled cells are later determined.

These kinds of experiments demonstrate that, in terms of normal fate, the crest is separated into discrete anteroposterior domains (Figure 8.12). Many different structures arise from the cephalic crest. At first, the majority of the head mesenchyme originates from it. This eventually gives rise to cartilage, bone, cranial ganglia, blood vessel pericytes and smooth muscle, and odontoblasts, or the cells that form teeth enamel. It also becomes general connective tissue. Two different migration pathways are visible in the trunk crest (Figure 8.13). Melanocytes are produced via the dorsolateral pathway, which runs between the somites and the epidermis. The ventral pathway becomes the dorsal root ganglia, sympathetic ganglia, adrenal medulla, and it also contributes to the cardiac crest, the heart's outflow tract, after passing through the sclerotome. Each dorsal root ganglion receives a contribution from the crest next to the posterior sclerotome segments on either side. This migration only happens through the anterior half of each sclerotome segment. Last but not least, the vagal and sacral regions of the crest contribute cells to the network of enteric ganglia.

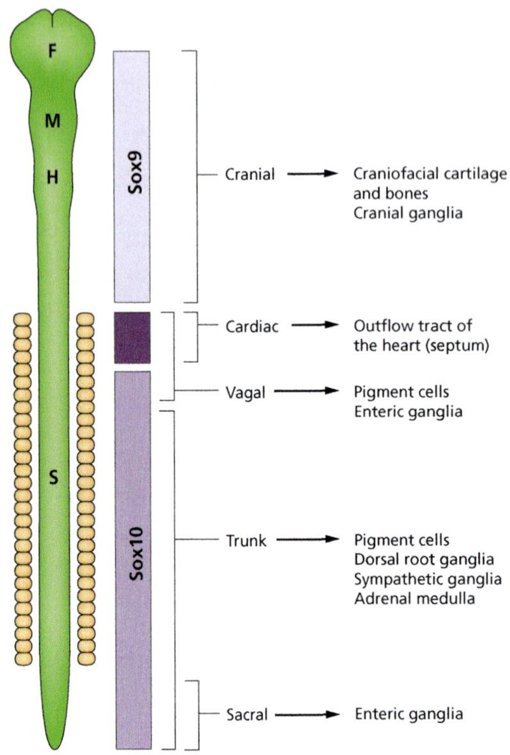

Figure 8.12. *Fate map of neural crest compiled from chick–quail orthotopic grafts. F, forebrain; M, midbrain; H, hindbrain; S, spinal cord.*

Source: https://dn790000.ca.archive.org/0/items/SlackJMWEssentialDevelopmentalBiology/Slack_J_M_W_Essential_Developmental_Biology.pdf.

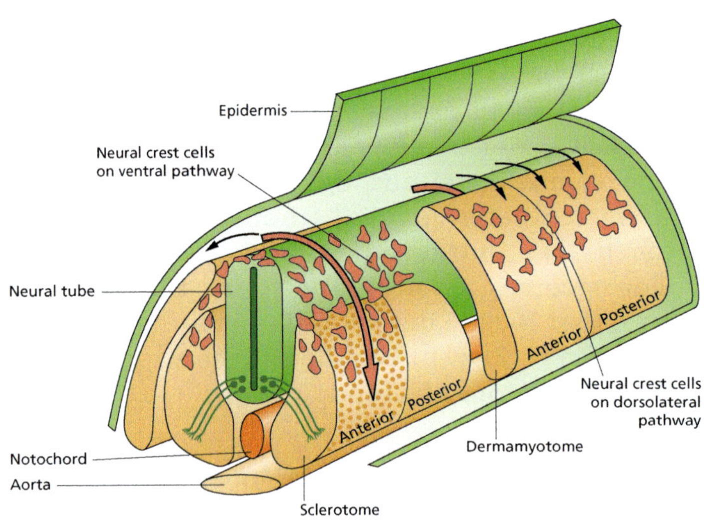

Figure 8.13. *Dorsolateral (black arrows) and ventral (red arrows) pathways of trunk neural crest migration.*

Source: J.M.W. Slack; Essential Developmental Biology; ISBN-13: 978–1-4051–2216–0.

8.4.1. Progressive Commitment of Crest Cells

The clonal descendants of individual cells within the neural folds can include neural crest, or epidermal cells, or any combination of the three. This implies that the environment in this area encourages the development of neural crest cells rather than the existence of a distinct neural crest rudiment. Neural crest can be generated from cells on both sides of the junction when the neural folds are removed and the neuroepithelium is joined directly to the epidermis. This suggests that the formation of the neural crest may result from an interaction between the epidermis and neuroepithelium. Although an essential component of this signal, BMP4 from the epidermis might not be enough to trigger neural crest formation on its own. Studies on Xenopus and chicks indicate that retinoic acid, Wnts, and FGFs may also be necessary.

The neural crest determination molecular pathway appears to have some characteristics of a linear sequence in addition to some feedback effects. The expression of the homeodomain transcriptional repressors Msx1 and 2 is the first reaction to BMP signaling. Next come Snail and Slug, two zinc-finger transcriptional repressors, and other transcription factors like Sox9 and 10. Experiments involving overexpression, inhibition, and epistasis provide evidence for the pathway. Although the Xenopus results are presented here, there is general consistency in the evidence from mouse knockouts and experiments involving retroviral infection of chicks. Excess Msx factors have been shown to induce the expression of snail in Xenopus, snail overexpression results in the expression of sox10, and sox10 overexpression results in the expression of slug. Neural crest formation can be driven by any of these factors, and injection of the appropriate dominant negative domain-swap constructs or antisense morpholinos will inhibit neural crest formation. According to the overexpression data, the sequence is Msx→ Snail→Sox10→Slug. It is most likely not strictly linear, though. Sox10 will induce Slug, and Slug will induce Sox10, indicating that the pathway contains some autocatalytic feedback. Moreover, Slug's dominant negative inhibitor will lower Snail and Sox10 expression. A glucocorticoid-inducible form of the dominant negative Slug will prevent cell migration if it is induced later, after the crest has formed, and it will prevent crest formation if it is induced early. This demonstrates that the properties of the crest cells are controlled by Slug function over an extended period of time. Initially expressed throughout the crest, Sox10 is quickly superseded in the cranial crest by Sox9 (Figure 8.12). It is well known that Sox9 plays a crucial role in cartilage differentiation from the mesoderm. The significance of Sox9 in the development of cartilage from the neural crest is demonstrated by a tissue-specific knockout in mice. The neural tube expresses the Cre recombinase in constitution wnt1-cre; floxed sox9 embryos, but not the mesoderm. Without influencing the development of the trunk skeleton, the consequent ablation of sox9 in the neural crest prevents the formation of the cranial skeleton.

A vast range of extracellular matrix elements are necessary for crest cell migration. The adhesion molecule N-cadherin is expressed by neural tube cells. Migratory crest cells lose this, but upon reaching their destinations, they re-express it. Antibodies that neutralize fibronectin or laminin, two components of the extracellular matrix, prevent the migration of crest cells. The posterior sclerotome specifically inhibits the trunk crest migration of the ventral pathway *in vivo*, allowing the anterior sclerotome to become the preferred migration route. The explanation is that ephrin B1 is expressed by the posterior sclerotome, while ephB3 is expressed by the crest cells. These combine in an aversive way, causing the crest cell processes to collapse upon contact with the ephrin B1-containing surface as a result of actin depolymerization facilitated by the RhoGTPase system. Chick embryos treated with an ephrin inhibitor allow the crest cells to migrate through both the anterior and posterior sclerotomes. The Eph system also prevents the crest's outward migration in rhombomeres 3 and 5 of the hindbrain.

Through the exchange of neural tube segments and the study of the fate of the crest cells that emerge from the graft, the overall anteroposterior commitment of the neural crest has been explored. Because only the cranial crest can produce skeletal tissues, this kind of experiment has demonstrated that, at least in higher vertebrates, the difference between the cranial and trunk crests emerges early. While cranial crest, when grafted to the trunk, forms the different types of neurons found in the trunk and also contributes to the skeletal structures of the body, trunk crest, even when grafted to the neural tube of the head, does not produce cartilage. The distinction may have more to do with the ability of trunk crest cells to migrate than with their actual competence to form cartilage, since these cells can form cartilage if they are grafted into the facial primordia rather than the hindbrain. Generally speaking, any area of the crest that is grafted to the proper location along the axis will develop into parasympathetic, sympathetic, or sensory ganglia. In the thoracic crest, for instance, adrenergic (sympathetic/adrenal medulla) neurons and cholinergic (parasympathetic) neurons are typically produced. However, the cell types that form are suitable for the new location of the graft if these regions are switched.

Labeling individual crest cells *in vivo* reveals that some of the labeled clones include multiple cell types, including pigment, sensory neurons, adrenal medulla, and glial cells. This suggests that those leaving the neural tube early may be pluripotent. Individuals who exit the neural tube later in life are more likely to become melanocytes, which is one type of terminal cell. This demonstrates that at least some of the crest's cells are pluripotent; to reflect their affinity for different kinds of stem cells, these cells have begun to be referred to as neural crest stem cells (NCSC) in recent years.

The environment that the cells migrate through does have a significant impact on the final differentiation of the cells (Figure 8.14). For instance, exposure to brain-derived neurotrophic factor (BDNF) from the neural tube is necessary for the development of dorsal root ganglia. The crest cells will not form the dorsal root ganglia if a barrier is placed between them and the neural tube; however, if they are given BDNF protein, they will. Dorsal root ganglia do not form in mice deficient in the bdnf gene. Environmental cues also determine whether smooth muscle, glia, or autonomic neurons will form. Trench crest cells in culture will preferentially form smooth muscle when exposed to TGFβ, autonomic neurons when exposed to BMP2 or –4, and Schwann cells when exposed to neuregulin (= glial growth factor). Although the factors may also contribute to differential survival of their own cell type, these effects are primarily educational. These factors, which include neuregulin in the nerve sheaths, TGFβ in the major cardiac outflow vessels, and BMP2/4 in the dorsal aorta and gut, are expressed in the specific areas of the embryo through which the cells migrate. The effects of mouse knockouts are as expected; for instance, a neuregulin knockout decreases the number of Schwann cells, and a TGFβ gene knockout impedes the development of the heart. The majority of autonomic neurons are absent in the mash1 knockout, which encodes a bHLH transcription factor that is a homolog of the Drosophila Achaete-Scute family. The transcription factor gene mash1 is expressed by autonomic neurons. Endothelin 3 is the inducing factor that determines commitment to form the pigment cells and enteric ganglia. Mice deficient in either of these cell types lack this factor or its receptor. Making transgenic mice for an inducible form of the receptor and causing its activity at different times can demonstrate that the requirement for activity is in the time interval 9:5–12:5 days of gestation.

Apart from the data demonstrating the regulation of NCSC differentiation by environmental cues, there exist multiple experiments wherein crest cells exhibit differentiation into a single cell type despite being subjected to diverse environmental conditions. This implies the existence of a component of autonomous commitment to specific fates that develops gradually over time.

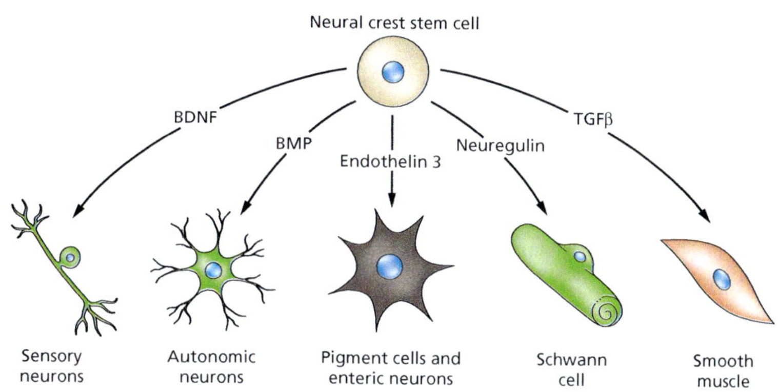

Figure 8.14. *Preferred differentiation pathways of trunk neural crest stem cells exposed to different inducing factors.*

Source: https://dn790000.ca.archive.org/0/items/SlackJMWEssentialDevelopmentalBiology/Slack_J_M_W_Essential_Developmental_Biology.pdf.

8.5. DEVELOPMENT OF NEURONAL CONNECTIVITY

8.5.1. The Growth Cone

Neurons send out their axons to innervate their targets. While sensory neurons in the dorsal root ganglia innervate a variety of peripheral receptors and sense organs, motor neurons in the ventral spinal cord innervate muscles or glands. The final pattern of connections is incredibly complex, so how do they figure out where to go? The specificity of connections is actually controlled by a whole hierarchy of distinct mechanisms. Even though they are all fairly low specificity, when combined, they produce an extremely precise outcome.

A growth cone is present at the tip of every developing axon (Figure 8.15). This structure resembles a migrating fibroblast in certain ways. Filopodia are continuously extended and retracted, and it actively crawls across the substrate. The growth cone's cytoskeleton is in a state of dynamic equilibrium. The microfilaments that support its lamellipodia and filopodia are continuously elongating at their plus (exterior) ends. Myosintype motor proteins are simultaneously pulling the filaments back into a centrally depolarized domain. The microtubule bundles that keep the axon structurally intact terminate in this area. Myosin motor activity is reduced, actin polymerization rate is increased, and microtubule elongation is increased when growth cone advance stimuli are applied. In order to lengthen the axon, new material is laid down as the growth cone advances. Although the proximal axon and cell body are immobile, they are essential to the process because they perform a large portion of the synthesis of the new materials needed for elongation. Owing to the growth cone's relatively large size, different regions may be exposed to distinct signals, and a unilateral stimulus will cause the growth direction to change. The small GTP exchange proteins mediate the connection between signal receptors and the cytoskeleton. When RhoA is active, growth cone collapse occurs, but Rac and Cdc42 are active and encourage elongation.

The neurotrophins are the substances that are most effective in stimulating axonal extension. These include neurotrophin 3 (NT3), neurotrophin 4/5 (NT4/5), nerve growth factor (NGF), and brain-derived neurotrophic factor (BDNF). The receptors are known as receptor tyrosine kinases, or Trk (pronounced "track") proteins. NT3 primarily binds TrkC, while BDNF and NT4/5 bind TrkA, and NGF binds TrkA. This gives the system some selectivity. Growth cone elongation is made possible by the tonic function of this, which activates RhoA. In the presence of neurotrophin, this activation is halted.

Furthermore, laminin and fibronectin, two extracellular matrix proteins, frequently create an environment that is favorable for elongation. If the neurotrophins are only applied to the cell body, they can still sustain survival; however, the growth cone is where they must be applied to achieve elongation. It is believed that the changes in gene expression are caused by endocytotic vesicles that contain active receptors and are transported retrogradely to the cell body.

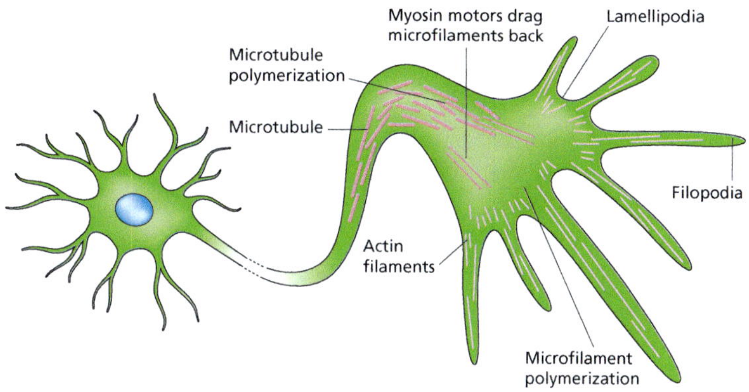

Figure 8.15. *A growth cone at the tip of a developing axon.*

Source: J.M.W. Slack; Essential Developmental Biology; ISBN-13: 978–1-4051–2216–0.

8.5.1.1. Guidance Molecules

Certain factors promote the guidance of growth cones, while others impede it. The guidance factors that are accountable for contact attraction, contact repulsion, long-range attraction, and long-range repulsion fall into four distinct categories. The primary method for identifying them is through *in vitro* assays (Figure 8.16). It is easy to test whether the contact factors will encourage axon outgrowth from a neural tube explant by coating them on a culture dish. Since growth cones *in vivo* will constantly be choosing between different environments, providing a choice of alternative substrates instead results in a more discriminating assay. Using this kind of test, ephrins that had growth-cone-repelling activity, for instance, were separated from the optic tectum. It is necessary to demonstrate that long-range factors are diffusible and that growth cones are able to detect a concentration gradient. Two explants are positioned in a collagen gel to achieve this. A long-range chemoattractant secreted by the target explant is required if axons grow straight from one to the other. It takes the movement of the source and a change in the axons' growth direction to provide definitive evidence of chemoattraction.

Extracellular matrix elements like laminin and fibronectin as well as adhesion molecules on other neurons like N-CAM, NgCAM, or N-cadherin are the factors that attract contact. The factors that repel contact are ephrins, which attach to Eph group receptors, and certain extracellular matrix elements like tenascin. Netrins are diffusible extracellular factors related to laminin, semaphorins, and slits that provide long-range guidance. Axon growth is generally attracted to netrins while semaphorins and slits are repulsive, though this does depend on the competency of specific cell populations.

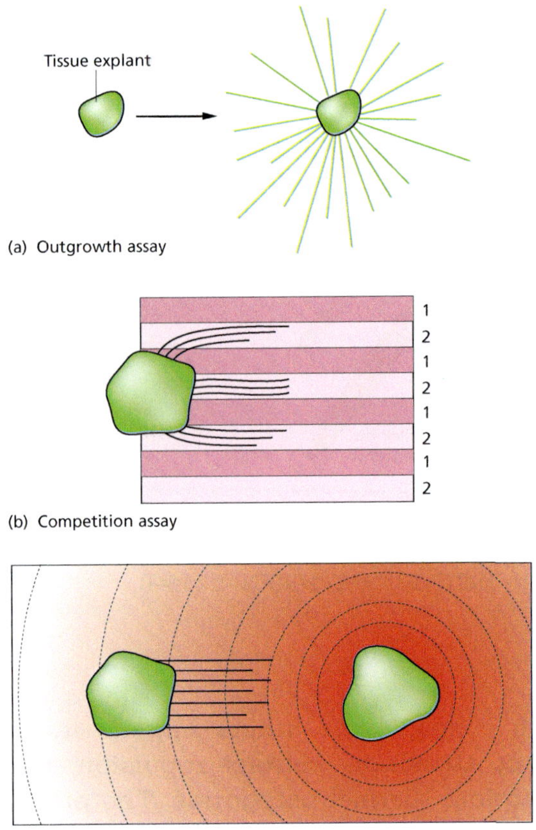

Figure 8.16. *Assays for attraction and repulsion. (a) Simple assay for outgrowth of fibers by plating a tissue explant on a coated surface, or by treating it with factors. (b) Competition assay. Two substrates, 1 and 2, are coated onto the thin alternating strips so that the axons growing out of the explant can choose between them. (c) Chemoattraction assay. The fibers from one explant grow towards the second explant which is emitting a chemoattractant.*

Source: https://dn790000.ca.archive.org/0/items/SlackJMWEssentialDevelopmentalBiology/Slack_J_M_W_Essential_Developmental_Biology.pdf.

DCC ("deleted in colon carcinoma") refers to receptors for netrins, neuropilins and plexins to semaphorin receptors, and robos (encoded by homologs of the Drosophila roundabout gene) to slit receptors. While some semaphorins are transmembrane molecules that function at short range through cell contacts, others are secreted and function as long-range factors.

8.5.1.2. Pathways

The limb regions where the spinal nerves fuse into the brachial or lumbosacral plexus and then branch again into a few principal nerves supplying the limb are the primary study areas for the initial pathways. The dorsoanterior sclerotome that encircles the neural tube is the first point of growth for the spinal nerves that innervate the legs. The posterior and ventral sclerotomes contain a repulsive factor that inhibits growth, just like neural crest cells do. The plexus region's mesenchyme provides a growth-promoting environment, but the nerves must pass through two openings in the developing pelvic girdle in order to function. More holes will allow the nerves to proliferate through them. The nerves will still find their proper, normal destinations even if a small amount of the neural tube is rotated or displaced. However, if it is moved significantly, the nerves from an area that does not typically innervate the limb may grow into it and form a pattern that resembles normalcy. These studies demonstrate that the existence of axon-follow able pathways through the tissues must be the most basic form of control. These pathways could just be empty spaces, or they could be areas rich in appropriate substrates like laminin, or they could be areas devoid of factors that repel.

Insects have been the subject of the most discerning research on this scale of connectivity. But it's crucial to keep in mind that the neural development of insects and vertebrates differs significantly. While many neurons in insects originate in peripheral organs and send axons into the central nervous system, most neurons in vertebrates arise in the central nervous system and send axons out to peripheral organs. Research on the formation of these pathways has demonstrated that the entire pathway is segmented into manageable chunks measuring roughly 100 µm. Growth cones will form along the pathway over this distance. When they reach a decision point—which could be an already-existing neuron or a shift in the substrate's characteristics—they will halt and extend their processes in all directions to investigate their environment and find the best path for the subsequent growth segment. "Pioneer neurons" form these first connections when the embryo is still relatively small, which results in a pathway with a small overall length. The development along pre-existing axonal tracts, also known as fascicles, is primarily responsible for later pathways. These give subsequent axons a "highway" to grow along. Axon retention within a fascicle is contingent upon the local balance between repulsion and attraction. The fascicle will remain intact if axons are more attracted to one another than to the substrate; if not, the fascicle will disintegrate. For instance, adding polysialic acid to N-CAM may cause defasciculation in vertebrates. As a result of the introduction of a high surface negative charge, the axons repel one another.

In the dorsal region of the spinal cord, there is a population of commissural neurons in the central nervous system. Their axons descend to the ventral midline, where the majority of them cross over to the opposite side and either make instantaneous contact with a neuron or spiral along the midline to eventually form a synapse at a different anteroposterior level (Figure 8.17). The floor plate's netrin secretion determines the direction of growth in the ventral direction. This is supported by the observation that both floor plate explants and cells transfected with netrin1 will stimulate the growth of commissural axons *in vitro*. Furthermore, a disordered pattern of commissural axon growth is observed in mice lacking netrin1, or its receptor DCC. Axons that cross at the midline do so because they express an NgCAM that can attach to a molecule that is similar on the floor plate cells.

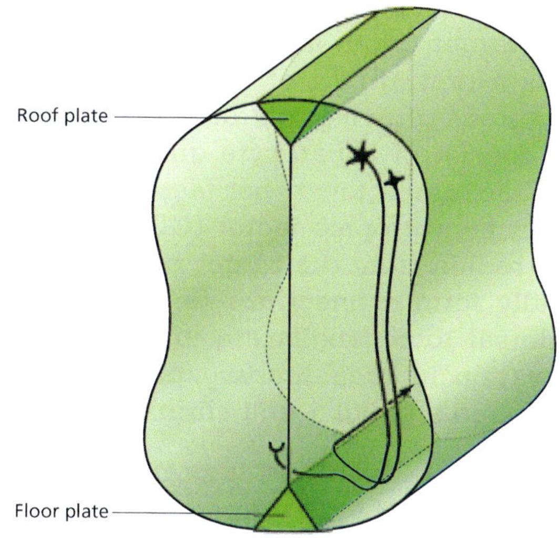

Figure 8.17. *Pathway of the axons from the commissural neurons.*

Source: https://dn790000.ca.archive.org/0/items/SlackJMWEssentialDevelopmentalBiology/Slack_J_M_W_Essential_Developmental_Biology.pdf.

When neutralizing antibody is added to NgCAM, the majority of axons grow longitudinally without crossing. Furthermore, the floor plate region has an active repulsive mechanism that includes the presence of the repulsive factor known as slit, which is secreted by midline glia. After the growth cones cross the midline, the slit receptor, robo, expresses itself. When the robo is activated by a slit, the growth cone is typically repeled from the floor plate and the guidance function of DCC is deactivated, but not the growth-promoting function. By doing this, it is made sure that the netrin signal keeps encouraging axonal elongation after the axons leave the floor plate and does not draw them back toward the midline.

8.5.2. Neurotrophins and Cell Survival

Vertebrate neurons select their final peripheral targets based on the neurotrophins secreted by those targets. For instance, the sympathetic nervous system normally innervates the pineal gland and the external ear. This is lost in mice lacking neurotrophin 3, but by applying the factor in organ culture experiments, it can be recovered.

Neurotrophins are in charge of controlling how long neuronal populations in the central nervous system and dorsal root ganglia survive. After the initial connections to the target organ are made, a significant number of motor and sensory neurons die. This is a typical part of the development of neurons, especially in higher vertebrates. For neurons grown in culture, a number of growth factors can serve as survival factors. They consist of hepatocyte growth factor, the NGF group (NGF, BDNF, NT3, NT4/5), a few FGFs, and a few other factors, such as ciliary neurotrophic factor (CNTF) and glial-derived neurotrophic factor (GDNF). Neurons must take up enough of the appropriate neurotrophin from the target organ to survive. The majority of the neurons that would typically form connections to the target organ die off if it is removed. For instance, the dorsal root ganglia are uniformly sized at first. However, those that do not have innervation in their limbs experience more cell death, which causes them to shrink. A limb bud's ganglia will shrink and experience increased cell death if it is removed.

If an extra limb bud is grafted in, then more cells will survive in the ganglia that supply it, and they will remain larger (Figure 8.18). This mechanism is known as the "neurotrophic hypothesis."

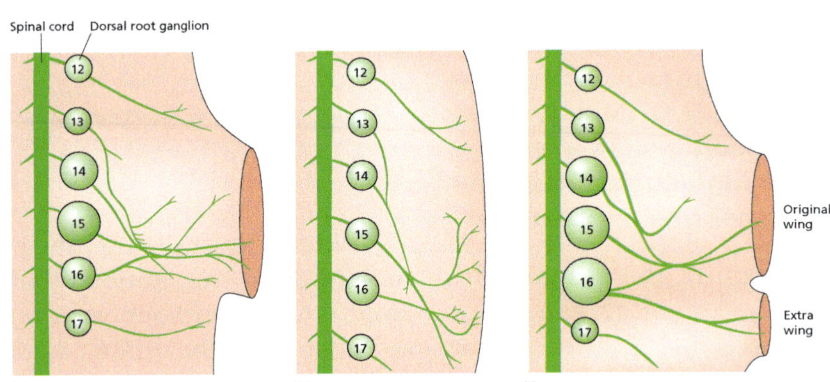

Figure 8.18. *Effects of neurotrophins on the dorsal root ganglia. (a) Normal brachial plexus in a chick embryo. (b) Reduced ganglion size after removal of the wing bud. (c) Increased ganglion size after grafting of extra wing bud.*

Source: J.M.W. Slack; Essential Developmental Biology; ISBN-13: 978–1-4051–2216–0.

NGF and other neurotrophins act on sensory and sympathetic neurons in culture in two different ways. They will support the survival of the cell bodies and encourage the growth of axons. The cell bodies continue to live if an experiment is designed so that the factor is only exposed to the growth cones and apical portions of the axons. This indicates that the factor can be absorbed at the growth cone and that retrograde transport is responsible for transferring its effects down to the cell body. It is possible to replicate the protective effect on dorsal root ganglia of grafting an additional limb bud into their vicinity by injecting NGF into the flank tissue. On the other hand, mice lacking neurotrophins or their receptors exhibit a marked reduction in sensory and sympathetic neurons. Notably, the deletion of NT3 has the greatest impact on the dorsal root ganglia.

8.5.2.1. Axonal Competition

There is a reorganization of connections following the formation of the first synaptic connections and the cell death-induced thinning of the central neurons. When multiple neurons initially project to a single multinucleate myofiber, the innervation of muscle is achieved by reorganizing the situation such that each myofiber is innervated by a single neuron but has a larger number of synapses (the specialized synapses on muscle are called neuromuscular junctions). This later phase, which depends on electrical activity, happens postnatally and is distinct from the first, rather rudimentary and diffuse pattern of connections. The principle that governs it says that if a neuromuscular junction has recently been active, then muscle excitation will strengthen it; if not, then muscle excitation will weaken it. As a result, every neuromuscular junction within a motor neuron will receive mutual reinforcement from its impulse, while every neuromuscular junction outside of it will be weakened. Until all myofibers have removed the neuromuscular junctions from all but one neuron, this process will continue on its own. A similar idea governs interactions between neurons within the central nervous system (CNS), where a connection made stronger by activity is referred to as a "Hebb synapse."

8.5.3. Neuronal Connections in the Visual System

One of the most intricate examples of neuronal specificity is the visual system. The inner surface of the optic cup gives rise to the retina. Every point on the retina receives light from a specific point in the visual field due to the camera-like optics of the vertebrate eye. The optic nerve, the second cranial nerve, is formed when axons from the retinal neurons travel down the optic tube. When the fibers on the optic tecta (or, in mammals, the lateral geniculate nuclei) form synapses, the fibers grow back into the midbrain. As a result, the surface of the tectum has a one-to-one, or topographic, relationship with the retina and, consequently, with the external visual field. Every point on the retina sends fibers to a specific point on the tectum. By illuminating a point in the visual field and recording electrical activity from the corresponding point on the tectum, this projection can be seen. Mammals have a projection from both retinas to both tecta, whereas lower vertebrates and birds have a complete crossing of optic fibers at the optic chiasma, allowing the right retina to project to the left tectum and vice versa. The projection's structure places the dorsal retina toward the ventral tectum and the anterior retina toward the posterior tectum (Figure 8.19). The anteroposterior axis of the retina is referred to as the nasal-temporal axis in the neuroscience literature; however, this usage is consistent with the rest of developmental biology.

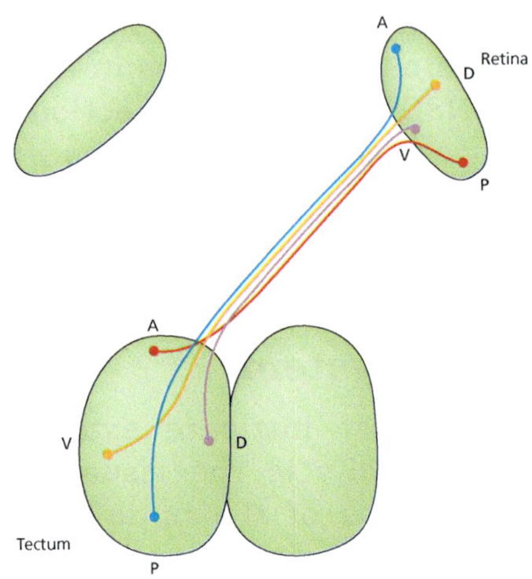

Figure 8.19. *Normal retinotectal projection in a lower vertebrate.*

Source: https://dn790000.ca.archive.org/0/items/SlackJMWEssentialDevelopmentalBiology/Slack_J_M_W_Essential_Developmental_Biology.pdf.

It is evident that a system of labeling cells in both the retina and the tectum, as well as a way to match the two sets of labels, are necessary for the specificity of connections. This idea is referred to as Sperry's "chemoaffinity hypothesis." Early in the embryonic development, the pattern in the eye is established. A Xenopus eye's projection to the tectum will continue to grow normally even if it is rotated before the tailbud stage. The projection will form but stay inverted if it is rotated later. If the optic nerve is severed in lower vertebrates, it will regenerate and the topography of the regenerated map can be examined. The connections restore the accurate projection during a typical regeneration. If the nerve is severed at the same time as half of the eye, the remaining half will protrude normally across the entire tectum. Similarly, when the nerve is severed, if half of the tectum is gone, the entire eye will project to the other half normally oriented. Several such experiments demonstrated that the labeling systems have a preferred hierarchy of connectivity but are quantitative. Additionally, they demonstrated how dynamic and self-rearranging the projection must be. This is a result of the retina's radial growth, which adds new cells from the margin's germinative zone. The tectum, on the other hand, adds cells at one end as it grows on the posterior-medial surface. Because of these various growth modes, the connections need to constantly reposition themselves to preserve the overall topographic projection.

It is now understood that a gradient of adhesivity throughout the retina and tectum serves as the foundation for the topographic projection. A tiny portion of the retina's dispersed cells will adhere to tectal explants with the same specificity as a typical *in vivo* neuronal projection (i.e., dorsal retina to ventral tectum and anterior retina to posterior tectum). At least some understanding of the molecular basis exists. Applying the stripe test (refer to Figure 8.16), it was demonstrated that the development of posterior retinal axons could be inhibited by an extract of posterior tectum. There was no repulsive activity in the anterior tectum, and the anterior retina did not respond to the tectal factor. After the factors were refined, it was discovered that they were adhesion molecules belonging to the ephrin family, exhibiting a gradient of expression ranging from high posterior to low anterior. On the retinal cells, there is a similar gradient of Eph-type receptors, with high expression in the posterior and low expression in the anterior. Hence, the ephrin system, which causes posterior fibers to adhere to the anterior tectum preferentially, is the molecular basis of the system, resulting in a short-range repulsion. The competition between fibers displaces the anterior fibers to the posterior tectum, resulting in the observed specificity, even though the anterior fibers could also adhere to the anterior tectum (Figure 8.19).

The transcription factor engrailed, which was first created by the action of the isthmic organizer on the midbrain, controls the expression of the ephrins. It also forms a posterior-antero gradient across the tectum. When engrailed is misexpressed in chick embryos through the use of a retrovirus, ephrin expression is upregulated and posterior retinal fibers are excluded from the regions of ectopic expression on the tecta.

CASE STUDY

INVESTIGATING THE DEVELOPMENT OF THE NERVOUS SYSTEM IN ZEBRAFISH EMBRYOS

Background

The nervous system's development is a complex and highly regulated process involving the differentiation and organization of neurons and glial cells. Understanding this process has significant implications for developmental biology and neurobiology. Zebrafish (Danio rerio) embryos are commonly used as model organisms in developmental biology due to their transparency, rapid development, and genetic similarities to humans.

Problem Statement

Researchers at the Developmental Neurobiology Institute (DNI) aim to investigate the molecular mechanisms underlying the development of the nervous system in zebrafish embryos. Specifically, they seek to understand how neural progenitor cells differentiate into specific types of neurons and glial cells and how these cells migrate and organize into functional neural circuits.

Solution

The research team designs a comprehensive study using advanced genetic, molecular, and imaging techniques to elucidate the processes involved in nervous system development. Their approach includes genetic manipulation, *in vivo* imaging, and transcriptomic analysis.

- **Implementation**

1. Experimental Design:
    - **Model Organism:** Zebrafish embryos are selected due to their transparent bodies and rapid development, allowing for real-time observation of nervous system development.
    - **Genetic Tools:** The team employs CRISPR-Cas9 gene editing to create knockout and knock-in zebrafish lines to study the roles of specific genes in neural development.
    - **Imaging Techniques:** High-resolution confocal microscopy and live-cell imaging are used to visualize the development of neural cells in real-time.
2. Gene Function Analysis:
    - **CRISPR-Cas9 Gene Editing:** The researchers use CRISPR-Cas9 to knock out genes suspected to be involved in neural differentiation and migration. They observe the effects of these knockouts on neural development.

- **Transgenic Lines:** Knock-in zebrafish lines expressing fluorescent markers under the control of neural-specific promoters are created to visualize specific neural cell types.

3. *In vivo* Imaging:
    - **Time-Lapse Imaging:** Time-lapse confocal microscopy is used to capture the dynamic processes of neural progenitor cell differentiation, migration, and organization in living embryos.
    - **Cell Tracking:** Advanced imaging software is utilized to track the movement and behavior of individual neural cells over time.
4. Transcriptomic Analysis:
    - **Single-Cell RNA Sequencing:** Single-cell RNA sequencing (scRNA-seq) is performed on neural progenitor cells and differentiated neurons to identify gene expression profiles and pathways involved in neural differentiation.
    - **Data Analysis:** Bioinformatics tools are used to analyze the scRNA-seq data, identifying key regulatory genes and signaling pathways.
5. Functional Studies:
    - **Behavioral Assays:** The functional consequences of gene knockouts are assessed using behavioral assays to determine the impact on neural circuit formation and functionality.
    - **Electrophysiology:** Electrophysiological recordings are performed to measure the activity of neural circuits in wild-type and mutant zebrafish embryos.

Results

The study reveals several key findings:

- **Gene Identification:** Specific genes and their regulatory networks are identified as crucial for the differentiation of neural progenitor cells into neurons and glial cells.
- **Cell Migration:** The team discovers the signaling pathways that guide the migration of neural cells to their correct positions in the developing nervous system.
- **Functional Impact:** Behavioral assays and electrophysiological recordings demonstrate that disruptions in identified genes lead to abnormal neural circuit formation and impaired neural function.

Conclusion

This case study highlights the intricate processes involved in the development of the nervous system. By using zebrafish embryos as a model organism and employing cutting-edge genetic and imaging techniques, the research team at DNI advances our understanding of neural differentiation, migration, and circuit formation. These insights have broader implications for understanding neurodevelopmental disorders and potential therapeutic approaches.

CLASS ACTIVITY

EARLY DEVELOPMENTAL STAGES

Describe the process of neurulation and explain the significance of the neural tube in the development of the nervous system.

- Activity:
- Draw and label the stages of neurulation from the neural plate to the neural tube formation.
- Discuss in groups the possible defects that can arise if neurulation does not occur properly, such as spina bifida and anencephaly.

SUMMARY

- The vertebrate central nervous system (CNS) consists of the brain and spinal cord and is formed from the neural plate of the early embryo. The peripheral nervous system (PNS) consists of the cranial, spinal and autonomic nerves and their associated ganglia. It is formed from the neural crest, from epidermal placodes, and from axons growing from cell bodies in the CNS.
- Neurons carry out most of their protein synthesis in the cell body. Materials are transported along the axon by slow axonal transport, or, in the case of neurotransmitter vesicles, by fast axonal transport. This flow of materials is balanced in a mature neuron by retrograde transport of materials from the axon terminal to the cell body.
- The structure of the spinal cord is of an inner ependymal layer, a central region of gray matter, containing most of the cell bodies, and an outer layer of white matter containing mainly myelinated fibers. In the gray matter, the motor neurons are located ventrally, while the interneurons, which receive sensory input, and the commissural neurons, which connect one side of the spinal cord with the other, lie dorsally.
- The autonomic system with its numerous small ganglia and CNS connections is also formed from the neural crest. It is divided into a distinct sympathetic and parasympathetic system, also producing an extensive system of enteric neurons in the gut wall.
- Neural induction is due to inhibition of bone morphogenetic proteins (BMPs) by substances such as chordin released from the organizer. This induces a neuroepithelium with an anterior specification such that, in the absence of further signals, it will express transcription factors such as Otx and form forebrain structures.
- Manipulations of retinoic acid levels show particular effects on the hindbrain. Retinoic acid is produced from dietary vitamin A and the enzyme that makes it (retinaldehyde dehydrogenase, RALDH2) is found in the somites and lateral plate of the trunk region.
- The initial three-layered structure of ventricular, mantle, and marginal zones is maintained in the spinal cord, but in some parts of the brain it later gets much more complex. In the mammalian cerebral cortex, neurons from the mantle zone invade the marginal zone to establish six further layers of neocortex.
- The mitotic cells of the ventricular zone can both maintain themselves and generate progeny that differentiate into neurons and glia. Later, when this layer becomes known as the ependyma, it is still mitotic although cell divisions are then few and far between.
- The neural crest is a population of cells derived from the neural folds. They are present in the dorsal part of the neural tube at the time of closure but soon migrate away from the neural tube into the surrounding tissues.

- The migration of crest cells depends on a wide variety of extracellular matrix components. Neural tube cells express the adhesion molecule N-cadherin. This is lost by migratory crest cells but re-expressed when they reach their destinations.
- The axons of neurons grow out and innervate their targets. Motor neurons in the ventral spinal cord innervate muscles or glands, and sensory neurons in the dorsal root ganglia innervate a variety of peripheral receptors and sense organs.

REVIEW QUESTIONS

1. What are the main differences between neurons and glial cells in terms of structure and function?
2. Describe the structure and function of the various types of glial cells found in the nervous system.
3. What signaling molecules and pathways are involved in the anteroposterior patterning of the neural plate?
4. Explain the process of neurogenesis and the factors that regulate the differentiation of neural progenitor cells into neurons.
5. What are the origins and fate of neural crest cells during embryonic development?
6. How do axons find their targets during the development of neuronal connectivity?

REFERENCES

1. Anderson, D. J. (1997). Cellular and molecular biology of neural crest cell lineage determination. *Trends in Genetics, 13*(7), 276–280. https://doi.org/10.1016/S0168-9525(97)01182-2.
2. Bally Cuif, L., & Wassef, M. (1995). Determination events in the nervous system of the vertebrate embryo. *Current Opinion in Genetics and Development,* 5(4), 450–458. https://doi.org/10.1016/0959-437X(95)80048-X.
3. Blumberg, B. (1997). An essential role for retinoid signaling in anteroposterior neural specification and in neuronal differentiation. *Seminars in Cell and Developmental Biology, 8*(4), 417–428. https://doi.org/10.1006/scdb.1997.0154.
4. Cayouette, M., & Raff, M. (2002). Asymmetric segregation of Numb: A mechanism for neural specification from Drosophila to mammals. *Nature Neuroscience,* 5(12), 1265–1269. https://doi.org/10.1038/nn973.
5. Henrique, D., & Schweisguth, F. (2003). Cell polarity: The ups and downs of the Par6/aPKC complex. *Current Opinion in Genetics and Development, 13*(3), 341–350. https://doi.org/10.1016/S0959-437X(03)00088-1.
6. Jessell, T. M. (2000). Neuronal specification in the spinal cord: Inductive signals and transcriptional codes. *Nature Reviews Genetics, 1*(1), 20–29. https://doi.org/10.1038/35049500.

7. Joyner, A. L. (1996). Engrailed, Wnt and Pax genes regulate midbrain-hindbrain development. *Trends in Genetics, 12*(1), 15–20. https://doi.org/10.1016/0168-9525(96)80006-6.

8. Kintner, C. (2002). Neurogenesis in embryos and in adult neural stem cells. *Journal of Neuroscience, 22*(2), 639–643. https://doi.org/10.1523/JNEUROSCI.22-02-00639.2002.

9. Le Douarin, N. M., & Smith, J. (1988). Development of the peripheral nervous system from the neural crest. *Annual Review of Cell Biology, 4*, 375–404. https://doi.org/10.1146/annurev.cb.04.110188.002103.

10. Lumsden, A., & Krumlauf, R. (1996). Patterning the vertebrate neuraxis. *Science, 274*(5287), 1109–1115. https://doi.org/10.1126/science.274.5287.1109.

11. Marquandt, T., & Pfaff, S. L. (2001). Cracking the transcriptional code for cell specification in the neural tube. *Cell, 106*(5), 651–654. https://doi.org/10.1016/S0092-8674(01)00422-6.

12. Marshall, H., Morrison, A., Studer, M., Pöpperl, H., & Krumlauf, R. (1996). Retinoids and Hox genes. *FASEB Journal, 10*(10), 969–978. https://doi.org/10.1096/fasebj.10.10.8863377.

13. O'Leary, D. D. M., & Nagakawa, Y. (2002). Patterning centers, regulatory genes, and extrinsic mechanisms controlling arealization of the neocortex. *Current Opinion in Neurobiology, 12*(1), 14–25. https://doi.org/10.1016/S0959-4388(02)00300-2.

14. Ragsdale, C. W., & Grove, E. A. (2001). Patterning the mammalian cerebral cortex. *Current Opinion in Neurobiology, 11*(1), 50–58. https://doi.org/10.1016/S0959-4388(00)00156-8.

15. Simeone, A. (2000). Positioning the isthmic organizer: Where Otx2 and Gbx2 meet. *Trends in Genetics, 16*(5), 237–240. https://doi.org/10.1016/S0168-9525(00)02084-0.

16. Wilson, S. I., & Edlund, T. (2001). Neural induction: Toward a unifying mechanism. *Nature Neuroscience, 4*(11), 1161–1168. https://doi.org/10.1038/nn756.

CHAPTER

9 Development of Mesodermal Organs

LEARNING OBJECTIVES

After studying this chapter, you will be able to:

- Describe the development of somitogenesis and myogenesis;
- Understand about the kidney;
- Learn about germ cell and gonadal development;
- Know about limb development.

KEY TERMS FROM THIS CHAPTER

Dermatome	Dermomyotome	Epithelium
Lateral plate	Mesenchyme	Myoblasts
Myofibers	Myogenesis	Myotome
Myotubes	Neural crest	Notochord
Organ culture	Paraxial mesoderm	Progenitor cells
Sclerotome	Segmental plate	Splanchnic mesoderm
Stem cells	Tissue engineering	Ureteric bud

INTRODUCTION

The mesoderm of higher vertebrates divides early into four zones, ranging from medial to lateral. The notochord is located in the middle, followed by the somites-forming paraxial mesoderm, the kidneys, adrenal glands, and gonads-forming intermediate mesoderm, and finally the body wall, limb buds, blood vessels, and heart-forming lateral plate mesoderm. The arrangement is the same in lower vertebrates, but it goes from dorsal to ventral because the yolk mass is surrounded by the mesoderm in these animals.

The coelom divides the lateral plate into an inner splanchnic mesoderm, which forms the heart and mesenteries, and an outer somatic mesoderm, which later forms the limb buds. The vertebrae form from somites, the limb buds and associated lateral plate give rise to the bones and girdles of the limbs, and the neural crest forms the majority of the skull. These three areas are where the skeleton is formed.

Despite the formation of some comparable tissue types, such as muscle, gonads, connective tissues, and blood cells, invertebrate mesoderm does not exhibit the same regional subdivision as vertebrate mesoderm.

In this chapter, the general convention applied elsewhere is followed in sections primarily dealing with mice (kidney, gonad, and germ cells), and the mouse gene and protein nomenclature convention is followed in those sections.

9.1. SOMITOGENESIS AND MYOGENESIS

The process by which the paraxial mesoderm segments into pairs of somites, or balls of mesoderm, within the trilaminar embryo body is referred to as somitogenesis. In humans, the caudal addition of somite pairs occurs at a rate of one pair every 90 minutes, with an average of 44 pairs forming after the first pair appears on day 20.

A somite pair is created by adding a somite on either side of the notochord (axial mesoderm). Not in the head region, the segmentation starts cranially (head end) and extends caudally (tailward), adding a somite pair at regular intervals along the way. Due to the sequential nature of the process, the number of visible somite pairs is used to determine the age of embryos from many different species.

The developing somite's outer cellular shell is defined by a transition from mesenchymal to epithelial cells, while the core cells continue to be organized as mesenchymal cells. Each somite has a temporary fluid-filled area known as the somitocoel during the early stages of somite development, which is subsequently eliminated by cell proliferation. Later, neural crest cells also penetrate and combine with somatic cells.

One of the three germinal layers that emerges during the third week of embryonic development is the mesoderm. It is created by a procedure known as gastrulation. The lateral plate mesoderm, paraxial, intermediate, and axial plate mesoderms are the four major parts. The notochord develops from the axial mesoderm. The paraxial mesoderm forms the somitomeres, which give rise to the subcutaneous skin tissue known as dermatomes and the cartilage and bone known as sclerotomes. These somites then organize into somites in the occipital and caudal segments, giving rise to the mesenchyme of the head (Ruppert et al., 2004). The notochord, neural tube, and epidermis are among the surrounding structures that provide signals for somite differentiation (Kimelman & Bjornson, 2004). The lateral plate and the paraxial mesoderm are joined by the intermediate mesoderm. It eventually differentiates into urogenital structures, which include the adrenal cortex, gonads, kidneys, and the ducts that connect them. The circulatory system's heart, blood vessels, and blood cells, as well as the mesodermal parts of the limbs, are all products of the lateral plate mesoderm (Scott & Gilbert, 2010).

The muscle (smooth, cardiac, and skeletal), the tongue muscles (occipital somites), the pharyngeal arches muscle (mastication, facial expressions muscles), connective tissue, the skin's dermis and subcutaneous layer, bone and cartilage, dura mater, the endothelium of blood vessels, red blood cells, white blood cells, microglia, the dentin of teeth, the kidneys, and the adrenal cortex are among the mesoderm derivatives (Dudek & Ronald, 2009).

9.1.1. Normal Development of the Somites

The somite pattern is the most obvious sign of a segmental arrangement of the body pattern in vertebrates, along with the rhombomeres of the hindbrain. The paraxial mesoderm, also known as the segmental plate, is the mesoderm that flanks the notochord and gives rise to somites in anteroposterior sequence (Figure 9.1). Forkhead transcription factors Foxc1 and c2 expression sets this apart from the intermediate mesoderm. When these genes are overexpressed in chicks through retroviral infection, the width of the somite files increases, and the mice with double knockout of these genes are devoid of somites. The number of somites formed is frequently used as an accurate indicator of the stage of a vertebrate embryo because somite formation is progressive and readily visible. A characteristic of the species, the total number formed during development can differ greatly amongst animal groups, ranging from a few fish with approximately thirty to several hundred in snakes.

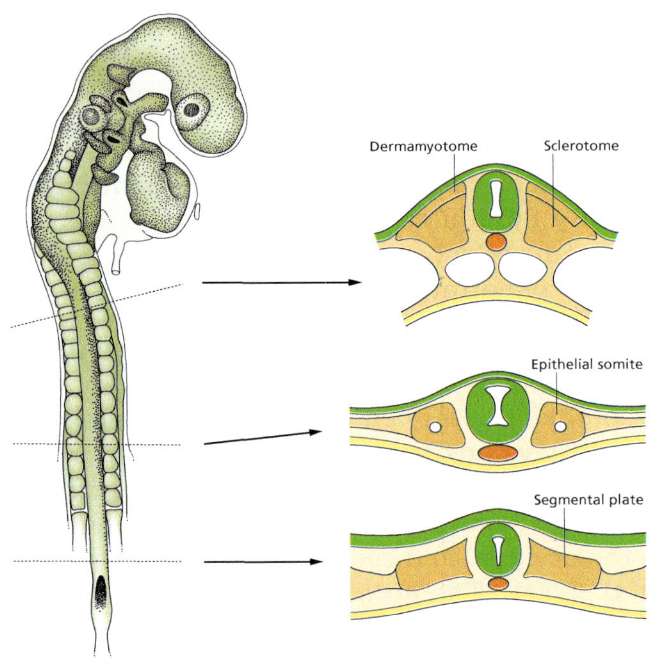

Figure 9.1. *Somitogenesis in the chick embryo. The process starts in the anterior and new segments are cut off at the posterior end of the file for a considerable time. The morphology is initially epithelial and then changes to an outer dense dermomyotome and an inner loose sclerotome.*

Source: Jonathan M.W. Slack. Essential Developmental Biology, 3rd Edition. Wiley-Blackwell. ISBN: 9781118387603.

Although all vertebrates form somites, the chick embryo has been used in the majority of experimental work, with some additional data derived from zebrafish, Xenopus, and mice. First discernible in the chick as loose cell associations, epithelial somites are vesicles of cells enclosing a cavity that eventually condense into somites. The changes in cell adhesion that propel the process are caused by an increase in the expression of fibronectin and N-cadherin with the formation of the epithelial somite. The epithelial somite is a transient structure because, on the medial side, it quickly experiences an epithelialto-mesenchymal transformation that forms the sclerotome, a mesenchymal condensation that eventually gives rise to the vertebrae and ribs (Figure 9.2). The term "syndetome" refers to the dorsal portion of the sclerotome, which also forms tendons. The vertebrae are first formed of cartilage, which is subsequently replaced by bone, just like other parts of the skeleton outside the skull. As a result of the posterior half of one somite combining with the anterior half of the next to form the sclerotome of each vertebra, the vertebrae end up half a segment out of phase with the original somite repeating pattern. This process, referred to as "resegmentation," is akin to the interaction between Drosophila's parasegments and definitive segments.

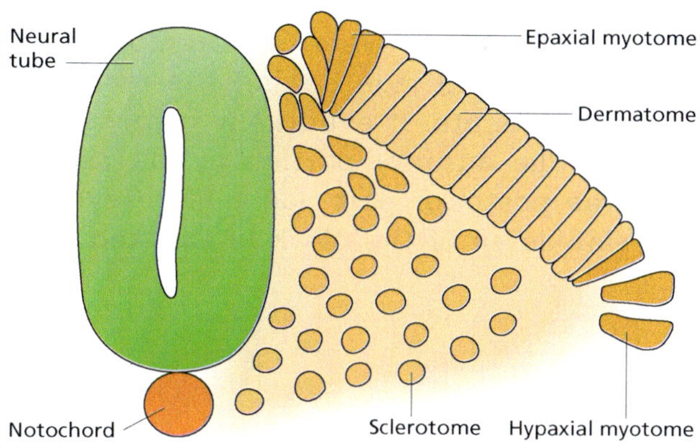

Figure 9.2. *Differentiation of the amniote somite.*

Source: Jonathan M.W. Slack. Essential Developmental Biology, 3rd Edition. Wiley-Blackwell. ISBN: 9781118387603.

The dermomyotome is a plate that is formed by the lateral portion of the epithelial somite. This further splits into the myotome, which forms skeletal muscle, and the dermatome, which forms the dermis of the dorsal midline. The segmental muscles of the main body axis are formed by the epaxial myotome, or region near the midline, and the ventral body wall, limbs, and diaphragm are formed by the hypaxial myotome, or more lateral region.

9.1.2. Segmentation Mechanism

The somites' segmental repeating pattern results from the interaction of a spatial gradient and a molecular oscillator, or clock. A single clock cycle initiates the formation of a single somite, with the gradient dictating the somites' formation in an anterior-posterior sequence. The clock's duration varies depending on the species; for example, zebrafish take 30 minutes, chicks 90 minutes, and mice 120 minutes. The clock in the chick begins to run in the early (H&H 3) embryo, and the first somite forms during the third cycle. The Notch pathway's periodic expression of its targets, the Hes genes and Lunatic fringe, can be seen as the clock's visual representation. The Hes genes are homologs of Drosophila hairy, a pair-rule gene also involved in segmentation, and they encode the transcription factors bHLH. Lunatic fringe is a gene that codes for a glycosyl transferase and is homologous to Drosophila fringe. The creation of compartment boundaries in imaginal discs is aided by Drosophila fringe.

These genes' mRNA levels oscillate at a periodicity that matches the rate at which new somites form. This oscillation is synchronous throughout the majority of the presomite plate, suggesting that cell communication is taking place to keep neighboring cells in the same phase of the cycle. Before it was eventually realized that the static *in situ* hybridization patterns were rapidly changing over time, it appeared that the patterns were highly variable between specimens, leading to the discovery of the periodic nature of this gene expression. Comparing the *in situs* of the left and right sides would allow one to see the phase shift if the embryos were cut longitudinally and fixed at different times. The oscillation slows down and lags behind the rest of the presomite plate in the area next to the last-formed somite. It eventually comes to an end in the area that will become the next somite. The anterior portion of the newly formed somite has a low level of mRNAs, while the posterior part has a high level, due to the progressive rather than instantaneous slowdown of the clock in this region (Figure 9.3).

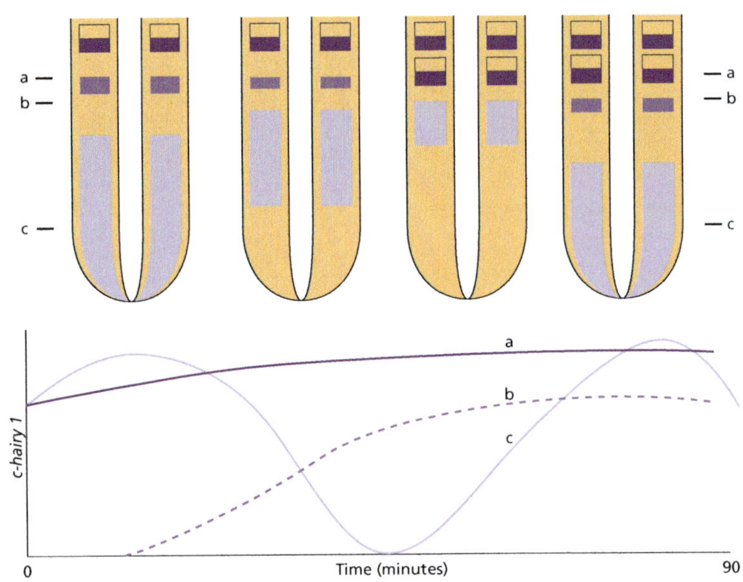

Figure 9.3. *Operation of the somite oscillator over one cycle of somite formation in the chick. The diagrams show the expression pattern of Hes at four times in the cycle, and the graphs show how the level of transcript varies at the points a, b, and c.*

Source: Jonathan M.W. Slack. Essential Developmental Biology, 3rd Edition. Wiley-Blackwell. ISBN: 9781118387603.

The Hes factors' autoregulation of transcription is essential to the oscillator mechanism. Since these proteins are highly unstable and will repress their own transcription, they break down quickly in the absence of fresh synthesis. When the product level is low, transcription is permitted. This leads to the accumulation of protein, which represses transcription. However, because the protein is unstable, this level eventually drops, causing transcription to increase once more (Figure 9.4a). The primary proof for this negative-feedback mechanism is the oscillator's cessation upon Hes synthesis inhibition through mutation (in mice) or morpholino treatment (in zebrafish). Zebrafish oscillations most likely result from this mechanism alone, wherein neighboring cells remain synchronized thanks to the Delta-Notch system. The oscillation in amniotes is longer in duration and incorporates the Delta-Notch system as a fundamental part of the oscillator. Hes proteins inhibit Delta transcription in this more intricate cycle, which lowers Notch signaling in nearby cells. Conversely, Notch signaling raises Hes transcription, which causes Hes proteins to decrease in nearby cells. This results in increased Delta transcription, which raises Hes transcription once more by increasing Notch signaling in the original cell (Figure 9.4b). In mice, loss of delta causes the oscillations to be suppressed, but in zebrafish, loss of delta causes the oscillations to continue but eliminates the synchrony in nearby cells. This shows that the Delta-Notch system in zebrafish just helps to keep the population of presomite plate cells in the same phase, but in mice it is an integral part of the oscillator itself.

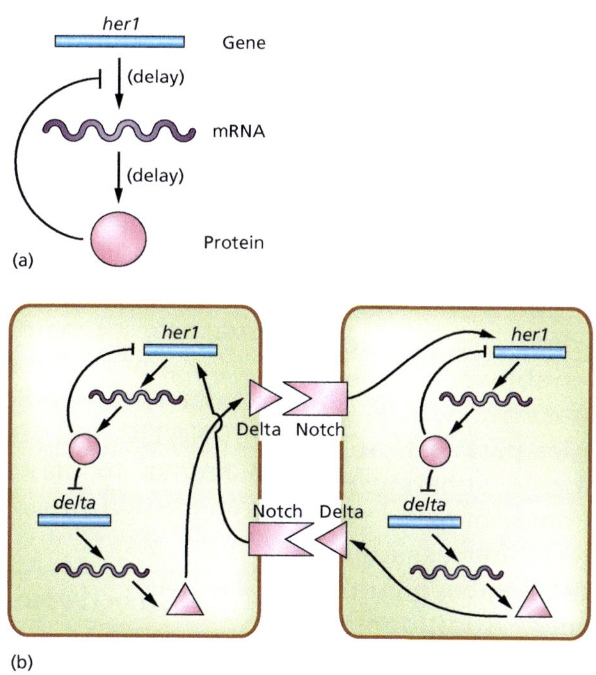

Figure 9.4. *(a) The simple feedback and delay model for the oscillator. (b) Coupling between cells by the Delta–Notch system.*

Source: Jonathan M.W. Slack. Essential Developmental Biology, 3rd Edition. Wiley-Blackwell. ISBN: 9781118387603.

Lunatic fringe and Hes gene expression are both controlled by the Notch pathway. Lunatic fringe has the ability to alter the Delta-Notch signaling system's constituent parts so that Notch activation only happens at the border between a region that expresses Lunatic fringe and one that does not. While the subsequent cell differentiation of the somite derivatives is relatively normal, the somitogenesis-related phenotypes of the mice lacking these components are similar, disrupting the segmentation pattern.

A gradient must interact with the oscillator's operation in order to produce a sequence of segments in space. The gradient portion of the system, which dictates other facets of the posterior body pattern, is the posterior-to-anterior gradient of fibroblast growth factors, primarily FGF4 and FGF8. Over time, the gradient progressively decays, resulting in the same FGF concentration being felt at increasingly posterior positions. Due to their shared involvement in controlling genes that regulate somite segmentation, such as the transcription factor gene Mesp2, the clock and gradient work together to produce segments. Mesp2 is upregulated by the clock components when the FGF level drops below a predetermined threshold. This causes the posterior part of each prospective somite to stabilize a steep gradient of Fringe expression, which in turn activates the posterior part of Notch. The expression of additional substances, such as ephrins, which contribute to the modifications in cell adhesion necessary to form the epithelial somite, is subsequently driven by notch signaling. Additionally, Mesp2 inhibits Snail's expression. Snail's product typically inhibits cell adhesion molecules like cadherins and integrins. As a result, the mesenchymal-to-epithelial transition and the upregulation of these cell adhesion molecules occur, which causes an epithelial somite to form. *In situ* hybridization allows for the visualization of a gradient of Fgf8 mRNA in zebrafish, chicks, and mice, while immunostaining reveals a corresponding gradient of phosphorylated ERK. When SU5402, an FGF signaling inhibitor, is added, segmentation is thrown off. Large somites are the result of this, as each cycle a larger than normal region receives FGF signaling below threshold, which is subsequently integrated into the somite. On the other hand, smaller-than-normal somites develop when FGF protein is administered via electroporation or a slow-release bead.

A "wavefront" is frequently mentioned in descriptions of somitogenesis, although practically no one understands the meaning of this term. The wavefront, which represents the threshold concentration of the FGF gradient and moves posteriorly as the gradient decays, is essentially the moving boundary of the mesenchymal-toepithelial transition.

Since the clock and gradient process for somitogenesis are similar in all vertebrates, it is assumed that humans also use this mechanism. It's interesting to note that certain congenital scoliosis types (spinal curvature, occasionally associated with other segmental abnormalities) are caused by mutations in genes, such as DELTA-LIKE 3, LUNATIC FRINGE, and MESP2, which are known to play important roles in the segmentation process.

9.1.3. Subdivision of the Somite

The somite's internal anterior–posterior subdivision, which was established during segmentation, is still significant during subsequent development. Because of its high ephrin level, the posterior sclerotome repels both neural crest cells and spinal nerves, which is why they preferentially grow through the anterior sclerotome of each segment.

The sclerotome, dermatome, and myotome are produced by the somite's division along the transverse axes, and their development is dependent upon interactions with the surrounding tissues (Figure 9.5). This can be demonstrated by the fact that even after the freshly formed epithelial somites have undergone microsurgical rotation or inversion, subdivision will still proceed normally with respect to the axes of the entire body. By demonstrating that a particular induction failed to occur when the presomite tissue was isolated and that it did occur at the site of contact when the signaling tissue and presomite tissue were cultured together, each unique induction was identified. The notochord and the neural tube's ventral portion induce the sclerotome. The notochord and dorsal neural tube induce the epaxial portion of the myotome, and the lateral plate mesoderm restricts the spread of myotomal induction. The neural tube induces the dermatome.

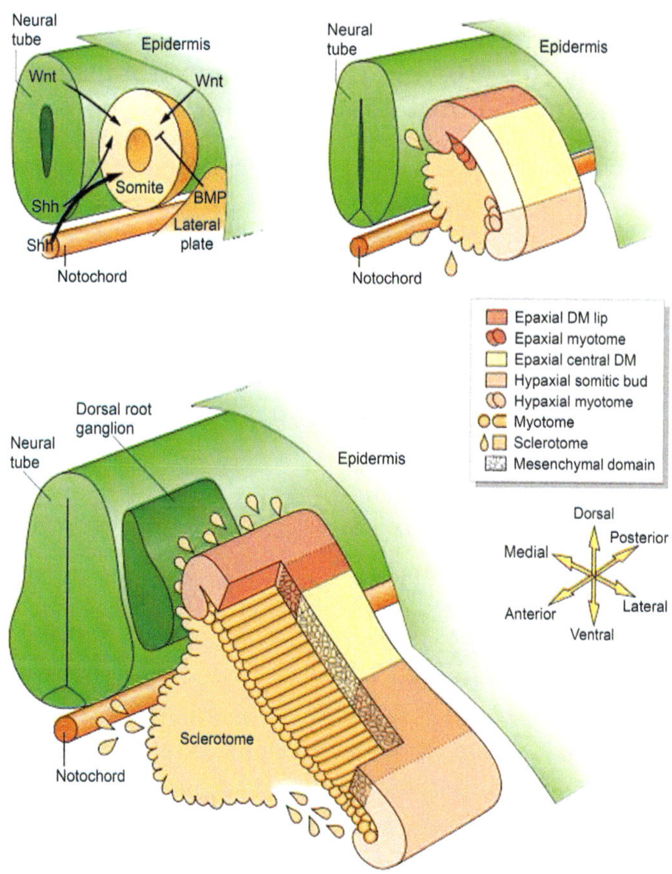

Figure 9.5. *Subdivision of somite by signals from surrounding tissues.*

Source: Jonathan M.W. Slack. Essential Developmental Biology, 3rd Edition. Wiley-Blackwell. ISBN: 9781118387603.

Sonic Hedgehog (Shh) is the main signal for sclerotome induction. This is necessary for the neural tube's ventrodorsal patterning and is expressed in the notochord and floor plate. Any region of the epithelial somite can produce cartilage when cells expressing Shh are present. The transcription factor genes Nkx3.2, Pax1 and 9, Scleraxis, and genes encoding myogenic inhibitors of the Id class are all upregulated during the early events. Since most sclerotome derivatives are absent in the Shh-knockout mouse, it is evident that Shh is essential for sclerotome formation *in vivo*. During typical development, the notochord remains intact and becomes the nucleus pulposus within the intervertebral discs.

As previously indicated, a process known as "resegmentation" causes each vertebra to eventually form from the posterior sclerotome of one somite joined with the anterior sclerotome of the next most posterior somite. The combination of Hox genes expressed at that level determines the regional characteristics of the vertebrae (cervical, thoracic, lumbar, sacral, and caudal) (Figure 9.6). Although the number of vertebrae in each of these body regions varies between the mouse and the chick, the Hox codes for each region are the same (Figure 9.7). For instance, the transition from cervical to thoracic vertebrae is indicated by the anterior boundary of Hoxc6 expression, the transition from lumbar to sacral vertebrae by Hoxd10, and the transition from sacral to caudal vertebrae by Hoxd12 expression. Changes in vertebral character can be predicted by transgenic overexpression and Hox gene knockout.

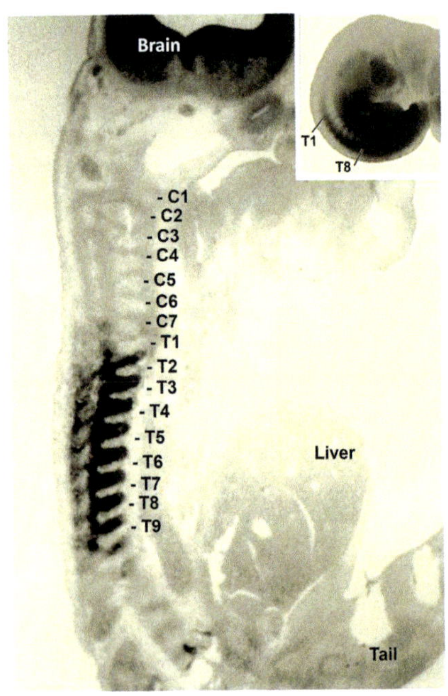

Figure 9.6. *Expression of Hoxc6 in the future thoracic vertebrae of a mouse embryo. C1–7 cervical vertebrae, T1–9 thoracic vertebrae.*

Source: Jonathan M.W. Slack. Essential Developmental Biology, 3rd Edition. Wiley-Blackwell. ISBN: 9781118387603.

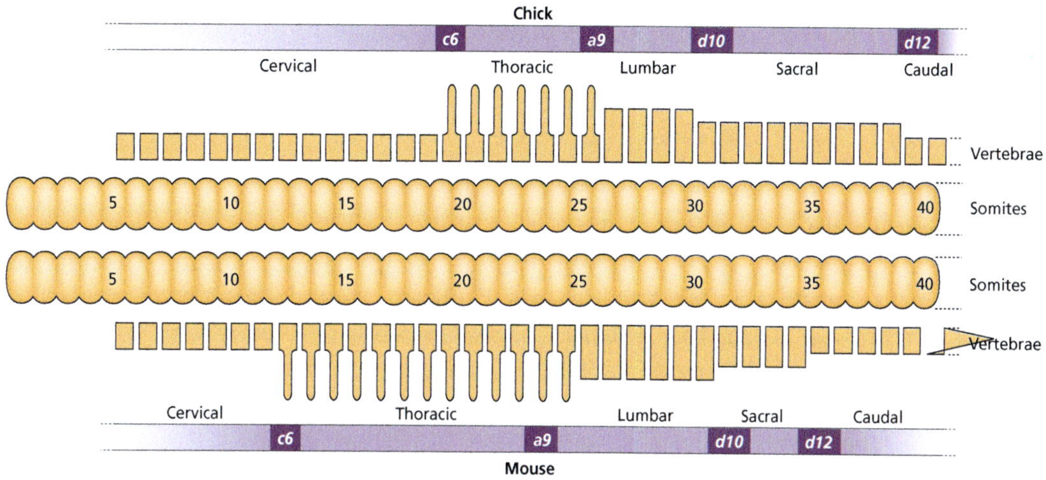

Figure 9.7. *Formation of vertebrae from the somites of the chick (above) and the mouse (below). Each vertebra is formed from two half somites, and the regional character of vertebrae correlates with the expression domains of Hox genes. c6, a9, d10, d12 indicate the anterior boundaries of these Hox genes in the two species.*

Source: *Jonathan M.W. Slack. Essential Developmental Biology, 3rd Edition. Wiley-Blackwell. ISBN: 9781118387603.*

The myotome's formation is a more intricate process. Before the somite has divided into the dermomyotome and sclerotome, some pioneer muscle cells originate from the dorsal medial region. Myoblasts then proliferate from cells at the dermomyotome's medial (epaxial) and lateral (hypaxial) edges, migrating around the edges of the dermomyotomal plate to join the myotome that is growing underneath. Following their migration into the myotome, these cells undergo post-mitosis and acquire anteroposterior orientation. While the epaxial and hypaxial regions have somewhat different signaling environments, they both produce myoblasts. This is just one of many instances in organogenesis where distinct rudiments give rise to the same cell type via initially disparate molecular pathways that ultimately converge on the same genes—in this case, the myogenic transcription factors of the bHLH class that are discussed below.

Early exposure to Shh during the presegmental stage is necessary for myogenesis in the epaxial region, and this is followed by a Wnt signal from the dorsal neural tube. As a result, Pax3, which is first expressed throughout the epithelial somite, is repressed and myogenic genes like Myf5 are inducted. The lateral plate mesoderm's removal exhibits a similar result. When BMP4 is supplied by one of the common overexpression techniques (cell pellet, slow-release bead, or retroviral overexpression), the lateral plate will express BMP4, which will suppress myogenesis. The balance between the effects of the BMP signal from the lateral plate and the Wnt signal from the dorsal neural tube appears to determine the spatial extent of the epaxial myotome. The medial region's induction of Noggin expression, which prevents the lateral plate-derived BMP4, is one of the Wnt signal's functions.

Since the primordium is located much farther away from the neural tube and outside the range of the Shh and Wnt signals, the induction of the hypaxial myotome occurs *via* a different pathway. The primary inductive signal is thought to consist of Wnt7A from the dorsal epidermis combined with a permissive level of BMP from the epidermis. The transcription factors Lbx1 (LIM class) and c-Met, which are subsequently necessary for myoblast migration to the limb buds, are expressed by the hypaxial myotomes.

The dorsal neural tube sends out a Neurotrophin 3 signal, which causes the dermatome to form. This factor's neutralizing antibody can stop the epithelial-to-mesenchymal transition, which is necessary for the dermis to form. The lateral and ventral regions of the dermis originate from the lateral plate mesoderm, whereas the dermatome produces the dorsal dermis. Dermo1, a bHLH protein, which inhibits the expression of the myogenic transcription factors, is partially responsible for its characteristics.

9.1.4. Myogenesis

Skeletal muscle, smooth muscle, and cardiac muscle comprise vertebrate muscle. All skeletal muscle originates from the corresponding regions of the unsegmented head mesoderm, which form the muscles of the jaw and eyes, or from the myotomes of the somites, which form the axial muscles, limb muscles, and body wall muscles. Both the lateral plate mesoderm and the hypaxial portion of the somites give rise to smooth muscle, while the early heart's myocardium, which is derived from anterior splanchnic mesoderm, gives rise to cardiac muscle.

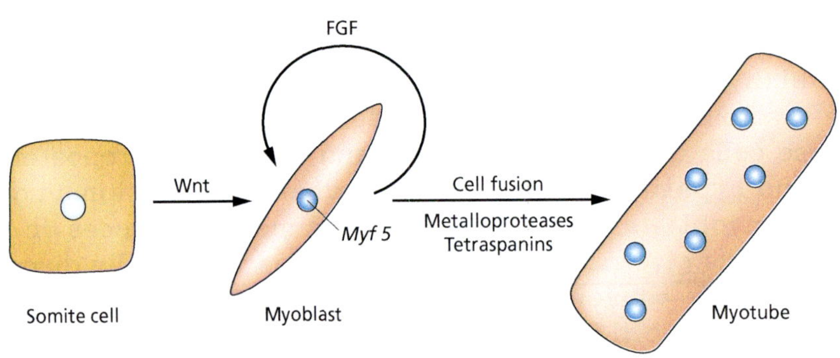

Figure 9.8. *Factors controlling myogenesis.*

Source: Jonathan M.W. Slack. Essential Developmental Biology, 3rd Edition. Wiley-Blackwell. ISBN: 9781118387603.

The well-known voluntary muscle that makes up most of an animal's mass that is vertebrate is called skeletal muscle. It is made up of multinucleate myofibers with myofibrils made of myosin and actin, two contractile proteins. Skeletal muscle differentiates by first committing its cells to become myoblasts, followed by a phase of migration and/or multiplication, and finally a break in myoblast division and fusion to form multinucleate myofibers (Figure 9.8). Proliferation stops in a normal developmental stage once myoblasts have moved from the dermomyotome to the underlying myotomes.

Although they should properly be referred to as "myocytes" once postmitotic, the term "myoblast" is still used.

9.1.4.1. Myoblast Fusion

Through the use of chimeric mice, it was demonstrated that multiple myoblasts can fuse to form a single myofiber *in vivo*. Muscle consists of both the two distinct forms and an intermediate form of isocitrate dehydrogenase, if a chimera is created from embryos from strains expressing different electrophoretic variants of the enzyme. This is a result of the enzyme's dimerization. Depending on the embryo from which they originated, single cells can only produce AA or BB dimers; however, because both forms are found in the same cytoplasm, fused cells can also produce the heterodimer AB.

Research on the mechanism of myoblast fusion has been conducted using *in vitro* work on mammalian models of myogenesis as well as genetics on Drosophila.

Thirty multinucleate fibers, each containing a distinct muscle, emerge in Drosophila embryos. A founder cell initiates each, fusing with multiple fusion-competent myoblasts. The competent cells that are encoded by the gene sticks-and-stones bind a receptor to a cell-surface molecule called Kinof-irreC (Kirre), which is expressed by the founder myoblasts. Fusion will be impossible if one of the components stops working. This is significant because fusing myoblasts in zebrafish express homologs known as Kirre-like, and blocking these with morpholinos will likewise stop fusion, suggesting a likely shared mechanism.

The majority of research on mammalian muscle fusion has used mouse C2C12 cells, a myoblast cell line that is simple to study in tissue culture. When FGF is present, these cells continue to grow; however, when it is not, they stop and begin to fuse together to form myotubes, which resemble the myofibers that are found *in vivo*. Myogenesis is inhibited by the transcriptional repressor Msx1, whose expression is sustained by FGF. The idea that fusion depends on a variety of molecules is supported by the fact that fusion can be inhibited by certain antibody treatments or accelerated by transfection of the right genes. Metalloproteases of the ADAM class, also referred to as meltrins, comprise one group and are also implicated in the process of mammalian fertilization. Furthermore, caveolin, a cell-surface molecule, and the protease calpain have been linked. The transcription factor NFATC2 is activated by an increase in intracellular Ca^{2+}, and mice lacking this gene develop smaller fibers with fewer nuclei.

9.1.4.2. Myogenic Transcription Factors

The myogenic protein family is essential to the differentiation of skeletal muscle. These transcription factors—also referred to as MyoD, Myf5, Myogenin, and MRF4 (Muscle regulatory factor 4, also known as Myf6)—are all members of the bHLH class. They were found because the genes could induce the cells to become myogenic when transfected into different kinds of tissue-culture fibroblasts. As the transfected cells fused to form multinucleate myotubes, they would begin to express genes for muscle proteins. Certain muscle protein genes, including those for acetylcholine receptor and creatine phosphokinase, have been demonstrated to be directly activated by MyoD. When turned on, it remains on because it also initiates its own transcription. Dimers known as bHLH factors typically form heterodimers with an E protein, which is another common bHLH protein. Additionally, there are inhibitory HLH proteins that function by sequestering the bHLH proteins in ineffective dimers. These proteins have the dimerization domain but lack the transcriptional activation domain. For instance, the HLHtype inhibitor Id protein, which is upregulated in the sclerotome in response to Shh, forms ineffective dimers with myogenic bHLH proteins and prevents the somite's portion from developing muscle.

While the other factor is also soon upregulated in both zones, in normal development Myf5 is initially upregulated in the epaxial myotome and MyoD in the hypaxial myotomes. Because the other gene will upregulate and take its place if one is removed, knockouts of either gene only slightly affect the development of muscles. But when both are knocked out at once, the result is a severely affected mouse that lacks skeletal muscle and myoblasts. The activity of Mrf4, which is located in the genome close to Myf5, is also lost in this double knockout. Normally, myogenin expresses itself throughout the myotome later than Myf5 and MyoD. There is a significant defect in the formation of skeletal muscle when it is knocked out, indicating that Myogenin plays a major role in the mechanisms underlying the effects of MyoD and Myf5.

Mammalian myofibers can survive for the duration of an animal once they have differentiated, and muscles can typically only grow larger by enlarging pre-existing fibers. A feedback mechanism involving the circulating inhibitor myostatin regulates growth. Despite the fact that myofibers are postmitotic, they are linked to muscle satellite cells, which are persistent progenitor cells. These tiny, mononuclear cells are found beneath each myofiber's basement membrane. They are characterized by the expression of the transcription factor Pax7 and come from the central region of the dermomyotome. Muscle satellite cells are in charge of the animal's growth in fiber quantity during growth as well as damage repair in adulthood.

9.2. THE KIDNEY

Since there aren't enough human kidneys available for transplantation, the kidney—which is the body's excretory organ—is one of the main targets for replacement through tissue engineering or another type of applied developmental biology. Because the development of the kidney during embryonic development involves reciprocal inductive interactions between an epithelium and a mesenchyme, it is representative of a broad class of organs. Kidney congenital anomalies are frequent and frequently result from mutations in genes essential to the kidney's development. For instance, loss of Fgfr1 can cause complete agenesis (absence); Pax2 mutations can cause hypoplasia (fewer nephrons than normal); and a contiguous defect affecting the Wt1 and Pax6 genes can cause WAGR syndrome (Wilm's tumor, aniridia, genitourinary abnormalities, and mental retardation).

A large number of nephrons make up the small, repetitious structure of the mature kidney. A renal corpuscle, together with a tubule and collecting duct connected, makes up a nephron. Fluid from the blood is filtered into the tubule by the renal corpuscle. As the fluid travels down the tubule, it is converted into urine. As ions are added or removed, water is moved in proportion. It eventually enters the collecting duct system, travels to the ureter, and eventually ends up in the urinary bladder of fish, amphibians, and mammals. The renal corpuscle itself is made up of a Bowman's capsule encasing a convoluted ball of capillaries called the glomerulus that is invested with kidney epithelial cells, or podocytes (Figure 9.9). Most of the experimental work on mammals has been made possible by the ease with which its development can be examined in organ cultures. An explanted rudiment can be used for a variety of procedures that cannot be performed on a mammalian embryo in utero because it can grow and differentiate for several days in culture. Tissues can be divided and reassembled, elements can be incorporated into or taken out of the culture medium, growth factors can be applied locally on slow-release beads, and certain antisense oligonucleotides or RNA interference can be used to prevent the expression of particular genes. The application of conditional knockout and mouse genetic labeling techniques has added a great deal of knowledge about *in vivo* mechanisms in recent times.

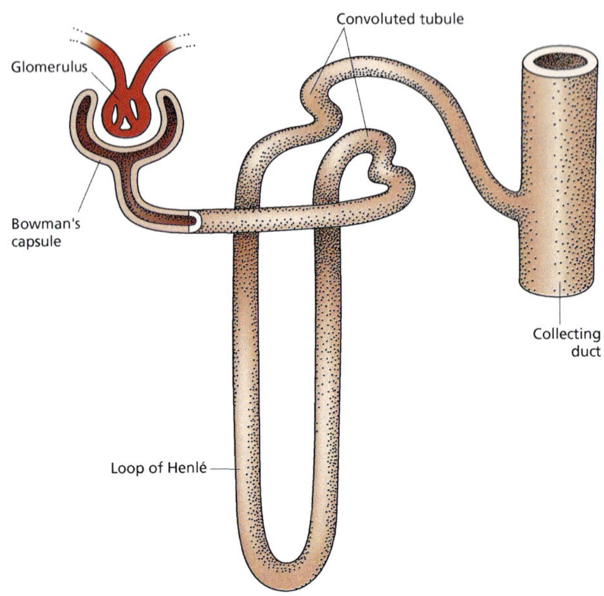

Figure 9.9. *Structure of a nephron.*

Source: *Jonathan M.W. Slack. Essential Developmental Biology, 3rd Edition. Wiley-Blackwell. ISBN: 9781118387603.*

9.2.1. Normal Development of the Kidney

The kidney and gonads originate from the intermediate mesoderm, which is lateral to the paraxial mesoderm that forms the somites. Amniotes, which include birds, mammals, and reptiles, have three kidneys: the pronephros, mesonephros, and metanephros, which are arranged anterior to posterior. The mesonephros may occasionally function during embryonic life, the pronephros is nonfunctional and vestigial, and the metanephros is the final kidney. Fish and amphibians are examples of lower vertebrates that lack the metanephros. The mesonephros forms the final kidney, while the pronephros may still function in larval life.

Two distinct components originate from the intermediate mesoderm, and certain stromal cells are also recruited from the neighboring paraxial mesoderm to form the kidney. The nephric duct, also known as the Wolffian duct, begins in the anterior at somite 10 in the chick and extends posteriorly to the cloaca. Pax2 and -8 expression is indicative of it, and nephric duct formation is inhibited by double knocking these genes. The duct elongates by intrinsic growth rather than by recruiting surrounding cells, as demonstrated by labeling it with DiI or by locally introducing the lacZ gene. Initially, the duct is solid and needs the surrounding epidermis to form a tube; however, a BMP slow-release bead can replace this necessity. The surrounding intermediate mesoderm contains a portion known as the nephrogenic mesoderm that has already committed to developing into a kidney. Tubules separate from the nephrogenic mesoderm and unite with it as the duct expands posteriorly. Nephrons in the mesonephros, in particular, are arranged in anteroposterior sequence, displaying a segmented arrangement. Unlike the other parts of the kidney, the metanephros' collecting system originates from the ureteric bud, an outgrowth of the main nephric duct. The source of this is a pseudostratified posterior segment of the nephric duct, which develops into the nephrogenic mesenchyme and begins to widely branch. The mouse has roughly ten generations of branching. Cyst formation results from the loss of genes associated with planar polarity, such as Wnt9b and Fat4, which impedes ureteric bud elongation. The collecting ducts' "principal" cells, which are in charge of reabsorbing water, and some "intercalated" cells, which control pH by secreting H^+ or HCO_3^-, make up the majority of these cells in their later differentiation. Intercalated cells proliferate at the expense of principal cells when Notch pathway components, such as the E3 ubiquitin ligase encoded by mindbomb1 (Mib1), are lost. This implies that the various cell types are produced by a Notch-based lateral inhibition mechanism, which is also the mechanism in charge of primary neurogenesis.

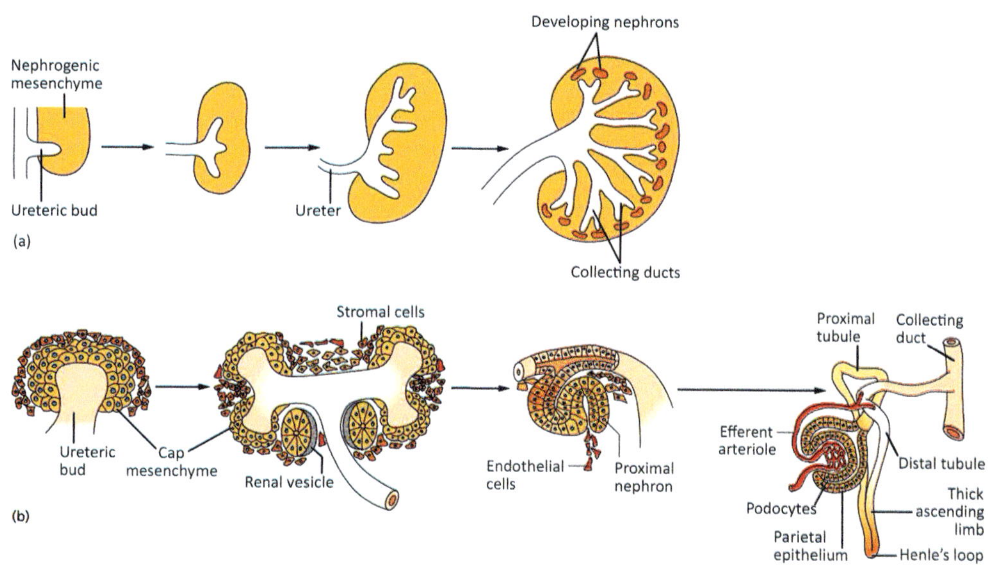

Figure 9.10. *Normal development of the metanephric kidney. (a) Growth and branching of ureteric bud into meta nephrogenic mesenchyme; (b) induction of a nephron.*

Source: Jonathan M.W. Slack. Essential Developmental Biology, 3rd Edition. Wiley-Blackwell. ISBN: 9781118387603.

The ureteric bud's branches unite with the nephrons that arise from the nephrogenic mesenchyme to form a compact, nonsegmental organ (Sariola & Sainio, 1997). An illustration of a mesenchymal-to-epithelial transition is this process. Around the tips of the ureteric buds, the mesenchyme first gathers and forms dense caps. These produce ovoid renal vesicles that eventually fuse into comma-shaped bodies at the tip of the bud, take on an S-shape, and lengthen into tubules (Figure 9.10). The homeobox gene Six2 is expressed in the condensed cap region, and the whole nephron is formed by this domain, as demonstrated by the labeling of mice with Six2-CreER x R26R. The mesenchyme first secretes a normal matrix that includes collagen I and III and fibronectin.

These products give way to laminin and collagen IV, which are typical elements of an epithelial basement membrane, during the transition. Simultaneously, E-cadherin replaces N-CAM on the cell surfaces. Renal vesicles form and differentiate prematurely when Six2 is lost. The Six2-positive cells in the mouse nephron are lost soon after birth, when nephron formation continues. Because they divide roughly four times during the second half of gestation while retaining the ability to form renal vesicles, these cells have been dubbed stem cells. Though the Six2-positive cells are better viewed as progenitor cells, their behavior does not meet the typical definition of a tissue-specific stem cell, which should have a much longer lifetime. In lower vertebrates, the kidney never stops producing new nephrons, as seen in zebrafish (Costantini & Kopan, 2010).

9.2.2. Tissue Interactions in Kidney Development

The nephric duct's ability to grow further is restricted if its tip is destroyed. Mesonephric tubules do not develop more posteriorly; instead, they develop normally at the level of the duct. This demonstrates that the duct is necessary for the nephrons to form. For the metanephros, the same holds true. Although the meta nephrogenic mesenchyme can be cultured *in vitro*, tubule formation cannot occur in the absence of the ureteric bud, indicating that the bud emits an inductive signal that is required for tubule formation. Because the kidney tubules are the only thing that the mesenchyme can form, this is an example of a permissive induction. Since the mesenchyme also affects the ureteric bud, the induction is actually a reciprocal process. The bud will not emerge from the nephric duct and will not be able to grow and branch in the absence of the mesenchyme. Thus at least two inducing factors are active: one originates from the mesenchyme and causes the bud to grow and branch, while the other originates from the bud and causes the mesenchyme to differentiate into tubules.

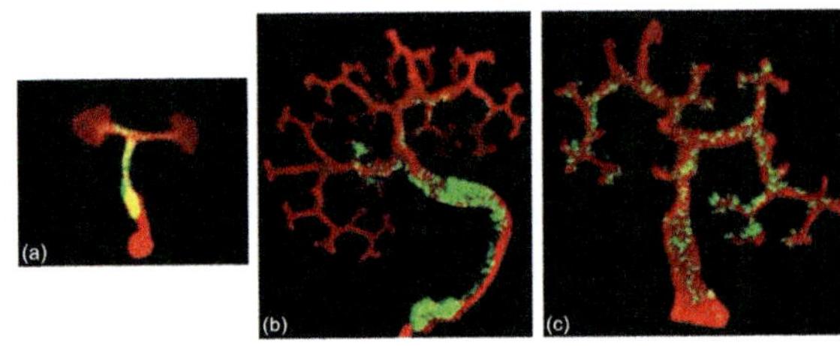

Figure 9.11. *Development of the collecting system of the kidney in chimeric embryos in which ES cells lacking the Ret gene were injected into normal blastocysts. The Ret⁻ cells are also GFP labeled. (a) A ureteric bud at E11.5; the Ret⁻ cells fail to colonize the tips. (b) The branching pattern after 3 more days of culture in vitro. The Ret⁻ cells are excluded from the terminal branches. (c) Control with Ret+ donor cells. Scale bars 250μm.*

Source: Jonathan M.W. Slack. Essential Developmental Biology, 3rd Edition. Wiley-Blackwell. ISBN: 9781118387603.

Glial-derived neurotrophic factor is the signal that originates from the mesenchyme (GDNF). Its receptor, RET, is expressed in the ureteric bud, and this is expressed in the mesenchyme (Figure 9.11). Tyrosine kinase RET functions via the ERK and PI3K pathways. Mice lacking the Gdnf or Ret gene develop kidney failure. When a GDNF slow-release bead is applied to a culture of nephrogenic mesenchyme derived from these embryos, the Gdnf knockout, which is deficient in the factor, experiences a restoration of duct branching, while the Ret knockout, which is incapable of responding to it, does not. On the other hand, the ureteric bud grows uncontrollably in a Ret gain-of-function mutant with constitutive receptor signaling activity. Therefore, this system passes the appropriate expression, activity, and inhibition tests three times. When cultured in contact with metanephrogenic mesenchyme, a variety of tissues other than the ureteric bud will induce tubules; in fact, the spinal cord was the inducer of choice in a large portion of the earlier experimental work on this issue. This is highly active and may function by secreting FGFs, which function similarly *in vitro*.

The identity of the ureteric bud's inducing factor that causes tubule formation has been the subject of much debate. With its expression in the ureteric bud, ability to cause condensations in the organ culture system, and ability to prevent the mesenchymalto-epithelial transition *in vivo*, WNT9B is currently leading the field. The upregulation of Wnt4, which is also necessary for kidney development, in the aggregating mesenchyme is an early effect of the induction. WNT4 appears to function in this context via the Wnt-Ca pathway, and tubule formation can also be triggered by calcium ionophore.

Expression of WT1, a zinc-finger transcription factor, determines the mesenchyme's capacity to undergo induction (Wilm's tumor 1). The complete failure of metanephric development is demonstrated by the Wt1 knockout mouse. Moreover, WT1 has later functions. After induction, its level rises in the mesenchyme, and it is essential for inhibiting Pax2, which is upregulated during the condensation stage and whose elimination is required to allow for nephron differentiation. Wilm's tumors are pediatric human tumors in which renal progenitor cells proliferate; somatic loss of the Wt1 gene is a contributing factor in some but not all cases. Rather than the early function of the mesenchyme's competence, this most likely represents loss of the late function, repression of Pax2.

The mesenchyme experiences massive cell death when the inducing signals are absent. The ureteric bud produces FGF2, which promotes aggregation as well as the survival of the mesenchyme. However, it won't encourage more advancement. Additionally expressed in the bud, BMP7 promotes tubule induction and preserves the survival of isolated mesenchyme. The major requirement appears to happen soon after the first induction of tubules, as evidenced by the severely dysplastic kidneys in the Bmp7 knockout animals. These kidneys contain numerous nephrons blocked at the comma or S stage of development. Tubulogenesis can also be induced *in vitro* by WNT4, which is expressed in the developing tubules themselves, and the knockout is stopped at the aggregation stage.

In summary, WNT9B from the ureteric bud induces the nephrons from the nephrogenic mesenchyme at first. WNT4 expressed in the mesenchyme itself is necessary for further development, and FGFs and BMP7 are likely necessary as survival factors.

9.3. GERM CELL AND GONADAL DEVELOPMENT

The gonads are crucial endocrine organs as well as the organs that store gametes. They come from two different embryos. The genital ridges, which develop from the intermediate mesoderm near the mesonephros, give rise to the somatic tissues of the gonads. Primordial germ cells (PGCs) are the source of the germ cells that make gametes (Wylie, 1999). These begin at a different location in the early stages of development in the majority of embryo types and migrate to the genital ridges. The morphology of the mature germ cells and the gonad varies based on the gender of the individual.

9.3.1. Germ Cell Development

The proximal portion of the egg cylinder is where PGCs begin to form in the mouse embryo. Using an injection of horseradish peroxidase (HRP), fate mapping demonstrated this. Since they cannot develop without this tissue, the extraembryonic ectoderm is responsible for inducing PGCs. In recombination experiments, extraembryonic ectoderm can induce PGCs from both proximal and distal epiblast. The signal is made up of -8b, -4, and BMP2. The extraembryonic ectoderm expresses these biologically active BMPs, and the knockouts exhibit varying degrees of failure in forming PGCs. A zone of tissue surrounding the posterior epiblast that expresses the transmembrane protein-encoding gene fragilis is the first result of induction. The allantois and PGCs are formed in this region. A zinc-finger transcription factor called blimp1 is upregulated by a small number of cells in the area. Several early developmental genes, including brachyury (T) and snail, are repressed by BLIMP1, but a group of genes linked to pluripotency, such as Oct4, Nanog, and Sox2, which are indicative of both primitive ectoderm and PGCs, are allowed to express themselves persistently. In all somatic tissue, the pluripotency genes Oct4 and Nanog normally undergo repression, whereas Sox2 remains active in the neuroepithelium. E7.5 forms about 40 PGCs, some of which are induced and some of which are multiplied by preexisting ones. Their high expression of the enzyme alkaline phosphatase makes them easily identifiable. Around E8, all transcription temporarily stops, but it does not last as long as it does in Drosophila or C. elegans.

The PGCs emerge from the base of the allantois at approximately E9–10, move up the mesentery, and arrive at the genital ridges, which give rise to the gonads, by E11.5 (Figure 9.12). Apart from Oct4, certain gene products that have been identified in Drosophila as essential for the development of germ cells are also needed in mice. These consist of the RNA helicase VASA and the RNA-binding protein NANOS. A chemokine expressed on the lateral plate mesoderm called SDF1 (stromal cell derived factor 1, or CXCL12) regulates the migration of PGCs. The PGCs express the G-protein coupled receptor that binds to it, CXCR4. The PGCs in the Cxcr4 knockout are unable to colonize the genital ridges. A number of factors are necessary for survival during migration, such as stem cell factor (SCF), which is encoded by the steel gene and binds to the KIT receptor expressed on germ cells.

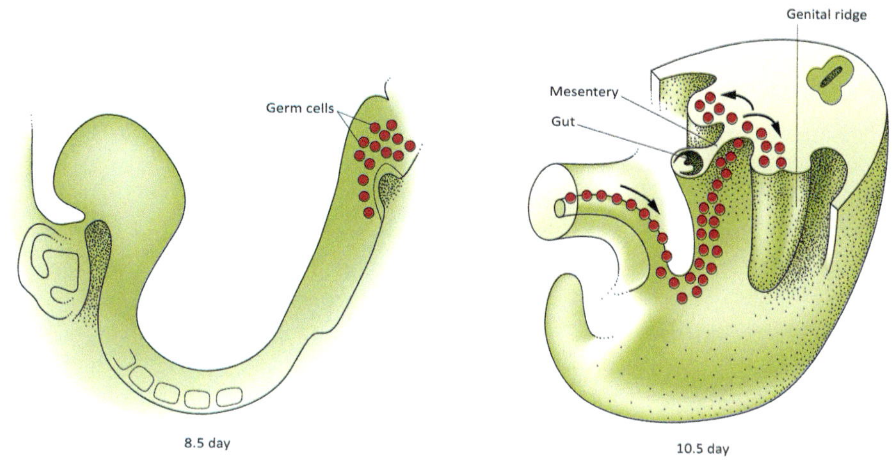

Figure 9.12. *Origin and migration of germ cells.*

Source: Jonathan M.W. Slack. Essential Developmental Biology, 3rd Edition. Wiley-Blackwell. ISBN: 9781118387603.

By cloning these germ cells by nuclear transplantation and analyzing the allele specificity of expression in the resultant embryos, it is possible to determine how soon the parental DNA imprints are removed from the gonads. De novo DNA methylation is then used to create the new sex-specific imprints.

9.3.2. Normal Development of the Gonads

Since the gonads of males and females differ from one another, sex determination is closely linked to the development of the gonads. The mechanism of sex determination is not at all well conserved across animal groups, unlike most other aspects of development, so the account that follows is limited to mammals. We know that gonadal development does not always require the presence of germ cells because the gonads can develop normally in a variety of mutant mice that are impaired in germ-cell migration.

The intermediate mesoderm gives rise to the gonads in close proximity to the kidneys. The mesonephros and gonad are formed by the medial and lateral portions of the intermediate mesoderm, respectively. Several gene products are needed in the early stages before sexual dimorphism manifests. These include SF1 (steroidogenic factor 1), a member of the nuclear hormone receptor family, and WT1, a zinc-finger transcription factor also necessary for kidney development. Genes encoding either of these can be knocked out to stop gonadal development. On both sides of the trunk, the genital ridge emerges at E9.5 and extends slightly into the coelom (Figure 9.13). The Limhomeodomain transcription factor LHX9 is expressed in the ridges, and no gonads form in the Lhx9 gene-knockout mouse. Cords of cells start to form from the coelomic lining epithelium and grow into the underlying mesenchyme slightly after the genital ridge appears.

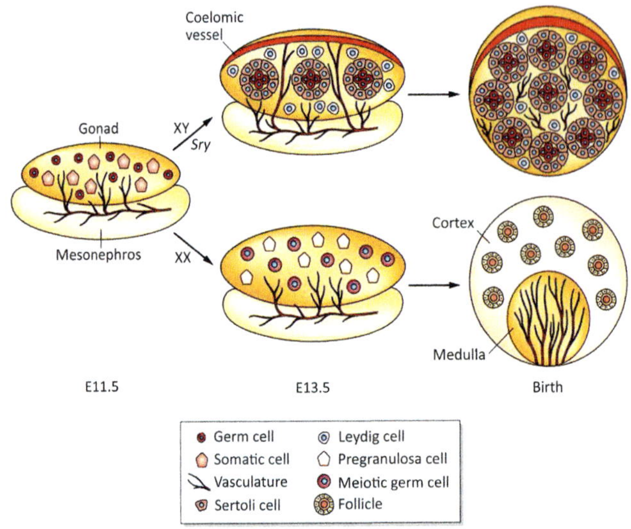

Figure 9.13. *Formation of gonads from the genital ridge.*

Source: Jonathan M.W. Slack. Essential Developmental Biology, 3rd Edition. Wiley-Blackwell. ISBN: 9781118387603.

During this time, two ducts that are crucial to the development of the sex organs also emerge. The nephric duct is also known as the Wolffian duct. After the metanephros begins to develop, it has no more excretory function; in the male, it eventually develops into the vas deferens. Alongside the Wolffian duct, the Müllerian (= paramesonephric) duct eventually gives rise to the female's oviduct, uterus, and proximal vagina. There is no discernible difference between the two sexes' differentiation up until this point (E12.5).

In males, the cords of coelomic lining cells that extend into the genital ridge create a complex network of Sertoli cell-based seminiferous tubules, which are integrated with the germ cells. Important components of the sex determination program, Sox9 and later Sry, are expressed by Sertoli cells. Leydig cells, which secrete testosterone, differentiate from the mesenchyme in between the tubules under their influence. The growing gonad shortens and encapsulates itself to form the testis as the tubules join the Wolffian duct. From E13.5 to birth, the germ cells are quiescent during the mitotic cycle. The Müllerian duct regresses in the interim. Testicular tubules are solid until postpartum, when they begin to hollow out and the spermatogonia (which are produced from germ cells) emerge (Figure 9.14). In a mature male, spermatogenesis is the process by which sperm are continuously produced from a population of stem cells.

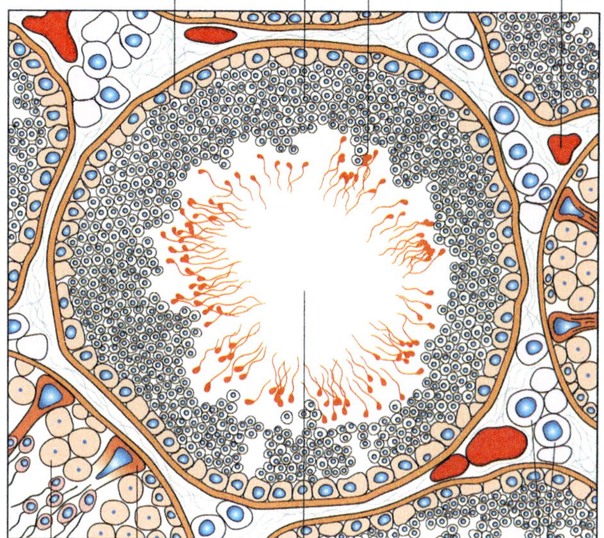

Figure 9.14. *Organization of a seminiferous tubule.*

Source: Jonathan M.W. Slack. Essential Developmental Biology, 3rd Edition. Wiley-Blackwell. ISBN: 9781118387603.

The cords of coelomic lining cells in females remain close to the surface as granulosa (= follicle) cells, adjacent to the germ cells, rather than growing very far into the mesenchyme. When the sex determination system is active, the follicle cells' upregulation of Wnt4 and Foxl2 is made possible by the lack of SRY. In addition to suppressing Sox9 expression, these also indirectly promote the development of mesenchymal thecal cells, which produce estrogen. The gonad then becomes encased in an ovary after that. At their posterior ends, the two Müllerian ducts fuse to form the proximal (i.e., Y-shaped in the mouse, this area is the deeper) portion of the uterus and vagina. They form the oviducts at the anterior end, the collecting funnels of which are situated in close proximity to the ovaries. The Wolffian duct degenerates in the interim.

After completing their last mitotic division, the oogonia—which are derived from the germ cells—become primary oocytes within the ovaries. Meiotic prophase begins at E13.5, and it is widely accepted that after this point, mammals do not go on to form new oocytes. Granulosa cells invest in each primary oocyte to create a primordial follicle. After reaching sexual maturity, each reproductive cycle triggers the activation of several follicles, which leads to the multiplication of granulosa cells and the expansion of the follicle containing the oocyte (Figure 9.15). It is possible for 8–12 oocytes to ovulate at once in a mouse. Every follicle that ruptures develops into a corpus luteum, which secretes progesterone to aid in the uterus's preparation for implantation.

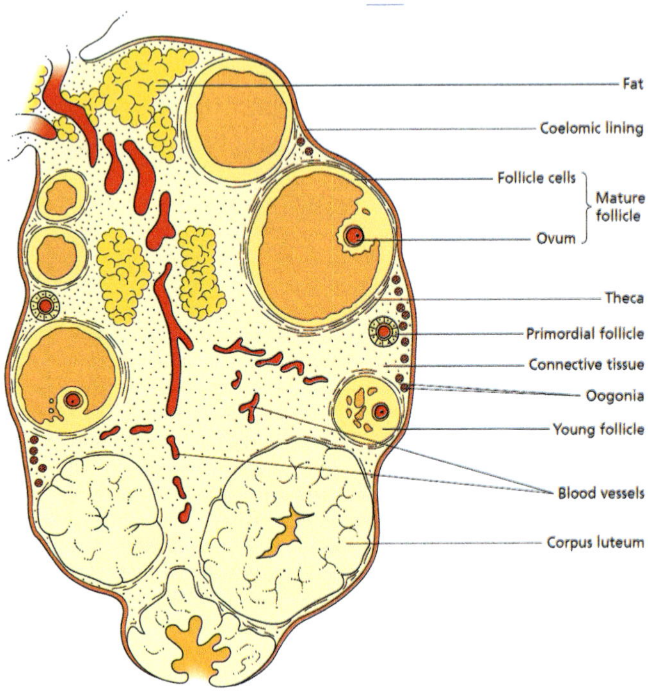

Figure 9.15. *A mammalian ovary.*

Source: Jonathan M.W. Slack. Essential Developmental Biology, 3rd Edition. Wiley-Blackwell. ISBN: 9781118387603.

9.3.3. Sex Determination

The chromosome composition ultimately determines whether the gonad develops into an ovary or a testis; in mammals, this is XY for males and XX for females. The Y chromosome, however, has a region that is not homologous to the X chromosome, which causes an imbalance in the situation. This contains the gene Sry (sex-determining region of Y), which is the crucial switch that regulates the sexual development pathway in mammals. It is the prototype of the Sox family of transcription factors and codes for a transcription factor belonging to the HMG (High Mobility Group) class. The following provides evidence for its function. An XY mouse that has Sry removed from its Y chromosome will mature into a female and even give birth to live eggs. An XX mouse will develop as a male even though it won't produce sperm if Sry is introduced as a transgene. Human chromosomal abnormalities that occur naturally have effects that are similar but not the same. Given that people with Turner's syndrome are female and those with Klinefelter's syndrome are male (XXY), it follows that having a Y chromosome must make a person male. Due to an abnormal recombination during meiosis, some phenotypically male XX individuals occasionally have a Sry gene on one X chromosome. On the other hand, it is occasionally discovered that female XY individuals have a mutated Sry gene. Sry is briefly expressed in the gonads of males just prior to the onset of sexual dimorphism (Figure 9.16). Its sole purpose is most likely to trigger the expression of Sox9, another HMG box family member. In contrast to Sry, Sox9 expression is linked to maleness in all vertebrates and endures throughout embryogenesis. It represses a set of genes necessary for female development, such as Wnt4 and Dax1, and upregulates a set of genes whose products are needed for male functions, such as Fgf9 and Amh (encoding anti-Müllerian hormone, AMH). Sertoli cells secrete AMH, also referred to as Müllerian inhibitory substance (MIS). The fact that the Amh knockout mouse still has a Müllerian duct indicates that it is a member of the TGFβ superfamily and plays a very specific role in the male Müllerian duct regression. The Sf1 gene plays a significant role in male development as well. It is expressed during early gonadal development, contributes to Sry upregulation, and is subsequently sustained by SOX9 during testis differentiation. In the Leydig cells, SF1 contributes to the upregulation of Amh and the genes involved in the testosterone synthesis pathway. The final stages of male development rely on the hormones that are in circulation. This is converted into the active form dihydrotestosterone, which causes the male external genitalia to differentiate. It is also the cause of the different secondary sexual traits that males exhibit, including facial hair and deeper voices.

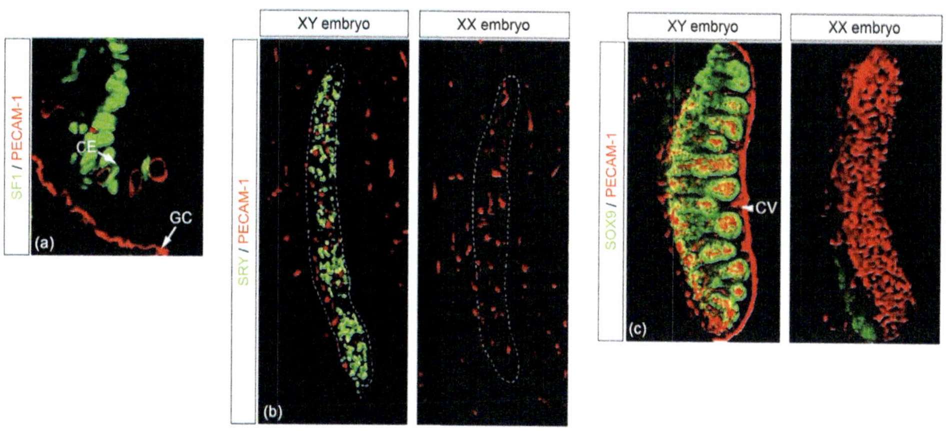

Figure 9.16. *Mouse gonads during the period of sex determination. (a) E10.5 the gonad of male and female are the same. Coelomic lining cells are immunostained green with antibody to SF1. Germ cells and endothelial cells of blood vessels are stained red. (b) At E11.5, Sry (green) is present in cells of the male but not female gonad. (c) At E12.5, Sox9 (green) is present in the male but not the female gonad and testis cords are forming. CE, coelomic epithelium; GC, germ cells; CV, coelomic vessel.*

Source: Jonathan M.W. Slack. Essential Developmental Biology, 3rd Edition. Wiley-Blackwell. ISBN: 9781118387603.

The female pathway begins with the upregulation of Wnt4, which is also necessary for kidney development. In the early genital ridges, Wnt4 is normally expressed. In males, it is repressed, but in females, it persists. In the event that Sry upregulation is postponed for any reason, WNT4 signaling is strong enough to inhibit Sox9 expression, favoring the female pathway. The ovary of the Wnt4 knockout is devoid of Müllerian ducts and contains fewer oocytes. Additionally, the cells secreting male-type steroids are not present. Even though the germ cells and gonads are from different cell lineages, the gonad's developmental path influences the germ cells' own. When germ cells in

XY animals migrate and fail to find the gonad, they stay in their ectopic location for a while and develop into oocytes instead of spermatogonia. A chimeric mouse made of both XX and XY cells exhibits the opposite condition. Because the Sry gene activates in the XY Sertoli cells and triggers the male program, which releases testosterone and AMH, these mice are male as long as they have more than approximately 35% of XY cells. The chimeric mouse develops a male morphology but has both testes and germ cells with a mixed XY and XX chromosome constitution because these hormones have the same effects on cells regardless of their chromosome makeup. Though they can develop into spermatogonia, the XX germ cells that enter the testis are unable to mature into sperm.

9.4. LIMB DEVELOPMENT

Developmental biologists are particularly interested in vertebrate limbs because of their visible nature, which makes human limb congenital defects particularly noticeable and upsetting. As such, one of the traditional subjects of organogenesis research is limb development. Despite some minor variations, the developmental mechanisms are highly conserved amongst vertebrate species. Tetrapods are animals with four limbs: a pair of forelimbs, which can be legs, arms, or wings, and a pair of hindlimbs. These animals include reptiles, birds, mammals, and amphibians. Snakes are among the few tetrapods that have never had limbs; they have lost them secondary to other evolutionary events (Christ & Brand-Saberi, 2002).

Most of the experimental work on microsurgery has been done on chick embryos; in more recent years, large-scale genetic testing on mice has been added. A window cut in the egg shell allows for the limb bud to be operated on in ovo in the chick embryo. To enable continued development, the egg can then be sealed in its shell and put back into the incubator. Genes can be electroporated, ectopically expressed using a replication-competent retrovirus, or administered from pellets of transfected cells or slow-release beads. However, it is not routinely possible to knock out genes in the chick and so the loss-of-function information has mostly come from the mouse.

9.4.1. Normal Development of the Limb

After the embryo reaches the phylotypic stage, limb buds emerge in the flank, each with an epidermis-overlaid mesenchyme core (Figure 9.17). The somatic mesoderm is the area of the mesoderm that gives rise to the limbs; it is located lateral to the intermediate mesoderm. The term "somatopleure" refers to the somatic mesoderm and the epidermis that covers it. An undifferentiated mass of loose mesenchyme encircled by epidermis makes up the early limb bud. An important structure that regulates outgrowth is the apical ectodermal ridge (AER), which is formed by thickening of the epidermis at the distal edge. All of the limb's skeletal structures and connective tissues, including the dermis, tendons, ligaments, cartilages, and sheaths that encase the muscles, are formed by the limb-bud mesenchyme. On the other hand, the somites give rise to the muscle myofibers. Making somite grafts from quail to chick at a stage prior to the limb bud beginning to grow out can demonstrate this. All of the myofiber nuclei in the developed limbs are of the quail type, whereas the other cell types in the limb are chick type (Figure 9.18).

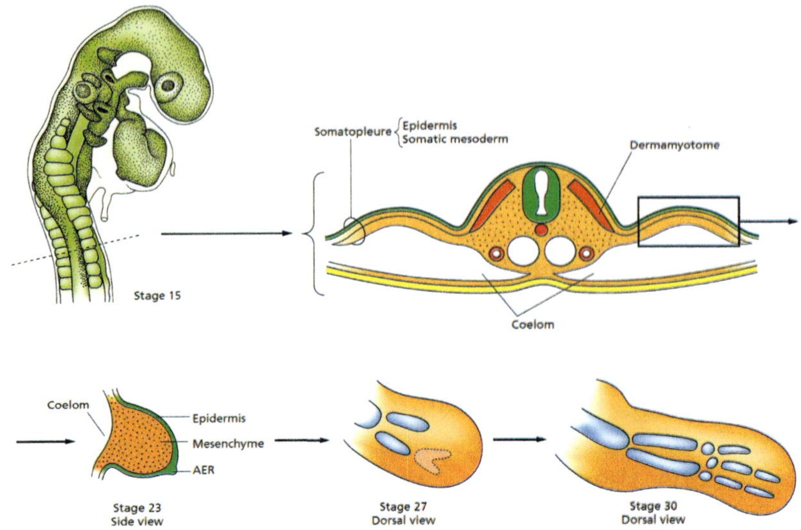

Figure 9.17. *Origin and growth of limb buds.*

Source: Jonathan M.W. Slack. Essential Developmental Biology, 3rd Edition. Wiley-Blackwell. ISBN: 9781118387603.

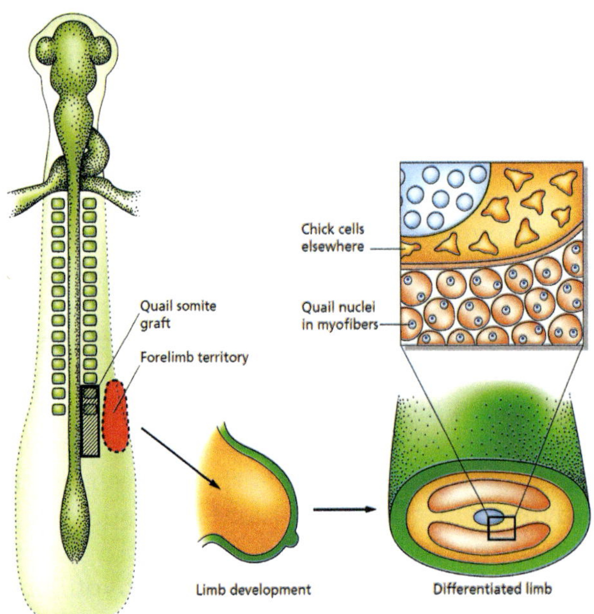

Figure 9.18. *Evidence that the limb muscles derive from the somites was obtained by grafting somite mesoderm from quail to chick and analyzing the resulting limbs.*

Source: Jonathan M.W. Slack. Essential Developmental Biology, 3rd Edition. Wiley-Blackwell. ISBN: 9781118387603.

The different muscles and cartilage components differentiate in a proximal to distal sequence, or outward from the body, as the limb buds lengthen and flatten during their outgrowth. Fate maps of the limb bud have been created by applying tiny marks of DiI and DiO. These maps demonstrate that during outgrowth, the proximodistal cell order is preserved, but the distal portions experience significantly greater elongation (Figure 9.19).

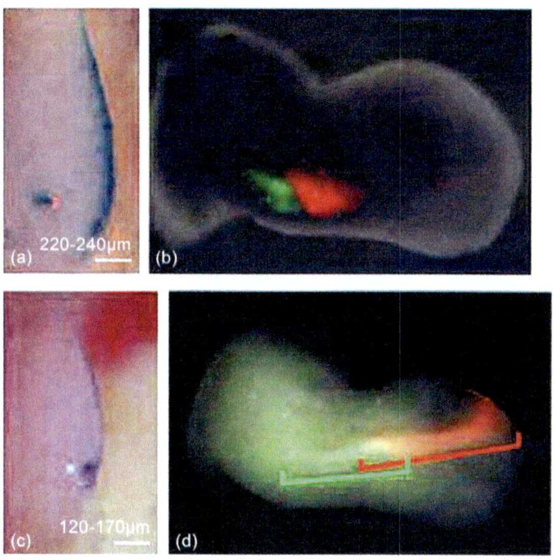

Figure 9.19. *Fate mapping of the chick limb bud. Small marks of DiI (red) and DiO (green) are applied at early bud stage (a, c) and the labeled cells are located 2 days later (b, d). The results show considerable elongation of the more distal parts of the early bud, and a lack of any proximodistal boundaries at early stages.*

Source: Jonathan M.W. Slack. Essential Developmental Biology, 3rd Edition. Wiley-Blackwell. ISBN: 9781118387603.

The humerus in the upper limb, the radius and ulna in the lower limb, the carpals in the wrist, and the metacarpals and phalanges in the digits make up the differentiated forelimb skeleton (Figure 9.20). The femur in the upper limb, the tibia and fibula in the lower limb, the tarsals in the ankle, the metatarsals, and the phalanges in the toes make up the hindlimb's skeleton. The proximal-distal axis, as described anatomically, extends from the humerus/femur to the phalanges. Terms for the upper limb, lower limb, and hand or foot, respectively, are stylopod, zeugopod, and autopod, which are frequently used in descriptions of proximodistal patterning. The first digit and the highest numbered digit form the anteroposterior axis. When the arm is extended with the palm facing ventrally, this corresponds to the anteroposterior axis of the entire body and is the same as the thumb-to-little finger axis of a human hand. Since there are rarely more than five digits on a vertebrate limb, the limb is referred to as pentadactyl. Nevertheless, the number is frequently lowered; for instance, the chick leg has four digits, but the chick wing only has three, which are oddly numbered 2, 3, and 4. The dorsal-ventral axis extends from the hand's back to its palm, connecting the upper and lower surfaces of the limb. Wholemount staining is a common method used to score experiment results and reveal the pattern of cartilage elements. Because all of the skeletal components are in one plane, this is a helpful guide for the proximal-distal and anterior-posterior patterns, but it hides the dorsal-ventral pattern. However, there are a number of additional characteristics that vary along the dorsal-ventral axis and can be utilized to examine odd patterns. These characteristics include the muscle pattern and the different kinds of epidermal specializations, such as nails, feathers, scales, and claws.

A differentiated limb is made up of many different cell types besides muscles and skeleton. The epidermis is where all skin specializations, including scales and claws, are formed. During the time of limb bud outgrowth, blood vessels develop from the capillary network that is already there in the mesoderm. The spinal cord sends nerves growing in, arriving when the limb bud is still undifferentiated but elongating. The brachial plexus innervates the forelimb, while the lumbosacral plexus innervates the hindlimb, formed by the union of the relevant spinal nerves. The neural crest is the source of pigment cells, just like it is in other parts of the body. The skeleton begins as cartilage and eventually transitions into bone. Ossification of the cartilage elements follows a preset sequence, usually beginning earlier in the larger skeletal elements and later in the smaller ones, unlike the early phase of limb development, which proceeds from proximal to distal.

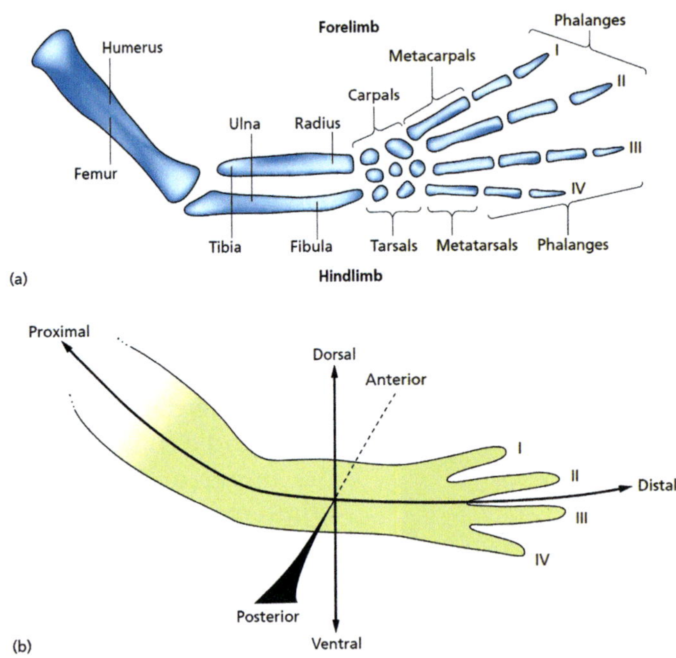

Figure 9.20. *Anatomy of the tetrapod limb. (a) Nomenclature of the skeletal elements in the forelimb and the hindlimb; (b) anatomical axes used for description of the limb.*

Source: Jonathan M.W. Slack. Essential Developmental Biology, 3rd Edition. Wiley-Blackwell. ISBN: 9781118387603.

9.4.2. Experimental Analysis of Limb Development

9.4.2.1. Limb Determination

The mesoderm first designates the limb rudiment as a territory. Transplanting the potential limb tissues to a location on the flank between the limb buds may demonstrate this. While the potential limb epidermis will not in this position, the mesoderm will cause the formation of a limb bud. But even the potential limb mesoderm cannot develop a limb until somite tissue from the same anteroposterior level is present at a stage prior to this, indicating that early somite induction is required for the acquisition of limb commitment.

By implanting a slow-release FGF bead, it is possible to induce the development of a new limb in the area of the flank between the normal limb buds (Figure 9.21). In response to such a bead implantation, the anterior portion of the flank gives rise to the forelimbs, and the posterior portion to the hindlimbs. Fgf10 is expressed in the developing limb mesoderm itself as well as in the early presomite plate. Furthermore, the lack of limb buds in the Fgf10 knockout mouse indicates that FGF10 is required for limb development. As a result, FGF10 meets the requirements for expression, activity, and inhibition to be the cause of the somatic mesoderm's initial induction of the limb-forming territories. Wnt signaling causes an upregulation of the Fgf10 gene. Wnt2b and Wnt8b are expressed in the forelimb-bud and hindlimb regions, respectively, and these factors can both cause Fgf10 expression and limb outgrowth. In addition, the Wnt inhibitor Axin has the ability to inhibit the development of limbs, and mice lacking either Lef1 or Tcf1, two elements of the Wnt signal transduction pathway, do not develop limb buds.

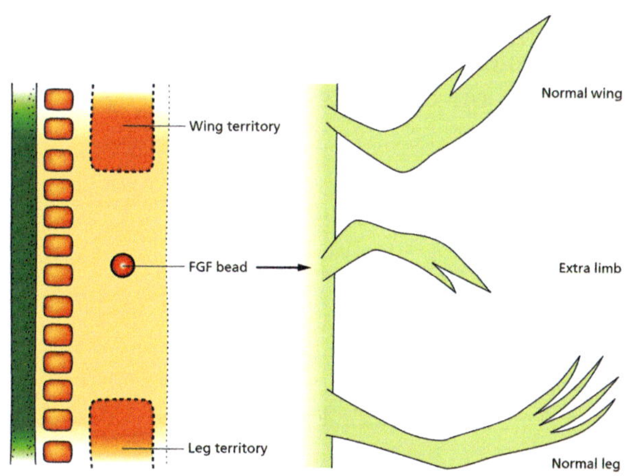

Figure 9.21. *Induction of a limb bud from the flank by an FGF bead. The induced limb has an inverted anteroposterior polarity compared to the normal ones.*

Source: Jonathan M.W. Slack. Essential Developmental Biology, 3rd Edition. Wiley-Blackwell. ISBN: 9781118387603.

T-box genes, which express Tbx5 in the early forelimb and Tbx4 in the early hindlimb, are responsible for differentiating the fore- and hindlimb. These genes are either directly or indirectly regulated by the combination of Hox genes that are active at different body levels. The partial conversion of the chick wing bud to a leg bud upon overexpression of Tbx4 suggests that T-box genes play a role in differentiating between the forelimb and hindlimb.

Myogenic cells migrate out of the somites a few hours before bud outgrowth begins, and somites of any level—not just those that normally supply the limbs—can contribute cells if they are grafted to the limb levels. Hepatocyte growth factor is a key element of the chemotactic signal for migration (HGF, = scatter factor). This is expressed in the mesenchyme of the prelimb-bud, and the myotomes express Met, its receptor. In mice lacking Met or Hgf, myoblasts do not migrate into the limbs, resulting in the development of muscle deficiency.

9.4.2.2. Proximal–Distal Outgrowth and Patterning

The development of the apical ectodermal ridge (AER) in the epidermis is the first noticeable event in the limb-bud process. By grafting an additional prelimb mesoderm beneath the flank epidermis, one can demonstrate how the mesoderm induces the AER. However, interaction between two epidermal territories—a lateral territory overlying the somatic mesoderm and a medial territory overlying the somites and intermediate mesoderm—is also necessary for its formation. The medial epidermis becomes the dorsal epidermis of the limb bud, and the lateral epidermis becomes the ventral epidermis of the limb bud, as well as forming the AER itself, according to fate mapping of the prospective limb epidermis using tiny marks of DiI (Figure 9.22). Transgenic mice that express a dominant negative BMP receptor under the control of an epidermis-specific promoter can avoid AER formation, which is dependent on BMP signaling. The transcription factor gene Engrailed-1 is expressed in the lateral epidermis and is likely involved in inhibiting the formation of AERs because a ventral AER is present in mice that lack the Engrailed-1 gene.

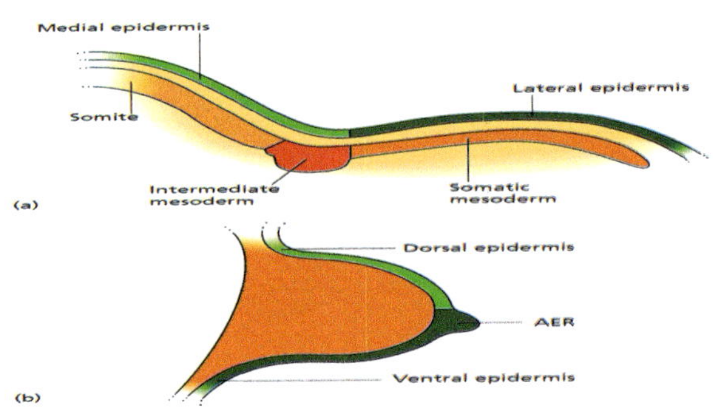

Figure 9.22. *Origin of dorsal and ventral limb-bud epidermis. The left section shows a pre-limb bud stage, the right one shows a stage 23 limb bud.*

Source: Jonathan M.W. Slack. Essential Developmental Biology, 3rd Edition. Wiley-Blackwell. ISBN: 9781118387603.

The AER endures as the limb bud grows outward. A distal truncation of the final pattern results from its surgical removal at any point (Figure 9.23). It is assumed that this represents a sequence of determination of successive proximal-to-distal levels of the pattern during limb-bud outgrowth, with more structures lost the earlier it is removed. For the AER to survive, it requires elements from the mesenchyme; otherwise, it will rapidly regress if grafted to a nonlimb-bud site. Consequently, AER and mesenchyme have a reciprocal relationship in which each is necessary for the other's function to occur. When outgrowing, the tip region, which is located approximately 250 μm from the

AER, does not differentiate. Msx1, Ap2, and Lhx2 are among the specific transcription factor genes expressed in this region. Lhx-engrailed, an inhibitory domain swap, can be overexpressed to prevent limb outgrowth, indicating that at least this gene is required for the continuous formation of structures. It is noteworthy that Lhx2 is a homolog of the apterous gene, which is necessary for Drosophila wing outgrowth.

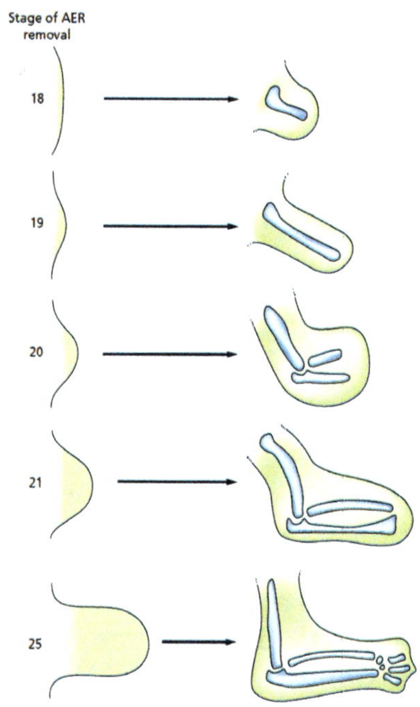

Figure 9.23. *Increasing extent of distal truncation the earlier the AER is removed.*

Source: Jonathan M.W. Slack. Essential Developmental Biology, 3rd Edition. Wiley-Blackwell. ISBN: 9781118387603.

Numerous fibroblast growth factor genes (Fgf 4, 8, 9, 17) are expressed by the AER itself. A slow-release FGF bead can replace the AER in the event that it is removed, supporting the distal parts' continued development and outgrowth. When Fgf8 is specifically knocked out using the Cre-lox system, the result is defects in the distal limbs, while when Ffg4 and -8 are knocked out together, the result is no limb development. Thus, the FGFs are the active factors released by the AER that meet the standard three requirements of expression, activity, and inhibition.

Sonic hedgehog (Shh) from the zone of polarizing activity (ZPA) and a secreted BMP inhibitor known as Gremlin contribute to the mesenchyme's AER maintenance activity. The mesenchyme of the limbs and the AER both express a number of BMPs. The AER will grow at this point if the BMP activity is locally inhibited, suggesting that the BMPs have a suppressive effect on the AER during outgrowth. BMPs are also necessary later on in the development of the limb for the cell death that takes place in between the developing digits. Duck-like webbed feet will result from stopping this in transgenic mice by driving Noggin expression using an epidermis-specific promoter.

The expression of different genes at different levels foreshadows the differentiation of proximodistal structures. The expression of the homeobox genes Meis1 and -2, which are triggered by retinoic acid originating from the main body, indicates the most proximal territory, the stylopod. These genes are causally involved in defining proximal character because overexpression of them can proximalize the more distal parts. Humans with mutations in the short stature homeobox gene, Shox, have shorter ulnas and radiuses. This gene is expressed in the middle section, or zeugopod. The Hoxa10, -11, and -13 genes also exhibit nested expression in the early limb bud; all three are expressed in the future autopodium (Figure 9.24a). Although it is commonly believed that this regulates proximal–distal differentiation, the precise function of Hox genes in limb patterning is still unknown because of the significant changes in expression patterns that occur during development. Individual Hox gene knockouts only slightly alter the limb phenotype. While they do not have homeotic effects, knockouts of entire paralog groups can result in stunning, segment-specific deletions (Figure 9.25). Thus, although Hox genes are undoubtedly required for limb development, their primary function may involve growth regulation as opposed to regional specialization. When it comes to signals, the gradients of retinoic acid from the body axis and FGFs from the AER may be enough to identify the various zones of gene upregulation.

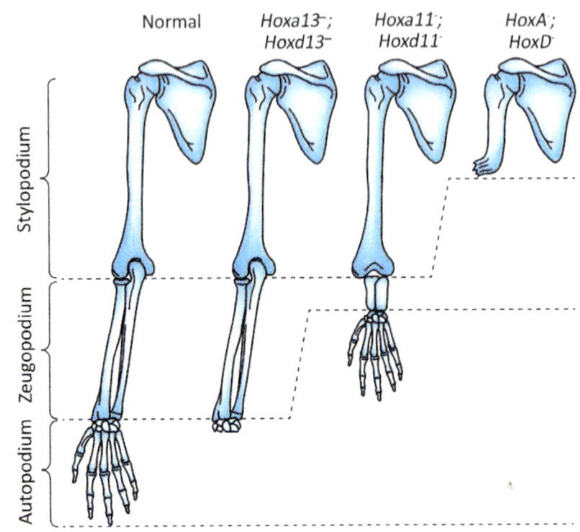

Figure 9.25. *Limb phenotypes resulting from loss of Hox genes of the a and d clusters. In the case shown on the right the entire Hoxa and d clusters have been deleted.*

Source: Jonathan M.W. Slack. Essential Developmental Biology, 3rd Edition. Wiley-Blackwell. ISBN: 9781118387603.

From an early stage, the joints that form between the various cartilage elements are prespecified. The future femur, ulna, and radius are all included in the upper limb's initial cartilage condensation.

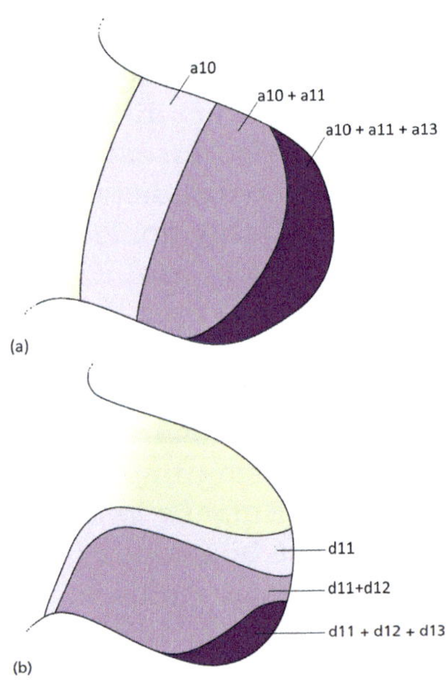

Figure 9.24. *Nested expression of Hox genes in the limb bud. (a) Hoxa genes; (b) Hoxd genes.*

Source: Jonathan M.W. Slack. Essential Developmental Biology, 3rd Edition. Wiley-Blackwell. ISBN: 9781118387603.

Because two cartilage condensations do not simply form because they are apposed, joints do not form as a result of the microsurgical ablation of the prospective elbow joint from this condensation. Numerous genes, including Wnt14 (now known as Wnt9a) and members of the BMP family, Gdf5, -6, and -7, are expressed in the potential joint regions. Joint defects are caused by loss of function of Gdf5 and -6. Extra joints will form as a result of Wnt9a's ectopic expression. Wnt9a functions by suppressing its own expression at long range and upregulating it at short range, resulting in the formation of a periodic pattern of expression that anticipates the location of joints.

9.4.2.3. Anteroposterior Patterning

At the posterior edge of the limb bud, the zone of polarizing activity (ZPA) regulates the anteroposterior pattern. This functions similarly to a diffusible morphogen source, which is maintained at a steady concentration within the ZPA and can diffuse into the surrounding tissues before being eliminated or destroyed there. An exponential concentration gradient will be established from posterior to anterior by such a mechanism. A second set of structures in a mirror-symmetrical arrangement to the first set is induced from the anterior tissue if a second ZPA is grafted to the anterior margin of the bud (a double posterior duplication; Figures 9.26 and 9.27). The six-digit 4–3–2–2–3–4 duplication structure can be changed to the three-digit 4–3–4. As the two ZPAs get farther apart, there are more structures included in the duplication. This is because the thresholds for the formation of more anterior-type structures will be activated if they are far apart, which will result in a lower central minimum of the U.

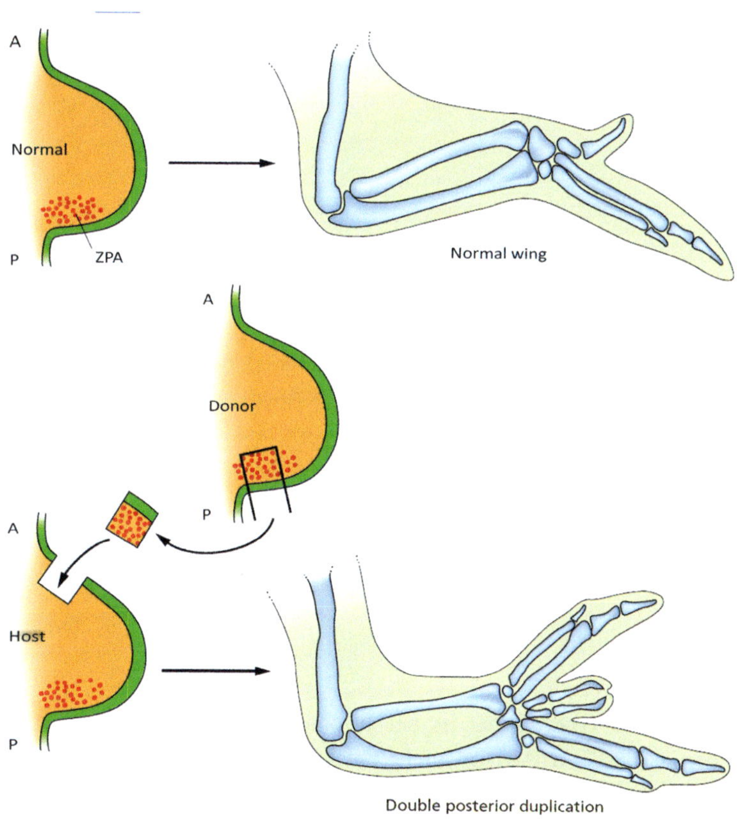

Figure 9.26. *The ZPA graft leads to the formation of a double posterior duplicated limb. Only the distal parts (digits) are usually affected because the more proximal parts are already determined when the graft is carried out.*

Source: Jonathan M.W. Slack. Essential Developmental Biology, 3rd Edition. Wiley-Blackwell. ISBN: 9781118387603.

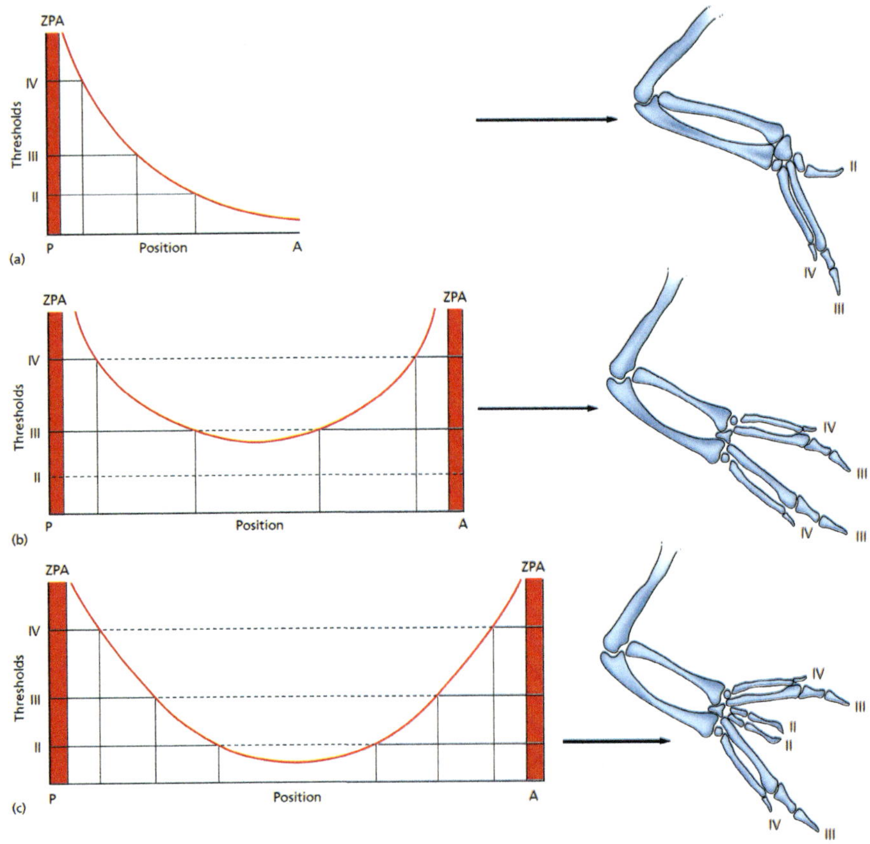

Figure 9.27. *How the gradient from the ZPA controls the pattern of digits: (a) normal development; (b) and (c) ZPA grafts. Because the separation of graft and host ZPAs is greater in (c), the minimum of the U-shaped gradient is lower and the duplication contains more elements than in (b).*

Source: Jonathan M.W. Slack. Essential Developmental Biology, 3rd Edition. Wiley-Blackwell. ISBN: 9781118387603.

Since the ZPA signal is the same in all vertebrate species, grafting mouse or lizard ZPA into the chick limb bud's anterior side will likewise cause a duplicate limb to form. Because the duplicate limb is made of chick tissue, every part of it has morphology more similar to that of a chick than of the donor species. When compared to normal limbs, limbs generated in the flank by FGF beads have anteroposterior polarity that is inverted (Figure 9.21). This is because the ZPA of the induced limb bud is on the anterior rather than the posterior side of the bud because it develops from the same region that forms the ZPA of the normal forelimb.

Similar to a ZPA graft, the implantation of a slow-release bead containing retinoic acid (RA) results in the induction of an additional set of posterior parts. The limb bud contains both retinoic acid and retinoiic acid receptors. Embryos of quail produced by depleting the mothers' vitamin A intake exhibit abnormalities in the limb bud.

However, transgenic mice carrying a retinoic acid reporter (comprising of a retinoic acid response element linked to lacZ) do not exhibit any retinoic acid activity at all, nor a gradient of retinoic acid in the limb bud itself, indicating that retinoic acid is not the endogenous ZPA signal. The *in vivo* function of retinoic acid signaling, which is present in the adjacent trunk of the body, is currently understood to be the induction of the Meis genes, which define the most proximal portion of the limb bud.

In reality, the Sonic hedgehog gene (Shh) produces the morphogen from the ZPA. Grafting experiments have shown that the expression of Shh occurs in the area known to have ZPA activity. A tissue culture-based bioassay used to measure the amount of active Shh in limb tissue explants reveals that the posterior side has approximately five times higher concentrations of Shh than the anterior. Similar to the ZPA, slow-release beads or cells that produce Shh can cause duplications. Shh appears to act in the chick in accordance with the predictions of the straightforward exponential gradient model. However, the mouse's mechanism is more intricate because the loss of digits occurs in a different order than expected—from posterior to anterior—when the Shh dose is gradually reduced.

The function formerly known as "AER maintenance factor," Shh can also sustain the AER and its Fgf expression. Retinoic acid has the ability to increase Shh expression, but sustained contact with the AER or an FGF bead is necessary for the ZPA to maintain Shh expression. This indicates that a positive feedback loop is composed of the Shh and FGF genes. Maintaining the ZPA is done by FGF from the apical ridge and by Shh from the ZPA. The Shh knockout mouse has deformed limbs that lack anteroposterior polarity and have severe distal defects. Grafting to the anterior side of normal limb buds reveals that, as would be predicted, there is no ZPA activity in the knockout limb buds. Loss of anteroposterior patterning and loss of AER support combine to produce the knockout phenotype. Retinoids can cause the initial upregulation of Shh expression, but the transcription factors dHAND and Gli3 appear to repress each other in a way that establishes Shh expression in normal development. Gli3 is in the Hedgehog signal transduction pathway, whereas dHAND is a bHLH factor that is first expressed in the entire limb mesenchyme. The expression of dHAND in the posterior of the future limb bud activates the expression of Shh, while Gli3 is expressed elsewhere. In normal development, these two factors reach an equilibrium. The fact that the expression of one factor expands in a wider domain in mice when the other is knocked out provides evidence for the mutual repression.

One of the three Gli-type transcription factors that are a part of the Shh signaling pathway is called Gli3. The Gli factors are stabilized against degradation by Shh signaling, which also permits them to enter the nucleus and control target genes. However, this is the only way that Gli2 acts. Target gene expression is repressed by Gli3's predominant repressiveness and Gli1's lack of activity. This explains why the Gli3-knockout mouse has polydactyly, or an excessive number of fingers. This is caused in part by the larger-than-usual dHAND and, consequently, Shh domain as well as the increased effectiveness of the Shh signal, which induces digits from a greater portion of the limb bud. Human polydactyly can also result from loss or reduction of Gli3, as demonstrated by Greig syndrome, for instance.

The expression patterns of the genes Hoxd9, -10, -11, -12, and -13 show nested territories that span the limb bud from posterior to anterior (Figure 9.24b). They can be applied with Shh or upregulated ectopically in this nested pattern using a ZPA graft. Many people believe that these genes encode distinct states of determination for various limb parts. The outcomes of mouse knockouts, however, contradict this theory since the loss of Hoxd cluster subsets typically results in distal deletions rather than homeotic transformations. The nested patterns of Figure 9.24 are only present at the stage indicated. The Hox genes have dynamic expression patterns that vary at different stages. Regarding the Hoxa genes, it is still unknown whether or not they actually code for states of anteroposterior determination and at what point in limb development the Hoxd genes are actually necessary.

9.4.2.4. Dorsoventral Patterning

The arrangement of muscles or epidermal specializations like feather germs and claws are the primary ways in which the dorsoventral pattern is observed. Early epidermis inversion can result in an inverted dorsoventral pattern of the internal limb tissues. This is accomplished by extracting the limb bud from the embryo, using trypsin to split the epidermis from the mesenchyme, recombining them with the epidermis inverted, and then using tiny platinum pins to reattach to the proper location on the embryo.

The dorsal epidermis was initially located above the somites, whereas the ventral epidermis covered the lateral plate, which accounts for the initial polarity of the epidermis (see Figure 9.22). The ventral epidermis is where Engrailed-1 is normally expressed. As we have seen, the Engrailed-1 knockout mouse also exhibits a ventral extension of the AER, indicating that Engrailed-1 activity limits the extent of the AER. The partial dorsalization of the ventral paw in the Engrailed-1 knockout mouse suggests that Engrailed-1 encodes the ventral state. Engrailed-1 is not upregulated and the resulting limbs have a double-dorsal appearance if the potential ventral epidermis is isolated from the underlying mesoderm by a barrier. This demonstrates that the dorsoventral polarity of the limb bud is ultimately caused by an inducing factor from the lateral mesoderm. BMP4, which is generated by the lateral mesoderm and contributes to the somites' regionalization, could be the signal.

Repressing Wnt7a expression, which is typically expressed in the dorsal but not the ventral epidermis, is one of Engrailed-1's roles. Wnt7a aids in the ZPA's maintenance of Shh expression, which obliquely supports AER maintenance. However, it plays a crucial part in dorsoventral specification as well. The Wnt7a knockout mouse has double ventral paws, while overexpression of Wnt7a results in double dorsal limbs. Therefore, there is strong evidence that Wnt7a is the signal from the epidermis that gives the underlying limb-bud mesenchyme its dorsal identity. In dorsal mesenchyme, Wnt7a activates the LIM-homeobox gene Lmx1. Since the Lmx1 zone is the only known example of a vertebrate compartment boundary that does not correspond to a discernible anatomical boundary, this has implications for cell adhesion. A compartment boundary is a line that, in the course of normal cell development, is not crossed by clones. Using clonal labeling techniques based on activating Cre recombinase in a very small percentage of cells with a low dose of tamoxifen, this specific boundary has been visualized. A ubiquitous promoter drives the CreER, and when the enzyme is active, it recombinates at the R26R reporter locus to produce a limited number of clones that are well-separated. By using this method, it is possible to observe that the ventral clones do not cross into the Lmx1 zone, and the dorsal clones do not cross out of it, despite significant cell dispersion during limb-bud outgrowth (Figure 9.28).

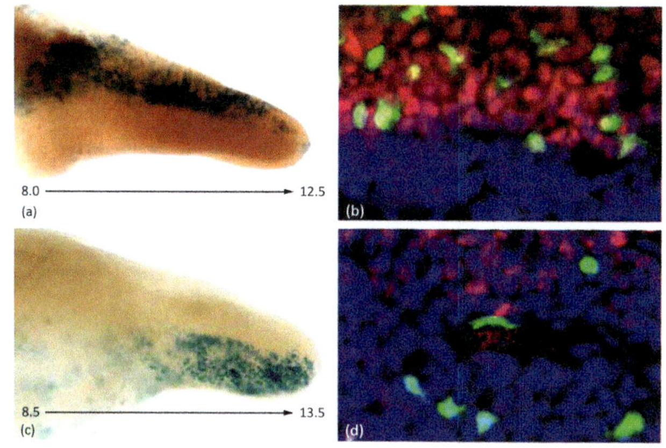

Figure 9.28. *Identification of a compartment boundary between dorsal and ventral mesenchyme in the mouse limb bud. Clones are induced early (E8.0 or 8.5) by a low dose of 4-hydroxytamoxifen to Pol2-CreER x R26R mice. In parts (a) and (c) clones are visualized in wholemounts by XGal staining, at E12.5. In parts (b) and (d), in section at higher power, clones are visualized by GFP fluorescence. Cells expressing the transcription factor LMX1B are immunostained red, nuclei are stained blue.*

Source: Jonathan M.W. Slack. Essential Developmental Biology, 3rd Edition. Wiley-Blackwell. ISBN: 9781118387603.

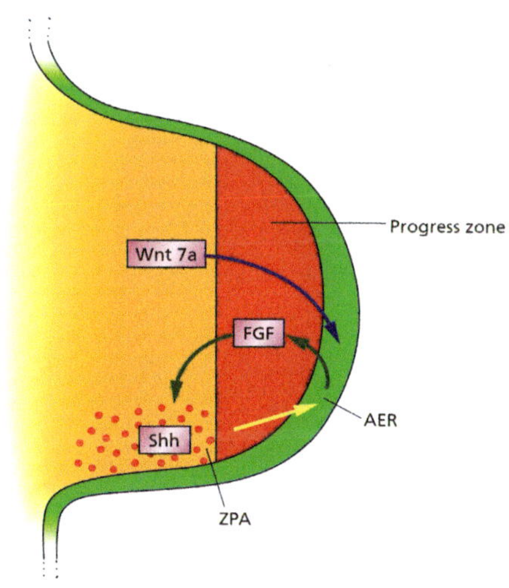

Figure 9.29. *The three signaling systems driving limb elongation and patterning.*

Source: Jonathan M.W. Slack. Essential Developmental Biology, 3rd Edition. Wiley-Blackwell. ISBN: 9781118387603.

The limb is not like most other organs in that it is formed by morphogenetic movements of cell sheets, but rather from a solid mass of tissue. It is asymmetrical in three dimensions. There is strong evidence that a distinct process controls the pattern in each of the three axes. Although the exact mechanism of the proximal–distal patterning is still unknown, it is thought to be dependent on the apical ridge's constant FGF supply and the proximal body region's retinoic acid signal. Shh gradients from the ZPA govern the anteroposterior pattern. Wnt7a from the dorsal epidermis regulates the dorsoventral pattern. The apical ridge needs both Shh and Wnt7a for its own survival and function (Figure 9.29), which ties development in the three axes together and indicates that the three processes cooperate to form an integrated organ.

ROLE MODEL

CHRISTIANE NÜSSLEIN-VOLHARD

Christiane (Janni) Nüsslein-Volhard (born 20 October 1942) is a German developmental biologist and a 1995 Nobel Prize in Physiology or Medicine laureate. She is the only woman from Germany to have received a Nobel Prize in the sciences.

Nüsslein-Volhard earned her PhD in 1974 from the University of Tübingen, where she studied protein-DNA interaction. She won the Albert Lasker Award for Basic Medical Research in 1991 and the Nobel Prize in Physiology or Medicine in 1995, together with Eric Wieschaus and Edward B. Lewis, for their research on the genetic control of embryonic development.

Early Life and Education

Nüsslein-Volhard was born in Magdeburg on 20 October 1942, the second of five children to Rolf Volhard, an architect, and Brigitte Haas Volhard, a nursery school teacher. She has four siblings: three sisters and one brother. She grew up and went to school in south Frankfurt, where she was exposed to art and music and thus was "trained in looking at things and recognizing things". Her great-grandfather was the chemist Jacob Volhard, and her grandfather was the known internist Franz Volhard. She is also the aunt of the Nobel laureate in chemistry Benjamin List.

After the Abitur in 1962, she briefly considered pursuing medicine, but dropped the idea after doing a month's nursing course in a hospital. Instead, she opted to study biology at Goethe University Frankfurt. In 1964 Nüsslein-Volhard left Frankfurt for the University of Tübingen, to start a new course in biochemistry. She originally wanted to do behavioral biology, "but then somehow I ended up in biochemistry (.) and molecular genetics because at the time this was the most modern aspect, and I was ambitious — I wanted to go where the leaders were. The old-fashioned botanists and zoologists were such dull people— there was nothing interesting there."

She received a diploma in biochemistry in 1969 and earned a PhD in 1974 for research into protein–DNA interactions and the binding of RNA polymerase in *Escherichia coli.*

Career

In 1975, Nüsslein-Volhard became a postdoctoral researcher in Walter Gehring´s laboratory at the Biozentrum, University of Basel. She was a specialist in the developmental biology of *Drosophila melanogaster'* (fruit fly) supported by a long-term fellowship from the European Molecular Biology Organization (EMBO). In 1977, she continued in the laboratory of Klaus Sander at University of Freiburg, who was an expert in embryonic patterning. In 1978, she set up her own lab in the newly founded European Molecular Biology Laboratory in Heidelberg with Eric Wieschaus, whom she had met in Basel. Over the next three years they examined about 20,000 mutated fly families, collected about 600 mutants with an altered body pattern and found that out of the approximately 5,000 essential genes only 120 were essential for early development. In October 1980, they published the mere 15 genes controlling the segmented pattern of the Drosophila larva.

In 1981, Nüsslein-Volhard moved to the Friedrich Miescher Laboratory of the Max Planck Society in Tübingen. From 1984 until her retirement in 2014, she was the director of the Max Planck Institute for Developmental Biology in Tübingen and also led its genetics department. After 1984, she launched work on the developmental biology of vertebrates, using the zebrafish (Danio rerio) as her research model.

In 2001, she became a member of the *Nationaler Ethikrat* (National Ethics Council of Germany) for the ethical assessment of new developments in the life sciences and their influence on the individual and society. Her primer for the lay-reader, *Coming to Life: How Genes Drive Development*, was published in April 2006.

In 2004, she started the Christiane Nüsslein-Volhard Foundation (*Christiane Nüsslein-Volhard Stiftung*) which aids promising young female German scientists with children. The foundation's main focus is to facilitate childcare as a supplement to existing stipends and day care.

Research

During the late 1970's and early 1980's, little was known about the genetic and molecular mechanisms by which multicellular organisms develop from single cells to morphologically complex forms during embryogenesis. Nüsslein-Volhard and Wieschaus identified genes involved in embryonic development by a series of genetic screens, generating random mutations in fruit flies using ethyl methane sulfonate. Some of these mutations affected genes involved in the development of the embryo. They took advantage of the segmented form of *Drosophila* larvae to address the logic of the genes controlling development. They looked at the pattern of segments and denticles in each mutant under the microscope, and were therefore able to work out that particular genes were involved in different processes during development based on their differing mutant phenotypes (such as fewer segments, gaps in the normal segment pattern, and alterations in the patterns of denticles on the segments). Many of these genes were given descriptive names based on the appearance of the mutant larvae, such as *hedgehog*, *gurken* (German: "cucumbers"), and *Krüppel* ("cripple"). Later, researchers Pavel Tomancal, Amy Beaton, et al., identified exactly which gene had been affected by each mutation, thereby identifying a set of genes crucial for Drosophila embryogenesis.

The subsequent study of these mutants and their interactions led to important new insights into early *Drosophila* development, especially the mechanisms that underlie the step-wise development of body segments. These experiments are not only distinguished by their sheer scale (with the methods available at the time, they involved an enormous workload), but more importantly by their significance for organisms other than fruit flies.

Her findings led to important realizations about evolution – for example, that protostomes and deuterostomes are likely to have had a relatively well-developed common ancestor with a much more complex body plan than had been conventionally thought.

Additionally, they greatly increased our understanding of the regulation of transcription, as well as cell fate during development.

Nüsslein-Volhard is associated with the discovery of *Toll*, which led to the identification of toll-like receptors.

As of 2023, Nüsslein-Volhard has an h-index of 104 according to Scopus.

Personal Life

Nüsslein-Volhard married in the mid-1960s while studying at the Goethe University Frankfurt, but divorced soon afterward and did not have any children. She lives in Bebenhausen, Germany. She has said that she loves to sing, play the flute and do chamber music. She published a cookbook in 2006.

CLASS ACTIVITY

Remind students that the kidneys are a filtering system for the blood. Explain that engineers design dialysis machines for people whose kidneys are not working properly. Have students write a few sentences about how the kidney functions on their journal sheets.

SUMMARY

- In higher vertebrates the mesoderm becomes partitioned at an early stage into four zones from medial to lateral. The notochord occupies the midline; then the paraxial mesoderm, which will become the somites; then the intermediate mesoderm, which will form the gonads, kidneys, and adrenals; then the lateral plate mesoderm, which forms body wall, limb buds, blood vessels, and heart. In lower vertebrates the arrangement is the same, but runs from dorsal to ventral since in these organisms the mesoderm surrounds the yolk mass.
- Invertebrate mesoderm does not show the same regional subdivision as vertebrate mesoderm, although some similar tissue types are formed including muscle, gonad, connective tissues, and blood cells.
- The term somitogenesis is used to describe the process of segmentation of the paraxial mesoderm within the trilaminar embryo body to form pairs of somites, or balls of mesoderm. In humans, the first somite pair appears at day 20 and adds caudally at 1 somite pair/90 minutes until on average 44 pairs eventually form.
- The mesoderm is one of the three germinal layers that appears in the third week of embryonic development. It is formed through a process called gastrulation. There are four important components, which are the axial, paraxial, intermediate, and lateral plate mesoderms.
- Vertebrate muscle is classified into skeletal muscle, smooth muscle, and cardiac muscle. All skeletal muscle is derived either from the myotomes of the somites, which form axial muscles, limb muscles, and body wall muscles, or from the corresponding regions of the unsegmented head mesoderm, which form the muscles of the eyes and jaw. Smooth muscle is formed from both lateral plate mesoderm and from the hypaxial part of the somites, and cardiac muscle arises from the myocardium of the early heart, derived from anterior splanchnic mesoderm.
- Skeletal muscle is the familiar voluntary muscle, making up a large proportion of the mass of a vertebrate animal. It consists of multinucleate myofibers, which contain myofibrils composed of the contractile proteins actin and myosin. Differentiation of skeletal muscle involves first the commitment of cells to become myoblasts, then a period of multiplication and/or migration, then a cessation of division and fusion of the myoblasts to form the multinucleate myofibers.
- The kidney is familiar as the excretory organ of the vertebrate body and because of the shortage of human kidneys for transplantation it is one of the principal targets for replacement through tissue engineering or some other form of applied developmental biology. The embryonic development of the kidney is representative of a large class of organs, as it involves reciprocal inductive interactions between an epithelium and a mesenchyme. Congenital abnormalities of the kidney are common and often arise from mutations in genes involved in its development.

- The gonads are both the repositories of the gametes and important endocrine organs. They have a dual embryonic origin. The somatic tissues of the gonads arise from the genital ridges, which are formed from the intermediate mesoderm in the vicinity of the mesonephros. The germ cells that produce the gametes are derived from primordial germ cells (PGCs).
- In the mouse embryo the PGCs originate from the proximal part of the egg cylinder. This was shown by fate mapping using horseradish peroxidase (HRP) injection. The PGCs are induced by the effect of the extraembryonic ectoderm since they do not develop in the absence of this tissue, and in recombination experiments extraembryonic ectoderm is able to induce PGCs from both proximal and distal epiblast.
- Male and female gonads are different from each other and so the development of the gonads is intimately associated with sex determination. Unlike most other aspects of development, the mechanism of sex determination is not at all well conserved between animal groups, so the account that follows applies only to mammals. In the mouse the gonads can develop normally in various mutants defective in germ-cell migration, so we know that gonadal development does not necessarily require the presence of the germ cells themselves.
- Vertebrate limbs are of special interest to the developmental biologist because their visible nature makes congenital defects affecting the limbs of humans particularly obvious and distressing. Accordingly, limb development is one of the classic topics of organogenesis research.
- Amphibians, reptiles, birds, and mammals are known as tetrapods because they have four limbs; a pair of forelimbs, which may be legs, arms, or wings, and a pair of hindlimbs. The few tetrapods that have no limbs, such as snakes, have lost them secondarily in the course of evolution.

REVIEW QUESTIONS

1. What do you understand by somitogenesis?
2. What is mesoderm?
3. What is called syndetome?
4. Name the two parts of the dermomyotome.
5. What do the kidneys develop from?
6. Name the five parts of nephron.
7. What is formed from intermediate mesoderm?

REFERENCES

1. Bryson-Richardson, R. J., & Currie, P. D. (2008). The genetics of vertebrate myogenesis. *Nature Reviews Genetics, 9*(9), 632–646. https://doi.org/10.1038/nrg2420.
2. Buckingham, M., & Relaix, F. (2007). The role of Pax genes in the development of tissues and organs: Pax3 and Pax7 regulate muscle progenitor cell functions. *Annual Review of Cell and Developmental Biology, 23*, 645–673. https://doi.org/10.1146/annurev.cellbio.23.090506.123642.
3. Buckingham, M., & Vincent, S. D. (2009). Distinct and dynamic myogenic populations in the vertebrate embryo. *Current Opinion in Genetics and Development, 19*(4), 444–453. https://doi.org/10.1016/j.gde.2009.06.007.
4. Christ, B., & Brand-Saberi, B. (2002). Limb muscle development. *International Journal of Developmental Biology, 46*(7), 905–914. https://doi.org/10.1387/ijdb.041836bc.
5. Costantini, F., & Kopan, R. (2010). Patterning a complex organ: Branching morphogenesis and nephron segmentation in kidney development. *Developmental Cell, 18*(5), 698–712. https://doi.org/10.1016/j.devcel.2010.05.004.
6. Dressler, G. R. (2006). The cellular basis of kidney development. *Annual Review of Cell and Developmental Biology, 22*, 509–529. https://doi.org/10.1146/annurev.cellbio.22.010305.104751.
7. Drew, U. (1993). *Color Atlas of Embryology*. Thieme.
8. Dudek, R. W. (2009). *High-yield embryology* (4th ed.). Lippincott Williams & Wilkins.
9. Goodfellow, P. N., & Lovell-Badge, R. (1993). SRY and sex determination in mammals. *Annual Review of Genetics, 27*, 71–92. https://doi.org/10.1146/annurev.ge.27.120193.000443.
10. Horsley, V., & Pavlath, G. K. (2004). Forming a multinucleated cell: Molecules that regulate myoblast fusion. *Cells Tissues Organs, 176*(1), 67–78. https://doi.org/10.1159/000077511.
11. Johnson, R. L., & Tabin, C. J. (1997). Molecular models for vertebrate limb development. *Cell, 90*(6), 979–990. https://doi.org/10.1016/S0092-8674(00)80382-0.
12. Kenderew, J. C., & Lawrence, E. (1994). Mesoderm induction. In *The Encyclopedia of Molecular Biology* (pp. 541). John Wiley & Sons.
13. Kimelman, D., & Bjornson, C. (2004). Vertebrate mesoderm induction: From frogs to mice. In C. D. Stern (Ed.), *Gastrulation: From Cells to Embryo* (pp. 363–380). CSHL Press.
14. Kuure, S., Vuolteenaho, R., & Vainio, S. (2000). Kidney morphogenesis: Cellular and molecular regulation. *Mechanisms of Development, 92*(1), 31–45. https://doi.org/10.1016/S0925-4773(99)00246-5.
15. Martin, G. R. (1998). The roles of FGFs in the early development of vertebrate limbs. *Genes & Development, 12*(11), 1571–1586. https://doi.org/10.1101/gad.12.11.1571.
16. McLaren, A. (2003). Primordial germ cells in the mouse. *Developmental Biology, 262*(1), 1–15. https://doi.org/10.1016/S0012-1606(03)00230-8.

17. Pappaioannou, V. E. (2004). Early embryonic mesoderm development. In R. P. Lanza (Ed.), *Handbook of Stem Cells* (Vol. 1, pp. 393–407). Gulf Professional Publishing.

18. Pownall, M. E., Gustafsson, M. K., & Emerson, C. P. (2003). Myogenic regulatory factors and the specification of muscle progenitors in vertebrate embryos. *Annual Review of Cell and Developmental Biology, 18*, 747–783. https://doi.org/10.1146/annurev.cellbio.18.031622.104913.

19. Rudnicki, M. A., & Jaenisch, R. (1995). The MyoD family of transcription factors and skeletal myogenesis. *BioEssays, 17*(3), 203–209. https://doi.org/10.1002/bies.950170305.

20. Ruppert, E. E., Fox, R. S., & Barnes, R. D. (2004). Introduction to Bilateria. In *Invertebrate Zoology* (7th ed., pp. 217–218). Brooks/Cole.

21. Sariola, H., & Sainio, K. (1997). The tip top branching ureter. *Current Biology, 9*(18), 877–884. https://doi.org/10.1016/S0960-9822(06)00314-5.

22. Scott, G. (2010). *Developmental Biology* (9th ed.). Sinauer Associates.

23. Shah, M. M., Sampogna, R. V., Sakurai, H., Bush, K. T., & Nigam, S. K. (2004). Branching morphogenesis and kidney disease. *Development, 131*(6), 1449–1462. https://doi.org/10.1242/dev.01037.

24. Sherman, L. S. (2001). *Human Embryology* (3rd ed.). Elsevier Health Sciences.

25. Wylie, C. (1999). Germ cells. *Cell, 96*(2), 165–174. https://doi.org/10.1016/S0092-8674(00)80578-3.

CHAPTER

10 Development of Endodermal Organs

LEARNING OBJECTIVES

After studying this chapter, you will be able to:

- Understand the overview of endoderm.
- Discuss on determination of the endoderm.
- Define the pancreas.

KEY TERMS FROM THIS CHAPTER

Amniotes	Animal embryo	Blastoderm
Cells migrating	Egg cylinder	Endocrine cell types
Endoderm	Eustachian tubes	Gastrulation process
Multihormone cells	Pancreatic cell lineage	Paneth cells
Posterior mesoderm	Septum transversum	Somatopleure
Splanchnic mesoderm	Splanchnic mesoderm	Splanchnopleure
Gut tube	Morphogenetic movements	

INTRODUCTION

The development of endodermal organs is a highly intricate process that begins with the formation of the endoderm during gastrulation in early embryonic development. The innermost of the three germ layers created during gastrulation is known as the endoderm. It creates the epithelial lining of the stomach and all of its progeny, which include the respiratory system, liver, and pancreas. The splanchnic mesoderm, which is the inner subdivision of the lateral plate after the coelomic cavity opens, forms the smooth muscle, connective tissue, and blood vessels that make up the outer coats of these organs. The term "splanchnopleure" refers to the combination of splanchnic mesoderm and endoderm, just as "somatopleure" refers to the combination of somatic mesoderm and epidermis.

10.1. OVERVIEW OF ENDODERM

The innermost of the three main germ layers in the very early embryo is called the endoderm. The mesoderm (middle layer) and ectoderm (outside layer) are the other two layers. The inner layer of the gastrula forms from cells that migrate inward along the archenteron and eventually develop into the endoderm. Initially made up of flattened cells, the endoderm eventually develops into columnar cells. It creates the epithelial lining for several different systems (Fukuda, Kimiko, Kikuchi, & Yutaka, August 2005).

All multicellular animals, with the exception of sponges, form two or three germ layers during the gastrulation stage. The terms diploblasty and triploblasty refer to the presence of two or three embryonic germ layers in the body, respectively. Ectoderm, endoderm, and mesoderm are all visible in triploblasty, whereas only ectoderm and endoderm are in diploblasty the endoderm layer present in both invertebrate and vertebrate embryos. This endoderm gives rise to the lower respiratory tract, the digestive tract, and other parts of the body.

10.1.1. Formation of Endoderm Layer

The embryo develops into a gastrula with several layers of cells from a blastula with a single layer of cells covering it. We refer to this stage of embryonic development as gastrulation. The term "germ layers" refers to these several layers that formed during gastrulation. Even in the earliest stages of an animal embryo, the cells have the potential to develop into endoderm. It is observed that the surface of the blastula is next to the blastocoel when studying the primitive endoderm cells in mice. The early blastocyst's fluid-filled cavity will grow into an extra-embryonic membrane. Extraembryonic endoderm has two main subtypes: parietal endoderm, which will form Reichert's membrane, and visceral endoderm, which surrounds the egg as a protective membrane. The endoderm cells start out on the blastula's surface and eventually grow into an internal structure. The developing embryo cells undergo a dramatic rearrangement during the process of gastrulation, resulting in the development of the correct position in the germ layers.

10.1.1.1. Formation of the Gut tube in Amniotes

Most of the work has been done with the chick embryo, with some help from experiments on early-stage Xenopus and zebra fish as well as mouse knockouts. The hypoblast cells are displaced to the blastoderm's periphery by the endoderm, which forms in the chick during gastrulation as a lower layer of definitive endoderm that emerges from the primitive streak. The area below the blastoderm is where the gut lumen begins. In a manner similar to the evagination of all three germ layers, the gut tube is created when the body folds away from the blastoderm (Figure 10.1). The way the head fold forms is similar to what would happen if a tiny finger were to be pushed up from below to deform the blastoderm and then sideways to lengthen the projection. As a result, the endoderm in this area develops into a foregut tube, and the anterior portion of the body is lifted off the blastoderm.

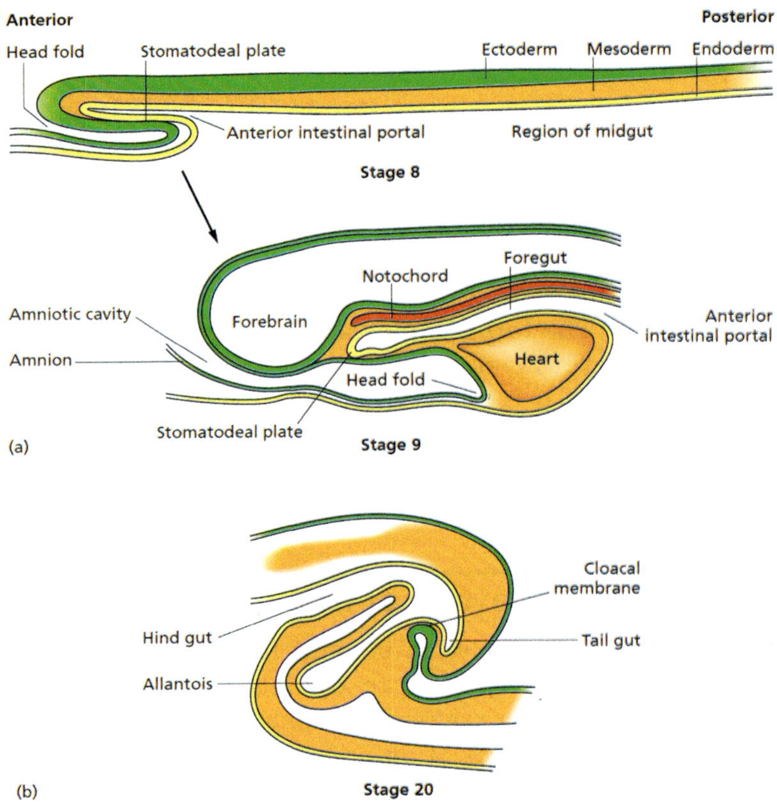

Figure 10.1. *Formation of the fore- and hindgut in the chick embryo. (a) Formation of headfold and foregut. (b) Formation of hindgut (later stage).*

Source: https://dn790000.ca.archive.org/0/items/SlackJMWEssentialDevelopmentalBiology/Slack_J_M_W_Essential_Developmental_Biology.pdf.

The level where the tube opens out into the flat sheet of endoderm is known as the anterior intestinal portal, or the junction of the foregut and midgut. A comparable process occurs at the back end somewhat later, resulting in the formation of the hindgut, with the posterior intestinal portal serving as the junction between the midgut and hindgut. A cloaca forms at the posterior end, but it is not perforate until hatching. The mouth forms at the beginning of the third day through fusion of the ventral foregut with the overlying epidermis. The embryo's extremities grow larger due to morphogenetic movements and growth, which causes the remaining open endoderm to rapidly contract. After four days, all that's left of this is a Vitello intestinal duct that joins the yolk mass and the midgut. The splanchnic mesoderm from both sides fuses to form the ventral and dorsal mesenteries as the gut closes (Figure 10.2a, b). While the ventral mesentery disappears everywhere but in the area surrounding the heart, liver, and cloaca, the dorsal mesentery endures and contains the mature gut. The ventral body wall is formed when the epidermis and mesoderm close around the gut (Figure 10.2c). With the exception of the umbilical tube, which connects to the extraembryonic tissues enclosing the Vitello intestinal duct, vitelline blood vessels, and allantois, this eventually becomes complete. A ventral outgrowth of the hindgut is the allantois (Figure 10.1b). It emerges from the embryo on the second day into the amniotic cavity, where it joins forces with the chorion to form the chorioallantoic membrane. This membrane, which is richly vascularized and serves as a vital respiratory organ prior to hatching, is visible as soon as the late embryo's egg is opened. The umbilical tube incorporates the allantois's stalk and blood vessels. After four days of incubation, the chick's gut has completely developed into a tube and has started to coil asymmetrically, under the direction of the left-right asymmetry system. Regarding the size and shape of the different organ rudiments, it is also beginning to exhibit some regional pattern (Figure 10.3).

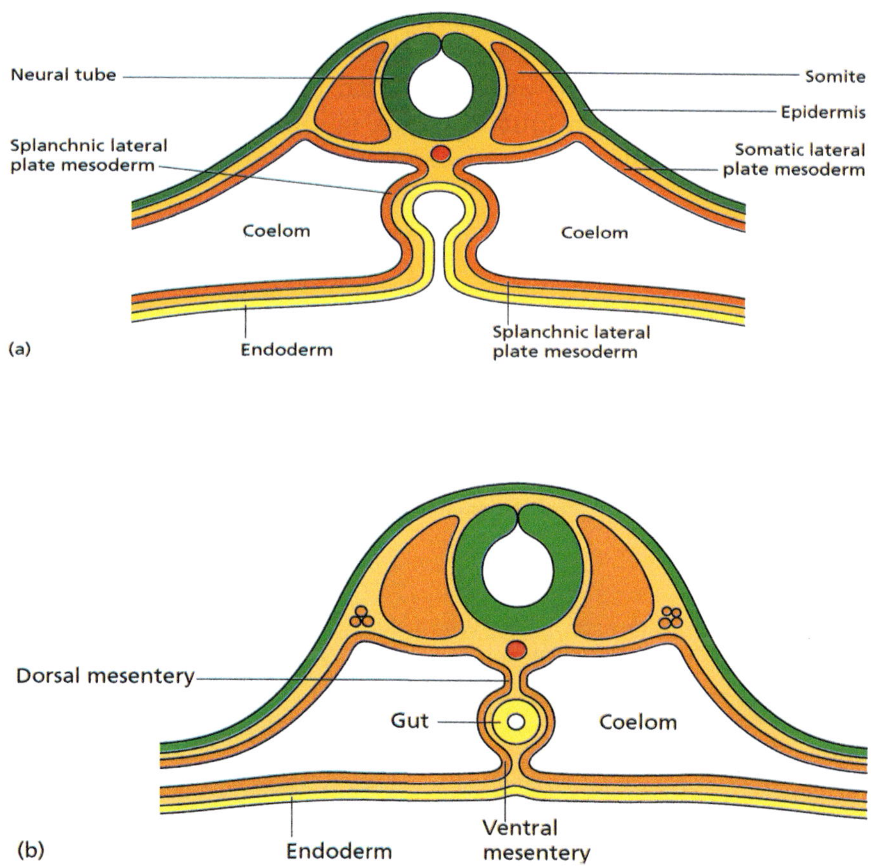

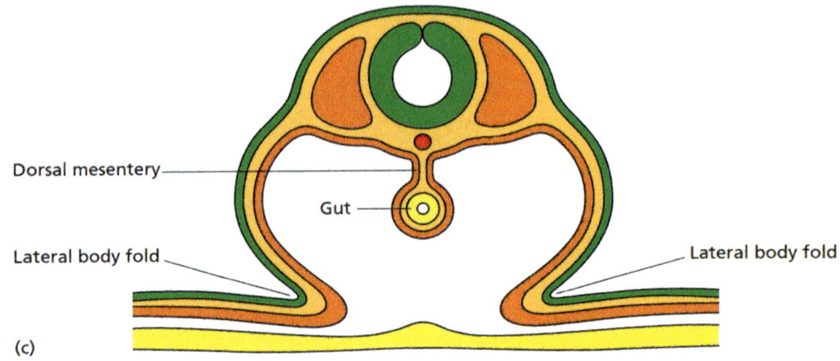

Figure 10.2. *Enclosure of the gut and ventral closure of the body in the chick embryo. (a)–(c) represent transverse sections between midgut and hindgut levels at three days of incubation.*

Source: https://dn790000.ca.archive.org/0/items/SlackJMWEssentialDevelopmentalBiology/Slack_J_M_W_Essential_Developmental_Biology.pdf.

In other higher vertebrates, the sequence of events is quite similar. In the mouse, the definitive endoderm begins as a ventral strip of tissue that borders the visceral endoderm on the exterior of the egg cylinder. With the help of the "turning" process, it integrates into the fore and hindgut pockets similarly to the chick. The placenta receives assistance from the allantois, and the gut's general structure is comparable.

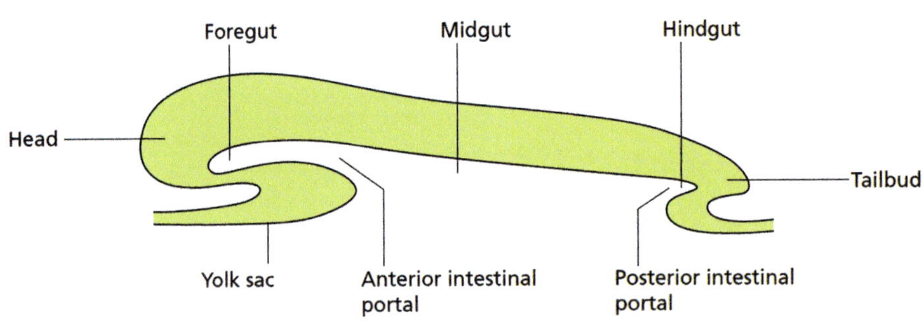

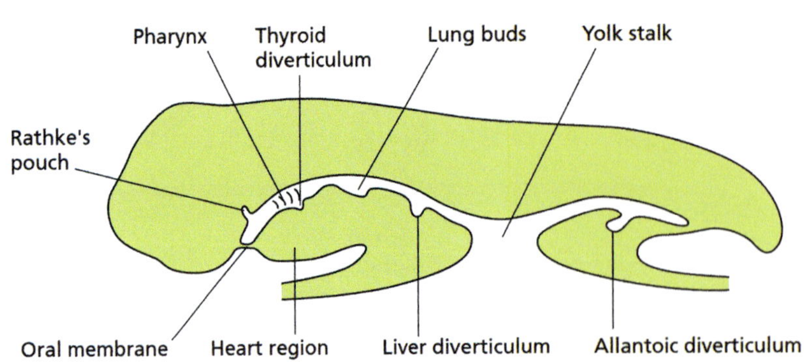

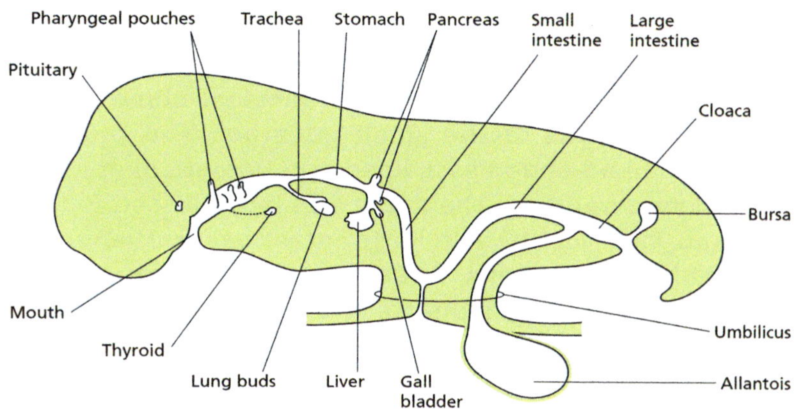

Figure 10.3. *Formation of the regions of the gut in an amniote embryo.*

Source: https://dn790000.ca.archive.org/0/items/SlackJMWEssentialDevelopmentalBiology/Slack_J_M_W_Essential_Developmental_Biology.pdf.

10.1.1.2. Organization of the Gut Tube

The pharyngeal pouches, which align with the branchial arches made of alternating rhombomeres, are formed in the region of the pharynx. The segmental arrangement of endodermal outgrowths is linked to four major pouches, as well as occasionally one or two more primitive ones (Figure 10.4). The Eustachian tubes and the middle ear cavities are formed by buds from the first pouch. The thyroid gland is located in the ventral midline, across from the second arch. Mammals have four pairs of pouches: the third pair forms the thymus, the fourth pair forms the parathyroids, and the second pair forms the tonsils. The paired buds that form the bronchi and lung tissue are produced by the pouches that form the laryngotracheal groove in the ventral midline, which later becomes the trachea. A large portion of the pharyngeal floor gives rise to the tongue, which is primarily made up of mesenchymal muscle from the head.

Figure 10.5 shows regional differentiation of the gut tube. The esophagus, which is initially lined with columnar and then later with stratified squamous epithelium, follows the pharyngeal region. Next is the stomach, where the epithelium develops into a glandular structure with a specialized function to secrete pepsin and hydrochloric acid. The pyloric sphincter, which leads from the stomach out into the small intestine, is located there. There, the columnar embryonic epithelium eventually differentiates into the crypt and villus arrangement. The junctions between the stomach and intestine and the esophagus and stomach are two of the very few locations in the mature animal where there is a sharp discontinuity of epithelial type that has arisen *in situ* from a common cell sheet. The pancreas and liver are connected to the duodenum, the first segment of the small intestine that forms a loop. Soon after this area of the foregut forms, the liver, a ventral outgrowth of the endodermal epithelium, spreads into the nearby ventral mesentery. In mammals, this area of the mesoderm is referred to as the septum transversum. A large dorsal bud and a small ventral bud—paired ventral buds in the chick—give rise to the pancreas. Later on in development, these move in unison and merge to form a single organ. The liver and ventral pancreas are intimately related, and the common bile duct is created when the ventral pancreatic and liver ducts combine. The jejunum and the ileum are the next two segments of the small intestine. The presence of outgrowths known as caeca, which are blind-ended sacs lined with intestinal epithelium, is frequently used to identify the junction between the small and large intestine (or colon). Although it lacks Paneth cells and villi, the mature large intestinal epithelium resembles that of the small intestine. The large intestine transforms into a rectum at its posterior end, and in mammals, the anal sphincter connects it to the outside.

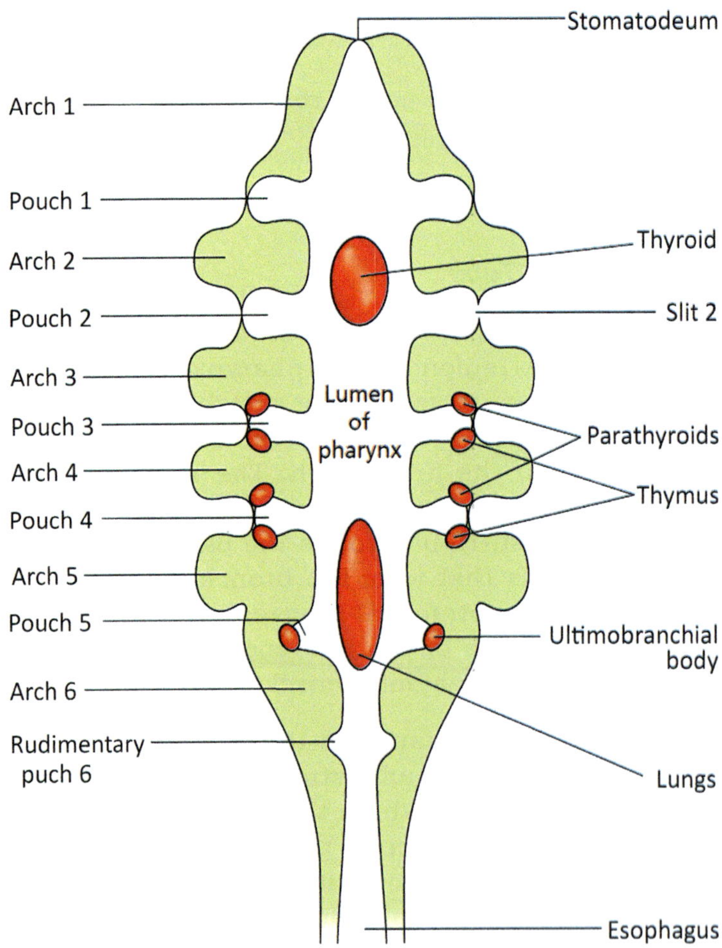

Figure 10.4. *Location of outgrowths of the endoderm in the pharyngeal region of the chick embryo.*

Source: *https://dn790000.ca.archive.org/0/items/SlackJMWEssentialDevelopmentalBiology/Slack_J_M_W_Essential_Developmental_Biology.pdf.*

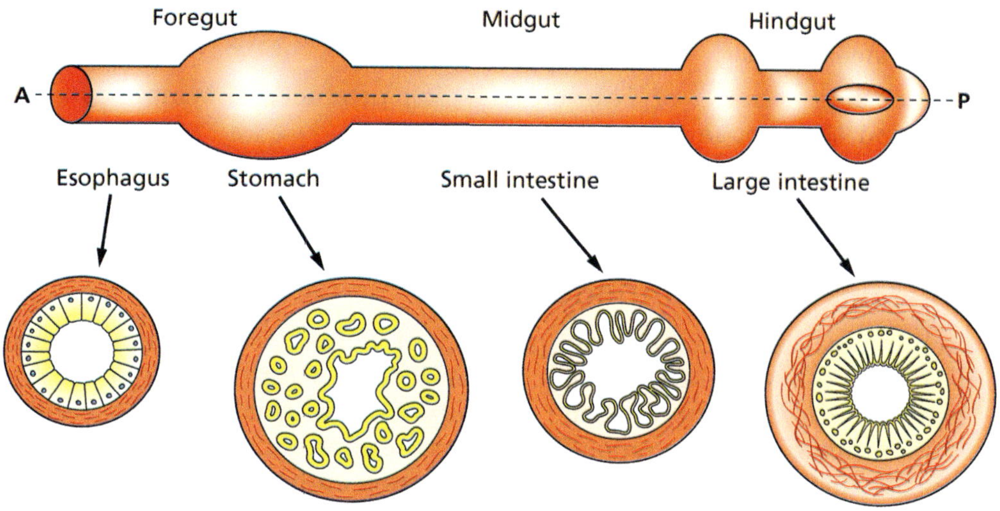

Figure 10.5. *Cytodifferentiation of regions of the chick gut tube.*

Source: *https://dn790000.ca.archive.org/0/items/SlackJMWEssentialDevelopmentalBiology/Slack_J_M_W_Essential_Developmental_Biology.pdf.*

The terms "foregut," "midgut," and "hindgut" are still used in the mature gut, but they no longer refer to the gut segments that are bounded by the anterior and posterior intestinal portals, which are changing during the gut tube's formation. Traditionally, the midgut is considered to extend to the junction of the small and large intestines, and the foregut is considered to extend as far as the pancreas and liver at late stages. Although the majority of amniotic vertebrates are covered by this account, most experimental work is done on mice or chicks, which are not exactly the same.

The following checklist is provided to keep track of the differences:

1. The mouse has tonsils in the second pharyngeal pouch, the chick does not.
2. The chick forms parathyroids from the third and fourth pharyngeal pouch, the mouse only from the third.
3. The mouse incorporates the ultimobranchial body into the thyroid as C (calcitonin) cells, the chick keeps them separate.
4. The mouse lungs have alveoli; the chick lungs have anastomosing channels running into large distal air sacs.
5. The chick has a specialization of the esophagus to form the crop, the mouse does not.
6. The chick has a division of the stomach into a glandular proventriculus and a muscular gizzard, the mouse has no gizzard.
7. The mouse embryo liver is a major hematopoietic organ while the chick liver is not.

8. The chick has paired ventral pancreatic buds; the mouse has one.
9. The mouse has one large cecum between small and large intestines, the chick has two small ones.
10. The chick has a bursa of Fabricius, which produces B lymphocytes, near the cloaca, and the mouse does not.
11. The chick has a cloaca while the mouse has separate urinary/ reproductive and alimentary orifices.

10.1.2. Fate Map of the Endoderm

Numerous techniques, such as orthopic grafting of quail tissue and tiny extracellular injections of DiI, have been used to map the endoderm's fate in the chick. The entire embryo is then cultured *in vitro* until the gut tube closes. Analogous research has also been conducted on mice and Xenopus. The anterior endoderm becomes the foregut, the middle section becomes the midgut, and the posterior endoderm becomes the hindgut, as would be expected from the broad conservation of anterior-posterior relationships (Figure 10.6a). The midline of the fate map represents structures such as the dorsal pancreatic bud because the early endoderm is an open sheet, and its midline becomes the dorsal midline of the gut tube. Similar to the liver, the ventral midline structures are produced from the edges of the definitive endoderm, which converge as the anterior and posterior portals approach one another. For this reason, ventral structures frequently have paired buds.

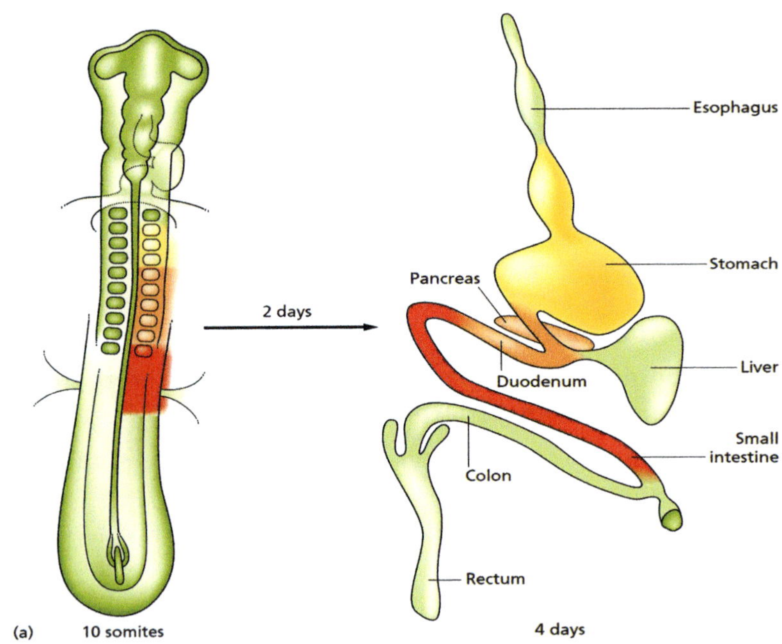

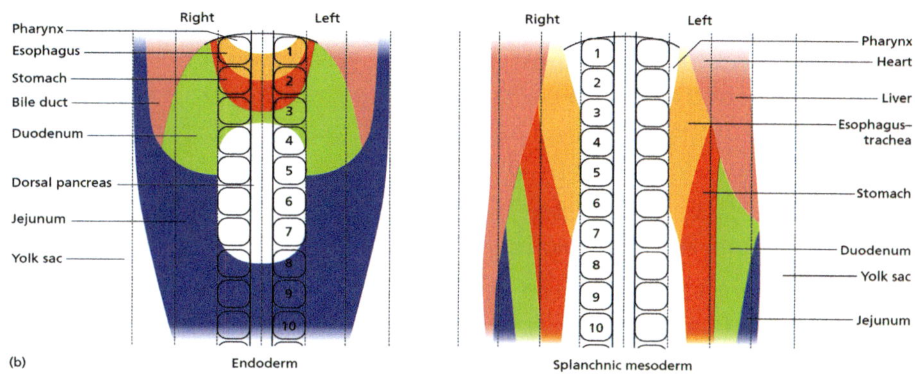

Figure 10.6. *(a) Fate map of the chick endoderm. (After Grapin-Botton.) (b) Mismatch between the fates of endoderm and splanchnic mesoderm at 1.5 days.*

Source: https://dn790000.ca.archive.org/0/items/SlackJMWEssentialDevelopmentalBiology/Slack_J_M_W_Essential_Developmental_Biology.pdf.

The relationship between potential endodermal regions and those in the splanchnic mesoderm is a crucial component of the fate map (Figure 10.6b). The mesoderm's prospective regions, which tend to be arranged longitudinally, differ significantly from those of the endoderm. This demonstrates that during the gut closure and ventral body wall closure processes, there is a significant relative movement of the two germ layers. The inductive interactions between mesoderm and endoderm are a key theme in gut development, and the fate map makes it abundantly evident that during the gut tube formation process, a specific endodermal region comes into contact with various mesodermal regions. This implies that as the endoderm interacts with various mesenchymes, its specification may alter.

10.1.3. Functions of Endoderm

The endoderm and mesoderm enter the embryo during invagination and begin forming the guts during the gastrulation process. The line of cells along the mesoderm is generated by the dorsal endoderm as the cells migrate into the inner region of the embryo. The gut cavity's precursor is the space that formed between the cells of the vegetal and dorsal endoderms. The reproductive system, digestive system, respiratory system, and other organ systems will all develop from the definitive endoderm.

The definitive endoderm is developed into the organs listed below:

- **Lungs:** The pair of lungs are part of the respiratory system, which purifies the incoming air. A pair of lungs work together and help to breathe. It mainly helps to oxygenate the fresh air to the body through the blood and extracts the carbon dioxide from deoxygenated blood.
- **Liver:** The liver is mainly found in all vertebrates. It detoxifies various metabolites,

and synthesizes proteins further, produces biochemicals necessary for digestion and growth.

- **Pancreas:** The pancreas plays an important role in converting the food particles to energize the body cell. The pancreas produces two major functions. One is an exocrine function, which helps the digestive system. And another is an endocrine function, which regulates blood sugar.
- **Female Vagina:** It is part of the female reproductive system. The vagina is an elastic, soft flexible muscular canal. It connects the uterus to the outside world.
- **Colon:** It is part of the digestive system, which is known as the large intestine. It absorbs the salt and water from the food materials and promotes digestion.
- **Stomach:** It breakdown the food by temporarily contracting and relaxing the stomach and promotes digestion. It temporarily stores the food.
- **Small Intestine:** It is a continuous tube running from the stomach to anus. It absorbs essential nutrients and water from the food.
- **Thymus:** It produces white blood cells called T-lymphocytes. It plays a major part in improving the immune system.
- **Prostate**: It produces the fluid with the sperm cells from testicles.
- **Thyroid:** It is the most important hormonal gland. It plays an important role in growth and metabolism.
- **Cecum:** The cecum connects the small intestine and large intestine. It simply acts as a storage unit for chyme which it receives from the ileum.

10.1.4. Migration through the Primitive Streak

FGF signaling appears to be involved in the migration of mesodermal cells through the anterior primitive streak and their condensation to form the chordamesoderm. The primitive streak expresses the Fgf8 protein, which keeps migrant cells away from the streak. Cells are still moving through the primitive streak inward at this time. The cells split into two layers as they enter the embryo. The hypoblast cells are pushed to the sides as the deep layer joins them along their midline. All of the embryo's endodemlal organs and the majority of its extraembryonic membranes are produced by these deep-moving cells; the remaining cells are hypoblasts and peripheral cells of the opaca region. Between this endoderm and the epiblast, a loose layer of cells is formed by the second migrating layer spreading outward. The mesoderm lining the extraembryonic membranes and the mesodermal parts of the embryo are produced by these cells in the middle layer. Wnt proteins appear to control the mesodermal precursors' further migration, while Fgf8-mediated chemo-repulsion appears to direct movement away from the streak. Wnt5a is not blocked in the more posterior areas, where it causes the cells to migrate widely in order to form the lateral plate mesoderm. However, Wnt3a opposes Wnt5a in the more anterior regions of the streak, preventing migration and causing the cells to form paraxial mesoderm. In fact, Wnt3asecreting pellets inhibit lateral migration and stop lateral plate mesoderm from forming when they are added to the posterior primitive streak.

10.1.5. Regression of the Primitive Streak and Epiboly of the Ectoderm

A new stage of development has now started. Hensen's node shifts from its location near the center of the area pellucida to a more posterior location as mesodermal ingression proceeds and the primitive streak begins to regress. The notochord and the embryo's dorsal axis are left behind by the regressing streak. Starting at the point where the ears and hindbrain form and extending caudally to the tail bud, the notochord is laid down in a head-to-tail orientation. Ectodermal precursors multiply and migrate to encircle the yolk by epiboly as the presumed mesodermal and endodermal cells inwardly migrate. Consequently, the mesoderm has positioned itself between these two areas, the endoderm has replaced the hypoblast, and the ectoderm has encircled the yolk as avian gastrulation approaches its conclusion. While many of the processes involved in avian gastrulation have been identified, our understanding of the molecular mechanisms underlying the execution of some of these processes is still in its early stages.

10.2. DETERMINATION OF THE ENDODERM

Xenopus and zebrafish have been the primary subjects of research regarding the early stages of endoderm formation. In Xenopus, at the 32-cell stage, the endoderm develops from tissue that roughly corresponds to the eight vegetal blastomeres. The transcription factor VegT's mRNA is what causes the endoderm to form. This stimulates the expression of additional transcription factors, such as GATA1–4, Mix1, Mixer, and Sox17α and β. While some of these can be activated directly, others need to be signaled between cells. In particular, nodal signaling is necessary for the expression of GATA4 and mixer. Zebrafish are more reliant on nodal signaling because the cyclops/squint double mutant, which is deficient in both nodal genes, does not form any endoderm at all. Loss of function results in the loss of the entire endoderm. Two genes, casanova (a member of the sox family) and spiel-ohne-grenzen (a POU domain homolog of oct4), are controlled by nodal. Similarly, nodal signaling-deficient mutants in mice fail to form the definitive endoderm. The numerous endodermal transcription factors exhibit a high degree of mutual reliance. If overexpressed in Xenopus animal caps, they will primarily activate the other factors, and endoderm development will typically be suppressed by the introduction of dominant negative constructs created with the engineered repression domain. Zebrafish and mice share a similar set of transcription factors expressed in their early endoderm, and mutants exhibit varying degrees of endodermal defects. For instance, mice lacking the sox17 gene completely lack the mid- and hindgut.

10.2.1. Regional Specification

Rather than designating a particular kind of epithelium or organ, the transcription factors previously mentioned appear to designate endoderm as a germ layer. Numerous additional transcription factors are expressed in the endoderm at different levels (Figure 10.7). Due to their mutant phenotype, some of these are known to be significant for regional specification. The pharyngeal pouches all express pax9 (paired domain), and the knockout mouse lacks an ultimobranchial body, parathyroid, or thymus. Both the tracheo-esophageal septum and the thyroid bud express Nkx2.1 (homeodomain, =TTF1). The mutant mouse lacks both a thyroid and lungs. The future pancreatic buds are included in the proximal intestine region where pdx1 (homeodomain, =XlHbox8 in Xenopus) is expressed, and the knockout mouse lacks a pancreas (see below). The majority of the intestine is located in the posterior region of the gut, which is where cdx2 (homeodomain, =cdxC in chick) is expressed. The complete loss of function in clones arising within the intestine of heterozygous animals can lead to transformation to an esophageal-like epithelium, suggesting that cdx2 distinguishes between anterior and posterior endodermal types of differentiation. However, the knockout of cdx2 is an early lethal one because it is also required in the trophectoderm. The three FoxA1–3 factors, which belong to the FoxA (forkhead domain) class, are arranged nestedly with staggered anterior boundaries. Even though foxA2 is expressed throughout the entire endoderm, its knockout results in a loss of the fore- and midgut. Interestingly, a number of these transcription factors function during the terminal differentiation of the concerned organs and are expressed again at a later stage. For instance, Pdx1 regulates not only the development of the entire pancreas but also its specific expression in β cells, where it aids in the regulation of insulin expression. In addition to regulating thyroid development overall, Nkx2.1 also regulates thyroid-specific gene expression, specifically thyroglobulin and thyroperoxidase.

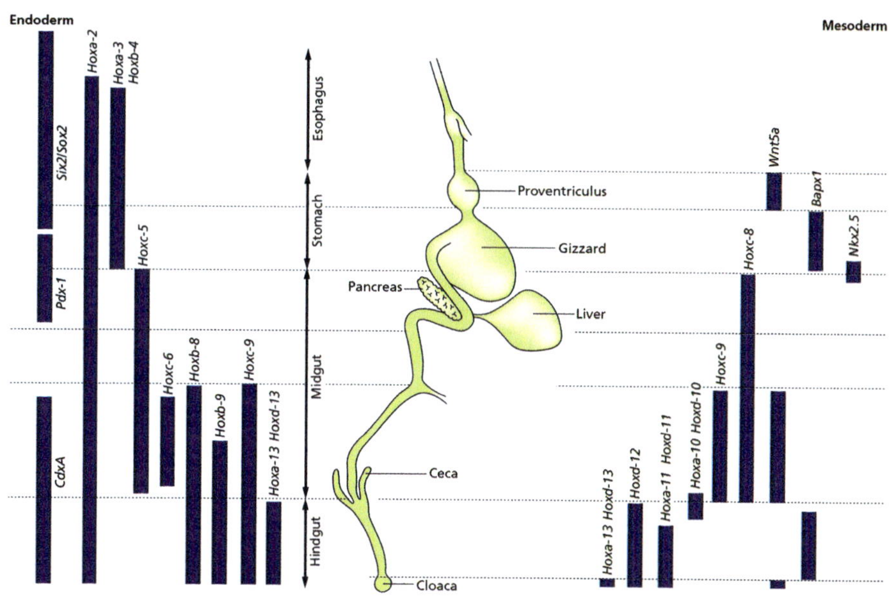

Figure 10.7. *Expression domains of transcription factor genes in the endodermal and mesodermal layers of the early chick gut.*

Source: https://dn790000.ca.archive.org/0/items/SlackJMWEssentialDevelopmentalBiology/Slack_J_M_W_Essential_Developmental_Biology.pdf.

Certain transcription factors may directly influence mesoderm differentiation because they are expressed regionally in the splanchnic mesoderm as opposed to the endodermal epithelium. For instance, loss of the ileocecal sphincter, a muscular structure formed from the mesoderm, will result from loss of the entire HoxD cluster (apart from d1 and d3). Others are most likely crucial for regulating the mesoderm's regional inductive signals, which regulate the endoderm's regional pattern (see below). For instance, the midgut epithelium develops into the large intestine rather than the small intestine when Hoxd13 is virally misexpressed in the midgut mesenchyme. The anteroposterior patterning system for the entire body during early development must be the original cause of the patterns depicted in Figure 10.7. This depends on a posteriorizing signal that involves retinoic acid, Wnts, and FGFs during gastrulation. The CDX family is a significant target gene group that in turn activates the more posterior Hox genes. Beginning at roughly stage 3, cdxA is expressed in the chick's posterior primitive streak. After stage 10, when the early expression domain that governs the overall body pattern fades, cdxA is expressed once more in the posterior endoderm in the area that corresponds to the majority of the future intestine.

10.2.2. Epithelial–Mesenchymal Interactions

The intricate and ongoing exchange of inductive signals between the endoderm and mesoderm is evident. Since the splanchnic mesoderm has developed into an undifferentiated layer of mesenchyme encircling the epithelial tube, it is now referred to as mesenchyme. Sonic Hedgehog is expressed throughout the whole endodermal tube (Shh). This regulates the mesenchyme's differentiation into the lamina propria, muscularis mucosa, submucosa, and smooth muscle layers of the mature gut by creating a radial gradient throughout it. Local induction of this pattern can be achieved by applying Shh from secreting tissue culture cells or through viral infection; its development is inhibited by the Shh inhibitor cyclopamine. Aside from this endodermal Shh signal, signals traveling from the mesoderm to the endoderm have received the majority of attention. The general consensus is that mesenchyme gives endodermal explants the permissive signal they need to mature and differentiate. Under certain conditions, it has been observed that the endoderm's differentiation can be reprogrammed to match the mesoderm's position of origin by combining mesenchyme and epithelium from distinct regions of the chick embryo's gut. The recombinants can be allowed to differentiate in organ culture or grafted to a neutral site in ovo for these experiments. The extracellular matrix associated with the mesenchyme may have a transient and nonspecific effect on the overall tissue architecture, which may lead to an exaggerated degree of reprogramming. However, when tissue architecture and cell differentiation are taken into account simultaneously, the following findings are observed. The chick embryo's gut displays "posterior dominance" from an early stage. This means that although anterior mesoderm can reprogram posterior endoderm; posterior endoderm cannot posteriorize anterior endoderm. Specifically, the future intestinal region that expresses CdxA seems committed from before day 10.5 of incubation, whereas the regions that are intended to form the esophagus and stomach are still capable of forming intestinal epithelium until approximately day 4.5 of incubation.

Treatment with various growth factors, such as activin A, FGF4, BMP7, and retinoic acid, can induce the expression of CdxA; however, mesenchyme is also necessary in these experiments, so the effects are not direct but rather function through an as-yet-unidentified mesenchymal signal. Unlike the stomach of a mammal, the stomach of a chick is separated into two distinct sections. The posterior gizzard has a squamous epithelial lining and a thick muscular wall, which are intended for churning and macerating the food. The anterior proventriculus has a glandular epithelial lining, similar to the stomach of a mammal. If one of these two epithelium regions is cultured with mesenchyme from the other, the two epithelium regions can still interconvert until roughly nine days of incubation. The proventriculus' mesenchyme expresses multiple BMPs, whereas the esophagus and gizzard do not. This indicates the significance of BMP as an inductive signal. When esophageal or gizzard epithelium is treated with BMP, proventriculus-like glands will differentiate; noggin prevents this from happening. Normal expression of the transcription factor Nkx2.3 (homeodomain = Bapx1) occurs in the mesenchyme of the gizzard. It stops BMP synthesis in the mesenchyme from being activated by Shh from the endoderm. When this factor enters the proventriculus mesenchyme, the surrounding epithelium will develop into a gizzard instead of a proventriculus. The BMP signaling regulates the expression of Nkx2.5, a transcription factor that regulates heart development, in a ring of mesenchyme at the distal end of the gizzard. This regulates the formation of the pyloric sphincter.

The development of the liver and how it differs from the ventral pancreas is another illustration that is starting to be understood. The liver develops from a ventral diverticulum of the foregut that multiplies into the ventral mesenchyme, which in mammalian embryos is known as the septum transversum. However, the neighboring cardiogenic mesoderm must send a signal for the liver diverticulum to form. FGF is this signal. FGFs are produced by the cardiogenic mesoderm; FGF inhibitors block the signal; and FGFs replicate the effects on hepatic endoderm that is isolated. It is possible to use the same signal again for a different purpose, even though FGFs are likely also involved in the early regionalization of the endoderm as posteriorizing factors. This interaction happens at the primitive streak stage, which is much earlier than the local interaction involved in the formation of the liver. Furthermore, it is likely that the liver-inducing signal consists of both BMP and a blood vessel component. The fact that normal hepatic endoderm will not develop normally in recombination experiments if the mesenchyme originates from an embryo devoid of blood vessels or BMP4 (flk1 knockout embryo) provides evidence for this. The later growth of the liver bud necessitates Wnt activity, as evidenced by the fact that overexpression of Wnt inhibitors will inhibit growth and viral overexpression of β-catenin will promote growth in the chick liver. There are undoubtedly a lot of additional variables besides Wnts that affect liver growth.

It is well known that FGF10 plays a role in the development of the intestinal cecum and lung buds. Both the fgf10 and fgfr2b knockouts are devoid of limb buds, the cecum, and the lungs. A lateral inhibition-type mechanism appears to be involved in the branching morphogenesis of the developing lungs, whereby newly formed tips generate FGF10 and inhibit the formation of neighboring tips.

10.3. THE PANCREAS

The significance of diabetes and pancreatic cancer for human health has led to a significant increase in research interest in the development of the pancreas in recent years. The developed pancreas performs two very different physiological roles. The production of digestive enzymes is one. The exocrine cells, which comprise the majority of the pancreatic mass, are where these are produced. They secrete their products into a duct system that runs into the duodenum and are grouped into acini (singular: acinus). The production of hormones that are discharged into the bloodstream is the second. Endocrine cells, which produce these, are primarily found in the islets of Langerhans, though some can also be found singly or in small groups. The vast majority of islet cells are insulin-secreting β-cells. Additionally, there are fewer PP cells that secrete pancreatic polypeptide, ε-cells that secrete ghrelin, δ-cells that secrete somatostatin, and α-cells that secrete glucagon. Endocrine cells make up roughly 10% of the embryo, but only 1–2% of the adult pancreas has them.

The mesenchyme is where the blood vessels and the relatively small amount of pancreatic connective tissue originate, whereas the endoderm gives rise to all of the pancreatic epithelial cell types. The spleen is also formed by the pancreatic mesenchyme. All vertebrates have pancreatic development that is largely similar. It begins as a large dorsal bud and a smaller ventral bud (paired ventral buds in the chick) that emerge from the duodenum. These buds enlarge and advance in the direction of one another, eventually merging. Every initial bud has a duct; in humans, the dorsal bud's duct becomes an accessory duct and the ventral bud's duct the principal duct. In zebrafish, endocrine cells predominate in the dorsal bud and exocrine cells predominate in the ventral bud. The majority of experiments have been conducted on mice, with a small amount also on chicks, Xenopus, and zebrafish. In mice, the ventral bud emerges roughly two days after the dorsal bud, which appears at 9.5 days. Unlike mature cells, endocrine hormone-containing cells are visible in the early buds and frequently produce multiple hormones. They are not, however, the progenitors of mature endocrine cells, as is now understood. During the 14.5–18.5 day period, when the buds are growing quickly, the number of endocrine cells increases and a large portion of the remaining pancreas differentiates into exocrine cells.

10.3.1. Induction of Pancreatic Buds

The range of pancreatic cell types produced by the dorsal and ventral buds is the same, but they appear to develop in distinct ways. The area where the notochord and gut roof meet is where the dorsal bud is visible. Shh and Ihh expression, which is expressed throughout the rest of the endoderm, is suppressed in this area. Administration of activin or FGF can replicate the effect of the notochord, suggesting that these could be the causative factors. Mice lacking the activin receptor exhibit numerous other defects in the gut along with small pancreases and fewer endocrine cells. Ectopic pancreatic buds from the Pdx1-expressing zone will form when the endoderm is exposed to the Shh inhibitor cyclopamine. But since cyclopamine application or contact with the notochord cannot cause an ectopic pancreas from more posterior endoderm, the notochord's function is regarded as permissive rather than instructive.

The liver prospective region will also develop into the ventral pancreas in the absence of the FGF signal. The ventral pancreas is formed from the area of the foregut floor that is adjacent to the liver. It appears that the FGF keeps Shh expression in the endoderm active when it would otherwise be inhibited. Normally, the cardiac mesoderm is too distant from the area that forms the ventral pancreas to receive the FGF signal. However, the ventral pancreas is absent in mice with hex knockouts, which are mutants for a homeodomain transcription factor. This is not because the hex factor codes for the pancreas; rather, hex is required for the cell migration and growth that typically lengthens the ventral foregut, preventing FGF from reaching the area that forms the ventral pancreas. Therefore, it seems that FGF may induce the dorsal bud, while its absence causes the ventral bud to develop. However, the inhibition of Shh expression in the endoderm is a common factor in the formation of both buds. Research on Xenopus and zebrafish shows that retinoic acid is also required early on in the process of pancreatic bud formation. BMS453 inhibitors stop buds from forming, and exogenous retinoic acid encourages the development of endocrine rather than exocrine processes.

The pancreatic mesenchyme must remain close to the buds in order for them to continue growing and differentiating into exocrine tissues. Once more, mesenchyme from different regions of the gut can be successfully substituted, and this is a permissive effect. Islet-1, a lim domain transcription factor, is expressed in the pancreatic mesenchyme, and the dorsal pancreas of islet-1-deficient mice does not grow. Nonetheless, the bud will develop normally if the endoderm from the knockout is cultured with mesenchyme from a normal mouse. FGFs are most likely among the elements from the mesenchyme. The mesenchyme expresses FGF10, FGFs can promote the growth of separated pancreatic endoderm, and pancreatic growth is decreased in those who lack FGF10 or the gene encoding its receptor, FGFr2b. Despite being permissive, mesenchymal factors can also influence the ratios of cell types that differentiate from the endoderm. Treatment with follistatin, an activin inhibitor, tends to suppress endocrine differentiation in organ culture experiments, while treatment with TGFβ tends to produce more endocrine cells. A signal from blood vessels also promotes endocrine development. A common environmental characteristic of the pancreatic buds is the close proximity of a major blood vessel: the vitelline veins near the ventral bud and the dorsal aorta near the dorsal bud. The proportion of endocrine cells can be increased in combinations of pancreatic buds and blood vessels, and decreased in cases where the vessels are removed. On the other hand, it appears that pancreatic endocrine cell development in zebrafish remains unaffected by the cloche (vegfr2) mutant's complete removal of blood vessels.

10.3.2. Pancreatic Transcription Factors

Pdx1 (lim-homeodomain) is crucial for the development of the pancreas and is expressed in the duodenum, including potential pancreatic bud territories. The knockout mouse only develops simple buds that do not differentiate or grow. Pdx1 is expressed throughout the early buds before being downregulated to a low level and remaining highly expressed in β-cells, where it functions as an insulin transcription factor. Insulin from the β-cells is lost in a postnatal knockout of Pdx1, which uses a tetracycline-regulated gene in place of the normal one. Pdx1 is not responsible for the formation of pancreas from other parts of the gut when misexpressed, despite its apparent role as the "master control gene" for the pancreas. However, Pdx1-VP16, an activated domain swap version, can reprogram the developing liver in Xenopus to become a pancreas. Pdx1 is expressed in a slightly earlier stage and in a smaller region than Hb9 (encoded by Hlxb9, homeodomain). The knockout mouse lacks a dorsal pancreas and has a normal ventral bud, but its differentiated β-cells do not mature properly because Hb9 is also required for them to mature.

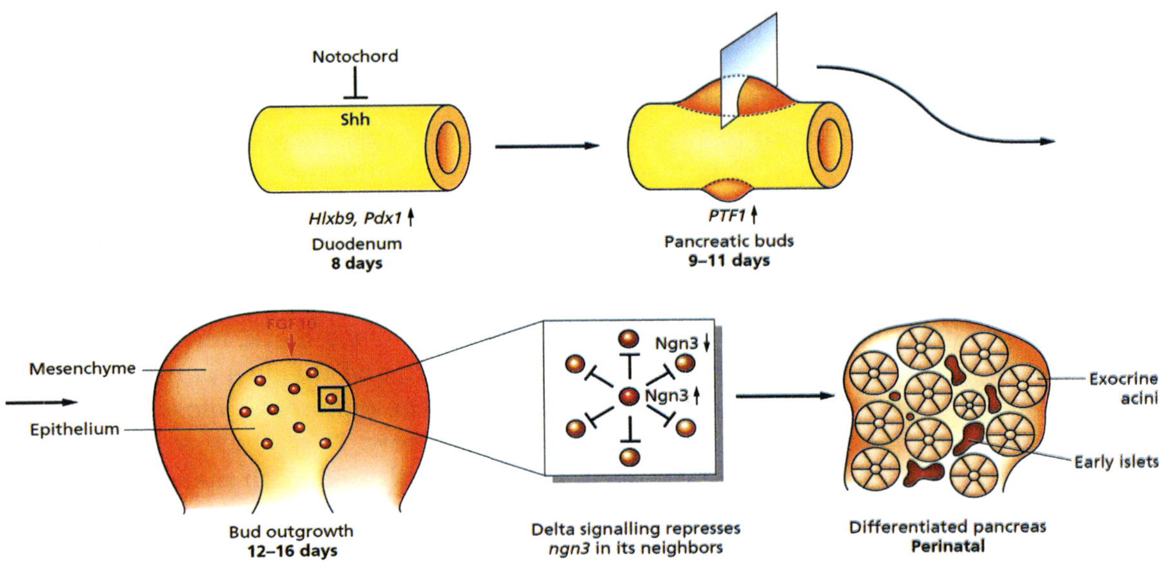

Figure 10.8. *Mechanism of pancreatic development.*

Source: https://dn790000.ca.archive.org/0/items/SlackJMWEssentialDevelopmentalBiology/Slack_J_M_W_Essential_Developmental_Biology.pdf.

To activate different exocrine genes, PTF1, a transcription factor with three subunits, is needed. The pancreatic buds themselves express it. The ventral bud is not able to form when the p48 subunit is deleted. The dorsal bud forms, but it does not have exocrine tissue and develops poorly. Endocrine cell development requires both Pax4 and Pax6 (paired box). Whereas the pax6 knockout mouse lacks α-cells, the pax4 knockout mouse lacks β- and δ-cells. β-cell development also necessitates the homeodomain factors Nkx6.1 and Nkx2.2, albeit at a later stage. The knockout mouse has a normal pancreas but no spleen at all. Hox11 (homeodomain), which is expressed in pancreatic mesenchyme, is not a part of the Hox clusters despite its name.

The pancreas is subject to the same lateral inhibition mechanisms that govern neurogenesis and intestinal epithelium cell differentiation. Early-stage endocrine cells stimulate the Notch pathway to suppress endocrine differentiation by their neighbors and produce Delta. Neurogenin 3 is the primary transcription factor that supports endocrine development (bHLH transcription factor). With a maximal expression at 13–15 points5 days, this is expressed in sporadic cells of the pancreatic buds that are not currently producing hormones. All endocrine cells disappear in mice lacking the ngn3 gene. A significant percentage of pancreatic cells, primarily α-cells, will become endocrine when ngn3 is overexpressed in transgenic mice using the Pdx1 promoter. Notch pathway knockouts of delta like 1, RBPJ⊠, and hes1 all increase the proportion of endocrine cells and have an effect akin to that of ngn3 overexpression. It is unknown what typically regulates the relative proportions of the various distinct endocrine cell types, but neurogenin 3 appears to be the primary driver of α-cell formation.

Figure 10.8 illustrates the pancreatic developmental trajectory inferred from the embryological experiments and the mouse knockouts.

10.3.3. Pancreatic Cell Lineage

In terms of morphology, behavior, and patterns of gene expression, pancreatic endocrine cells closely resemble neurons. Neural crest grafts from quail to chick revealed that they were not derived from the neural crest, as previously believed. To demonstrate that they are derived from endoderm, isolated endoderm explants can be cultured *in vitro* to produce a small number of endocrine cells. Alternatively, labeled endoderm can be reconstituted with mesenchyme to achieve good development in organ culture, and the resulting cell types can be shown to be all derived from endoderm. Additionally, retroviral labeling has demonstrated that a single endoderm cell is capable of producing both endocrine and exocrine progeny. Numerous Crelox lineage studies on the pancreas have been conducted in an effort to comprehend the lineage at a higher resolution. This kind of application involves the creation of a transgenic line driven by an interest promoter. Once mated to a reporter line, like R26R, the progeny should exhibit permanent labeling of the cells expressing the Cre promoter at any developmental stage. All types of pancreatic cells are labeled when the pdx1 promoter is used to drive the Cre. The addition of an appropriate agonist (typically the medication tamoxifen) can stimulate the activity of the Cre if it is fused to the estrogen receptor hormone binding domain (CreER). As a result, specific time windows can be studied because cells will only be labeled in conjunction with tamoxifen when they are also producing the CreER. If the CreER is induced after 8.5 days of development, exocrine and endocrine cells are labeled. However, the labeling of ducts is limited to the period between 9.5 and 11.5 days, indicating a unique mode of lineage formation at this stage. While the expression of ngn3 itself is only momentary, all endocrine cells are labeled when the neurogenin 3 promoter is used to drive the cre. These findings thus support the requirements for Pdx1 in the development of the pancreas and Ngn3 in the formation of endocrine cells. It is discovered that all pancreatic cells, not just the exocrine cells, are labeled when the p48 promoter is used to drive the cre. Thus, this factor may regulate exocrine differentiation in addition to its early role in pancreatic specification.

It was thought that multihormone cells, which express two or three hormones concurrently, could be the ancestors of all later endocrine cells because they were present in the early buds. However, the later formed α-cells are not labeled when the insulin promoter is used to drive cre, indicating that they had not previously expressed insulin. Similarly, in the presence of the glucagon promoter, the β-cells remained unlabeled, indicating a previous lack of glucagon expression. However, using the pancreatic polypeptide promoter labels β-cells, so it is possible that cells expressing this hormone are precursors. The question of whether β-cells are exclusively formed from pre-existing β-cells or from undifferentiated stem cells during postnatal life has also been addressed by Cre labeling. Nearly all of the β-cells can be labeled when the insulin promoter is used to drive CreER and the activity of the Cre is induced after birth. The fact that this label is then maintained indefinitely indicates that no new β-cells are being recruited from ducts or other islet cells that were not producing insulin at the time of labeling. Based on these experiments, a consensus pancreatic cell lineage is depicted in Figure 10.9.

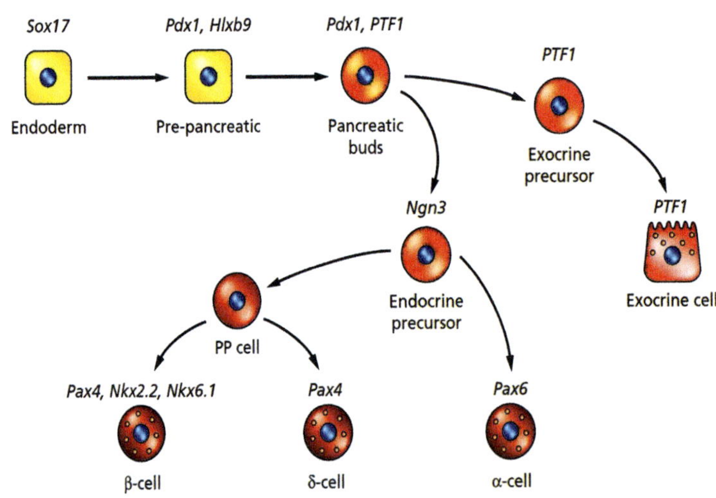

Figure 10.9. *Pancreatic cell lineage.*

Source: https://dn790000.ca.archive.org/0/items/SlackJMWEssentialDevelopmentalBiology/Slack_J_M_W_Essential_Developmental_Biology.pdf.

ROLE MODEL

KARL ERNST VON BAER: PRUSSIAN-ESTONIAN EMBRYOLOGIST

Karl Ernst von Baer (born February 17 [February 28, New Style], 1792, Piep, Estonia, Russian Empire—died November 16 [November 28], 1876, Dorpat, Estonia) was a Prussian-Estonian embryologist who discovered the mammalian ovum and the notochord and established the new science of comparative embryology alongside comparative anatomy. He was also a pioneer in geography, ethnology, and physical anthropology.

Baer, one of 10 children, spent his childhood with an uncle and aunt before he returned at the age of seven to his own family. His parents, of Prussian descent, were first cousins. After private tutoring Baer spent three years at a school for members of the nobility. In 1810 he entered the university at Dorpat to study medicine, receiving his medical degree in 1814.

Dissatisfied with his medical training, Baer studied in Germany and Austria from 1814 to 1817. The crucial year of his education was the academic year 1815–16, when his training in comparative anatomy at the University of Würzburg with Ignaz Döllinger introduced him to a new world that included the study of embryology.

In 1817 Baer began his teaching in Königsberg (now Kaliningrad, Russia), where he remained until 1834. In 1820 he married Auguste von Medem of Königsberg, by whom he had six children. Although Döllinger had suggested that Baer begin a study of chick development, he was unable to meet the expense of purchasing the eggs and paying an attendant to watch the incubators. This work was done instead by Baer's more affluent friend Christian Pander, who in 1817 described the early development of the chick in terms of what are now known as the primary germ layers—that is, ectoderm, mesoderm, and endoderm.

From 1819 to 1834 Baer devoted most of his time to embryology, extending Pander's concept of germ-layer formation to all vertebrates. In so doing Baer laid the foundation for comparative embryology. He made many important technical discoveries. In 1827 he described his discovery of the mammalian ovum (egg) in his De Ovi Mammalium et Hominis Genesi ("On the Mammalian Egg and the Origin of Man"), thereby establishing that mammals, including human beings, develop from eggs. He opposed the popular idea that embryos of one species pass through stages comparable to adults of other species. Instead, he emphasized that embryos of one species could resemble embryos, but not adults of another, and that the younger the embryo the greater the resemblance. This was in line with his epigenetic idea—basic to embryology ever since—that development proceeds from simple to complex, from homogeneous to heterogeneous.

One of the most important books in embryology is Baer's Über Entwickelungsgeschichte der There, in which he surveyed all existing knowledge on vertebrate development and from which he derived his far-reaching conclusions. He identified the neural folds as precursors of the nervous system, discovered the notochord, described the five primary brain vesicles, and studied the functions of the extra-embryonic membranes. This pioneering work established embryology as a distinct subject of research, at least in its descriptive aspects. He marked out the main lines of descriptive and comparative study that had to be accomplished before the modern approach—the causal analysis of development—could emerge.

In 1834 Baer moved to St. Petersburg, Russia, where he became a full member of the Academy of Sciences; he had been a corresponding member since 1826. His first duties were as librarian of the foreign division, but he eventually served the academy in a variety of administrative positions. He retired from active membership in 1862 but continued to work as an honorary member until 1867. After moving to Russia, Baer abandoned embryology. Particularly interested in the Russian north, he became a courageous explorer there; he was the first naturalist to collect specimens from Novaya Zemlya, which was then uninhabited. During his extensive travels throughout Russia, Baer developed a great scientific and practical interest in its fisheries. He made significant discoveries in geography, including one concerning the nature of the forces responsible for the configuration of riverbanks in Russia.

Baer's travels also increased his long-standing interest in ethnography. He contributed to the Academy at St. Petersburg by establishing an extensive skull collection. As a result of his interest in skull measurements, he called a meeting of craniologists in Germany in 1861, which led to the establishment of the German Anthropological Society and to the founding of the journal Archiv für Anthropologie. He was also responsible for the founding of the Russian Geographical Society and the Russian Entomological Society, of which he was the first president.

In his early days as an embryologist Baer had begun to consider possible relationships, in terms of kinship, between animals. In 1859, the year that Charles Darwin's Origin of Species appeared, Baer published a work on human skulls suggesting that stocks now distinct might have originated from one form; the ideas of the two men were formulated completely independently. Baer, however, was no strong adherent to the doctrine of transformation (the pre-Darwinian term for evolution). Although he believed that some very similar animals, such as goats and antelopes, might be related, he was vehemently against the concept expressed in the Origin of Species that all living creatures might have evolved from one or a few common ancestors. In his philosophical writings—and all his embryological writings were philosophical to some degree—Baer saw nature as a whole, even though not in terms of modern evolutionary theory. He viewed the development of organisms and of the cosmos in the same light, and his all-encompassing view of the universe brought together what might otherwise have seemed diverging threads in his thought.

CLASS ACTIVITY

MICROSCOPIC EXAMINATION OF ENDODERM-DERIVED TISSUES

Objective: Examine and identify endoderm-derived tissues under the microscope.

Materials: Microscope, prepared slides of endodermal tissues (e.g., lung tissue, intestinal lining).

Procedure:

1. Obtain prepared slides of various endoderm-derived tissues.
2. Examine each slide under the microscope.
3. Identify key features of each tissue type.
4. Draw and label the structures observed, noting their functions.

SUMMARY

- The endoderm of amniotes starts as a flat sheet of cells below the ectoderm and mesodermal layers. The gut tube is formed through the folding of the blastoderm to create a headfold and tailfold, which respectively contain pockets of fore- and hindgut. Subsequently the anterior and posterior intestinal portals move towards each other until they join. The gut tube and ventral body wall are closed by lateral body folds which meet each other on the ventral side of the embryo.
- The basic pattern of the gut is the same in all vertebrates with a pharynx, esophagus, stomach, liver, pancreas, small and large intestine. Each is lined with a characteristic epithelium derived from the endoderm.
- In the fate map the endoderm and mesenchyme of each part of the gut come from different positions in the embryo. They move together in the course of gut closure.
- Nodal signaling is necessary to activate expression of the early endodermal transcription factors which include Sox17 and mixer.
- Regional specification depends on inductive signals from the mesoderm. The posterior is determined early and the anterior later in development. Formation of the liver requires an FGF signal from the cardiac mesoderm.
- The pancreas consists of exocrine cells that secrete digestive enzymes into the intestine, and endocrine cells that secrete hormone, including insulin, into the bloodstream. Both cell types arise from the endoderm.
- The pancreas originates as dorsal and ventral buds from the duodenum. Suppression of Shh expression in the endoderm is necessary for bud formation and FGF is necessary for continued growth.
- Pdx1 is essential for the development of all pancreatic cell types. Neurogenin 3 is essential for the development of endocrine cells. Endocrine precursors expressing neurogenin 3 suppress endocrine development of their neighbors by Delta-Notch signaling.

REVIEW QUESTIONS

1. What are the primary organs that arise from the endoderm layer?
2. Describe the process of liver development from the endoderm.
3. At what stage of embryonic development does the endoderm form?
4. Explain the role of the yolk sac in the development of the gut tube.
5. How do the different regions of the gut tube (foregut, midgut, hindgut) arise during development?
6. Describe the fate of cells in the anterior versus posterior regions of the endoderm.
7. How does the endoderm interact with other germ layers during organogenesis?

8. Define the importance of proper cell migration through the primitive streak for embryonic development.
9. What molecular signals mediate epithelial–mesenchymal interactions during development?
10. Write the significance of pancreatic bud induction for the development of the pancreas.

REFERENCES

1. Cardoso, W. V. (2001). Molecular regulation of lung development. *Annual Review of Physiology, 63*, 471–494. https://doi.org/10.1146/annurev.physiol.63.1.471.
2. Duncan, S. A. (2003). Mechanisms controlling early development of the liver. *Mechanisms of Development, 120*(1), 19–33. https://doi.org/10.1016/S0925-4773(02)00349-4.
3. Fukuda, K., & Kikuchi, Y. (2005). Endoderm development in vertebrates: Fate mapping, induction and regional specification. *Development, Growth & Differentiation, 47*(6), 343–355. https://doi.org/10.1111/j.1440-169X.2005.00774.x.
4. Gilbert, S. F. (n.d.). *Endoderm*. Sinauer Associates. Retrieved from: https://www.sinauer.com (accessed on 5 September 2024).
5. Grapin-Botton, A., & Melton, D. A. (2000). Endoderm development from patterning to organogenesis. *Trends in Genetics, 16*(4), 124–130. https://doi.org/10.1016/S0168-9525(00)02011-3.
6. Gu, G., Brown, J. R., & Melton, D. A. (2003). Direct lineage tracing reveals the ontogeny of pancreatic cell fates during mouse embryogenesis. *Mechanisms of Development, 120*(1), 35–43. https://doi.org/10.1016/S0925-4773(02)00350-8.
7. Hebrok, M. (2003). Hedgehog signaling in pancreas development. *Mechanisms of Development, 120*(1), 45–57. https://doi.org/10.1016/S0925-4773(02)00348-2.
8. Herrera, P. L. (2002). Defining the cell lineages of the islets of Langerhans using transgenic mice. *International Journal of Developmental Biology, 46*(1), 97–103. https://doi.org/10.1387/ijdb.15266318.
9. Lammert, E., Cleaver, O., & Melton, D. A. (2003). Role of endothelial cells in early pancreas and liver development. *Mechanisms of Development, 120*(1), 59–64. https://doi.org/10.1016/S0925-4773(02)00347-0.
10. Murtaugh, L. C., & Melton, D. A. (2003). Genes, signals and lineages in pancreas development. *Annual Review of Cell and Developmental Biology, 19*, 71–89. https://doi.org/10.1146/annurev.cellbio.19.111301.153131.
11. Ober, E. A., Field, H. A., & Stanier, D. Y. R. (2003). From endoderm formation to liver and pancreas development in the zebrafish. *Mechanisms of Development, 120*(1), 5–18. https://doi.org/10.1016/S0925-4773(02)00351-7.
12. Roberts, D. J. (2002). Molecular mechanisms of development of the gastrointestinal tract. *Developmental Dynamics, 219*(1), 109–120. https://doi.org/10.1002/dvdy.10133.
13. Shivasani, R. A. (2002). Molecular regulation of vertebrate early endoderm

development. *Developmental Biology, 249*(2), 191–203. https://doi.org/10.1006/dbio.2002.0721.

14. Warburton, D., Schwartz, M., Tefft, D., et al. (2000). The molecular basis of lung morphogenesis. *Mechanisms of Development, 92*(1), 55–81. https://doi.org/10.1016/S0925-4773(99)00340-7.
15. Wells, J. M., & Melton, D. A. (1999). Vertebrate endoderm development. *Annual Review of Cell and Developmental Biology, 15*, 393–410. https://doi.org/10.1146/annurev.cellbio.15.1.393.
16. Yasugi, S. (1993). Role of epithelial–Mesenchymal interactions in differentiation of epithelium of vertebrate digestive organs. *Development Growth & Differentiation, 35*(1), 1–9. https://doi.org/10.1111/j.1440-169X.1993.00001.x.
17. Zaret, K. S. (2000). Liver specification and early morphogenesis. *Mechanisms of Development, 92*(1), 83–88. https://doi.org/10.1016/S0925-4773(99)00341-9.
18. Zaret, K. S. (2001). Hepatocyte differentiation: From the endoderm and beyond. *Current Opinion in Genetics & Development, 11*(5), 568–574. https://doi.org/10.1016/S0959-437X(00)00209-4.

Index

I

L

M

N

O

P

Q

R

S

T

U

V

X

Y

Z

Printed and bound by CPI Group (UK) Ltd, Croydon, CR0 4YY

18/09/2026

14965063-0006